Lecture Notes in Computer Science   13209

More information about this series at https://link.springer.com/bookseries/558

Vincent Andrearczyk · Valentin Oreiller ·
Mathieu Hatt · Adrien Depeursinge (Eds.)

Head and Neck Tumor Segmentation and Outcome Prediction

Second Challenge, HECKTOR 2021
Held in Conjunction with MICCAI 2021
Strasbourg, France, September 27, 2021
Proceedings

Editors
Vincent Andrearczyk
HES-SO Valais-Wallis University of Applied
Sciences and Arts Western Switzerland
Sierre, Switzerland

Valentin Oreiller
HES-SO Valais-Wallis University of Applied
Sciences and Arts Western Switzerland
Sierre, Switzerland

Mathieu Hatt
LaTIM, INSERM, University of Brest
Brest, France

Adrien Depeursinge
HES-SO Valais-Wallis University of Applied
Sciences and Arts Western Switzerland
Sierre, Switzerland

ISSN 0302-9743   ISSN 1611-3349 (electronic)
Lecture Notes in Computer Science
ISBN 978-3-030-98252-2   ISBN 978-3-030-98253-9 (eBook)
https://doi.org/10.1007/978-3-030-98253-9

This Springer imprint is published by the registered company Springer Nature Switzerland AG
The registered company address is: Gewerbestrasse 11, 6330 Cham, Switzerland

Preface

Head and Neck cancer (H&N) is among the deadliest cancers, with a large majority present in the oropharynx region. Treatment planning would benefit from automatic analysis of (FDG-)Positron Emission Tomography (PET)/Computed Tomography (CT) images, including automatic segmentation of the primary tumor and prediction of patient outcome. Building upon the success of tumor segmentation in HECKTOR 2020, we expanded the challenge to the new tasks of Progression Free Survival (PFS) prediction and augmented the dataset with new cases from a sixth center.

We proposed the second edition of the HEad and neCK TumOR segmentation and outcome prediction (HECKTOR 2021[1]) challenge to evaluate and compare the current state-of-the-art methods. We provided a dataset containing 325 PET/CT images (224 training and 101 test images) from six centers with manually delineated primary tumors and PFS time for training cases. Patient clinical data were also provided for all cases (with some missing data) including center, age, gender, TNM 7/8th edition staging and clinical stage, tobacco and alcohol consumption, performance status, HPV status, and treatment (radiotherapy only or chemoradiotherapy). The challenge was hosted by the International Conference on Medical Image Computing and Computer Assisted Intervention (MICCAI 2021[2]). The results were presented at a half-day event on September 27, 2021, as a designated satellite event of the conference.

Among the 42 teams who submitted their segmentation outcome to task 1, 21 teams submitted a paper describing their method and results. Of these, 20 papers were accepted after a single-blind review process with a minimum of two reviewers per paper. A total of 30 teams submitted their outcome predictions to task 2 and seven submitted to task 3, of which 17 and seven teams, respectively, reported their participation in a paper and all were accepted after review. Note that when participating in multiple tasks, the participants could decide whether to submit a single paper or multiple papers. The present volume gathers the total of 29 participants' papers together with our overview paper (which underwent the same reviewing process as the participants' papers).

We thank the committee members, the participants, the MICCAI 2021 organizers, the reviewers, and our sponsors Siemens Healthineers Switzerland, Aquilab, Bioemtech, and MIM. The papers included in this volume take the Springer Nature policies into account[3].

December 2021

Vincent Andrearczyk
Valentin Oreiller
Mathieu Hatt
Adrien Depeursinge

---

[1] https://www.aicrowd.com/challenges/miccai-2021-hecktor.

[2] www.miccai2021.org.

[3] https://www.springernature.com/gp/authors/book-authors-code-of-conduct.

Organization

General Chairs

Vincent Andrearczyk   University of Applied Sciences Western
Switzerland, Switzerland

Valentin Oreiller   University of Applied Sciences Western
Switzerland, Switzerland

Mathieu Hatt   University of Brest, France

Adrien Depeursinge   University of Applied Sciences Western
Switzerland, Switzerland

Program Committee

Sarah Boughdad   Lausanne University Hospital, Switzerland

Catherine Chez Le Rest   CHU Poitiers, France

Hesham Elhalawani   Cleveland Clinic Foundation, USA

Mario Jreige   Lausanne University Hospital, Switzerland

John O. Prior   Lausanne University Hospital, Switzerland

Martin Vallières   University of Sherbrooke, Canada

Dimitris Visvikis   University of Brest, France

Additional Reviewers

Daniel Abler

Manfredo Atzori

Carlos Cardenas

Pierre-Henri Conze

Pierre Fontaine

Alba Garcia

Mara Graziani

Vincent Jaouen

Oscar Jimenez

Niccolò Marini

Henning Müller

Sebastian Otálora

Annika Reinke

Ricky R. Savjani

Stephanie Tanadini Lang

Santiago Toledo

Alexandre Zwanenburg

Contents

Overview of the HECKTOR Challenge at MICCAI 2021: Automatic Head and Neck Tumor Segmentation and Outcome Prediction in PET/CT Images

Vincent Andrearczyk[1(✉)], Valentin Oreiller[1,2], Sarah Boughdad[2],
Catherine Cheze Le Rest[3,6], Hesham Elhalawani[4], Mario Jreige[2],
John O. Prior[2], Martin Vallières[5], Dimitris Visvikis[6], Mathieu Hatt[6],
and Adrien Depeursinge[1,2]

[1] Institute of Information Systems, School of Management, HES-SO Valais-Wallis
University of Applied Sciences and Arts Western Switzerland, Sierre, Switzerland
[2] Centre Hospitalier Universitaire Vaudois (CHUV), Lausanne, Switzerland
[3] Centre Hospitalier Universitaire de Poitiers (CHUP), Poitiers, France
[4] Department of Radiation Oncology, Cleveland Clinic Foundation,
Cleveland, OH, USA
[5] Department of Computer Science, University of Sherbrooke,
Sherbrooke, QC, Canada
[6] LaTIM, INSERM, UMR 1101, Univ Brest, Brest, France

Abstract. This paper presents an overview of the second edition of
the HEad and neCK TumOR (HECKTOR) challenge, organized as a
satellite event of the 24th International Conference on Medical Image
Computing and Computer Assisted Intervention (MICCAI) 2021. The
challenge is composed of three tasks related to the automatic analysis of
PET/CT images for patients with Head and Neck cancer (H&N), focusing on the oropharynx region. *Task 1* is the automatic segmentation of
H&N primary Gross Tumor Volume (GTVt) in FDG-PET/CT images.
Task 2 is the automatic prediction of Progression Free Survival (PFS)
from the same FDG-PET/CT. Finally, *Task 3* is the same as Task 2 with
ground truth GTVt annotations provided to the participants. The data
were collected from six centers for a total of 325 images, split into 224
training and 101 testing cases. The interest in the challenge was highlighted by the important participation with 103 registered teams and
448 result submissions. The best methods obtained a Dice Similarity
Coefficient (DSC) of 0.7591 in the first task, and a Concordance index
(C-index) of 0.7196 and 0.6978 in Tasks 2 and 3, respectively. In all
tasks, simplicity of the approach was found to be key to ensure generalization performance. The comparison of the PFS prediction performance
in Tasks 2 and 3 suggests that providing the GTVt contour was not crucial to achieve best results, which indicates that fully automatic methods can be used. This potentially obviates the need for GTVt contouring,

---

V. Andrearczyk and V. Oreiller—Equal contribution.
M. Hatt and A. Depeursinge—Equal contribution.

V. Andrearczyk et al. (Eds.): HECKTOR 2021, LNCS 13209, pp. 1–37, 2022.
https://doi.org/10.1007/978-3-030-98253-9_1

opening avenues for reproducible and large scale radiomics studies including thousands potential subjects.

Keywords: Automatic segmentation · Challenge · Medical imaging · Head and Neck cancer · Segmentation · Radiomics · Deep learning

1 Introduction: Research Context

The prediction of disease characteristics and outcomes using quantitative image biomarkers from medical images (i.e. radiomics) has shown tremendous potential to optimize and personalize patient care, particularly in the context of Head and Neck (H&N) tumors [58]. FluoroDeoxyGlucose (FDG)-Positron Emission Tomography (PET) and Computed Tomography (CT) imaging are the modalities of choice for the initial staging and follow-up of H&N cancer, as well as for radiotherapy planning purposes. Yet, both gross tumor volume (GTV) delineations in radiotherapy planning and radiomics analyses aiming at predicting outcome rely on an expensive and error-prone manual or semi-automatic annotation process of Volumes of Interest (VOI) in three dimensions. The fully automatic segmentation of H&N tumors from FDG-PET/CT images could therefore enable faster and more reproducible GTV definition as well as the validation of radiomics models on very large cohorts. Besides, fully automatic segmentation algorithms could also facilitate the application of validated models to patients' images in routine clinical workflows. By focusing on metabolic and morphological tissue properties respectively, PET and CT images provide complementary and synergistic information for cancerous lesion segmentation and patient outcome prediction. The HEad and neCK TumOR segmentation and outcome prediction from PET/CT images (HECKTOR)[1] challenge aimed at identifying the best methods to leverage the rich bi-modal information in the context of H&N primary tumor segmentation and outcome prediction. The analysis of the results provides precious information on the adequacy of the image analysis methods for the different tasks and the feasibility of large-scale and reproducible radiomics studies.

The potential of PET information for automatically segmenting tumors has been long exploited in the literature. For an in-depth review of automatic segmentation of PET images in the pre-deep learning era, see [19,25] covering methods such as fixed or adaptive thresholding, active contours, statistical learning, and mixture models. The need for a standardized evaluation of PET automatic segmentation methods and a comparison study between all the current algorithms was highlighted in [25]. The first challenge on tumor segmentation in PET images was proposed at MICCAI 2016[2] by Hatt et al. [23], implementing evaluation recommendations published previously by the AAPM (American Association of Physicists in Medicine) Task group 211 [25]. Multi-modal analyses of PET and CT images have also recently been proposed for different tasks,

---

[1] https://www.aicrowd.com/challenges/miccai-2021-hecktor, as of October 2021.
[2] https://portal.fli-iam.irisa.fr/petseg-challenge/overview#_ftn1, as of October 2020.

including lung cancer segmentation in [32,36,69,70] and bone lesion detection in [64]. In [5], we developed a baseline Convolutional Neural Network (CNN) approach based on a leave-one-center-out cross-validation on the training data of the HECKTOR challenge. Promising results were obtained with limitations that motivated additional data curation, data cleaning and the creation of the first HECKTOR challenge in 2020 [4,51]. This first edition compared segmentation architectures as well as the complementarity of the two modalities for the segmentation of GTVt in H&N.

In this second edition of the challenge, we propose a larger dataset, including a new center to better evaluate the generalization of the algorithms, as well as new tasks of prediction of Progression-Free Survival (PFS). Preliminary studies of automatic PFS prediction were performed with standard radiomics [18] and deep learning models [2] prior to challenge design. The proposed dataset comprises data from six centers. Five centers are used for the training data and two for testing (data from the sixth center are split between training and testing sets). The task is challenging due to, among others, the variation in image acquisition and quality across centers (the test set contains data from a domain not represented in the training set) and the presence of lymph nodes with high metabolic responses in the PET images.

The critical consequences of the lack of quality control in challenge designs were shown in [42] including reproducibility and interpretation of the results often hampered by the lack of provided relevant information and the use of non-robust ranking of algorithms. Solutions were proposed in the form of the Biomedical Image Analysis challengeS (BIAS) [43] guidelines for reporting the results. This paper presents an overview of the challenge following these guidelines.

Individual participants' papers reporting their methods and results were submitted to the challenge organizers. Reviews were organized by the organizers and the papers of the participants are published in the LNCS challenges proceedings [1,8,11,15,16,20,26,29,33,34,38,40,41,44,45,47,49,50,52–55,57,59–61,63, 66,68]. When participating in multiple tasks, participants could submit one or multiple papers.

The manuscript is organized as follows. The challenge dataset is described in Sect. 2. The tasks descriptions, including challenge design, algorithms summaries and results, are split into two sections. The segmentation task (Task 1) is presented in Sect. 3, and the outcome prediction tasks (Tasks 2 and 3) are described in Sect. 4. Section 5 discusses the results and findings and Sect. 6 concludes the paper.

2   Dataset

2.1   Mission of the Challenge

Biomedical Application
The participating algorithms target the following fields of application: diagnosis, prognosis and research. The participating teams' algorithms were designed

for either or both image segmentation (i.e., classifying voxels as either primary tumor, GTVt, or background) and PFS prediction (i.e., ranking patients according to predicted risk of progression).

Cohorts
As suggested in [43], we refer to the patients from whom the image data were acquired as the challenge cohort. The target cohort[3] comprises patients received for initial staging of H&N cancer. The clinical goals are two-fold; the automatically segmented regions can be used as a basis for (i) treatment planning in radiotherapy, (ii) further radiomics studies to predict clinical outcomes such as overall patient survival, disease-free survival, response to therapy or tumor aggressiveness. Note that the PFS outcome prediction task does not necessarily have to rely on the output of the segmentation task. In the former case (i), the regions will need to be further refined or extended for optimal dose delivery and control. The challenge cohort[4] includes patients with histologically proven H&N cancer who underwent radiotherapy treatment planning. The data were acquired from six centers (four for the training, one for the testing, and one for both) with variations in the scanner manufacturers and acquisition protocols. The data contain PET and CT imaging modalities as well as clinical information including age, sex, acquisition center, TNM staging, HPV status and alcohol. A detailed description of the annotations is provided in Sect. 2.2.

Target Entity
The data origin, i.e. the region from which the image data were acquired, varied from the head region only to the whole body. While we provided the data as acquired, we limited the analysis to the oropharynx region and provided a semi-automatically detected bounding-box locating the oropharynx region [3], as illustrated in Fig. 1. The participants could use the entire images if wanted but the predictions were evaluated only within these bounding-boxes.

2.2   Challenge Dataset

Data Source
The data were acquired from six centers as detailed in Table 1. It consists of PET/CT images of patients with H&N cancer located in the oropharynx region. The devices and imaging protocols used to acquire the data are described in Table 2. Additional information about the image acquisition is provided in Appendix 2.

---

[3] The target cohort refers to the subjects from whom the data would be acquired in the final biomedical application. It is mentioned for additional information as suggested in BIAS, although all data provided for the challenge are part of the challenge cohort.
[4] The challenge cohort refers to the subjects from whom the challenge data were acquired.

Table 1. List of the hospital centers in Canada (CA), Switzerland (CH) and France (FR) and number of cases, with a total of 224 training and 101 test cases.

Center	Split	# cases
HGJ: Hôpital Général Juif, Montréal, CA	Train	55
CHUS: Centre Hospitalier Universitaire de Sherbooke, Sherbrooke, CA	Train	72
HMR: Hôpital Maisonneuve-Rosemont, Montréal, CA	Train	18
CHUM: Centre Hospitalier de l'Université de Montréal, Montréal, CA	Train	56
CHUP: Centre Hospitalier Universitaire Poitiers, FR	Train	23
Total	Train	224
CHUV: Centre Hospitalier Universitaire Vaudois, CH	Test	53
CHUP: Centre Hospitalier Universitaire Poitiers, FR	Test	48
Total	Test	101

Table 2. List of scanners used in the various centers.

Center	Device
HGJ	hybrid PET/CT scanner (Discovery ST, GE Healthcare)
CHUS	hybrid PET/CT scanner (Gemini GXL 16, Philips)
HMR	hybrid PET/CT scanner (Discovery STE, GE Healthcare)
CHUM	hybrid PET/CT scanner (Discovery STE, GE Healthcare)
CHUV	hybrid PET/CT scanner (Discovery D690 ToF, GE Healthcare)
CHUP	hybrid PET/CT scanner (Biograph mCT 40 ToF, Siemens)

Training and Test Case Characteristics
The training data comprise 224 cases from five centers (HGJ, HMR[5], CHUM, CHUS and CHUP). Originally, the dataset in [58] contained 298 cases, among which we selected the cases with oropharynx cancer. The test data contain 101 cases from a fifth center CHUV (n = 53) and CHUP (n = 48). Examples of PET/CT images of each center are shown in Fig. 1. Each case includes a CT image, a PET image and a GTVt mask (for the training cases) in the Neuroimaging Informatics Technology Initiative (NIfTI) format, as well as patient information (e.g. age, sex) and center. A bounding-box of size $144 \times 144 \times 144$ mm^3 locating the oropharynx region was also provided. Details of the semi-automatic region detection can be found in [3].

Finally, to provide a fair comparison, participants who wanted to use additional external data for training were asked to also report results using only the HECKTOR data and discuss differences in the results. However, no participant used external data in this edition.

Annotation Characteristics
For the HGJ, CHUS, HMR, and CHUM centers, initial annotations, i.e. 3D contours of the GTVt, were made by expert radiation oncologists and were later

---

[5] For simplicity, these centers were renamed CHGJ and CHMR during the challenge.

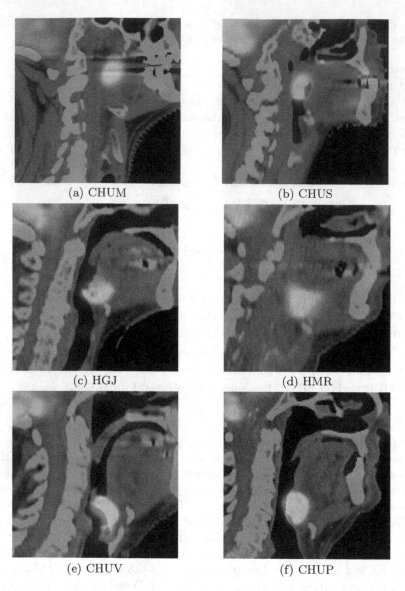

(a) CHUM   (b) CHUS

(c) HGJ   (d) HMR

(e) CHUV   (f) CHUP

Fig. 1. Case examples of 2D sagittal slices of fused PET/CT images from each of the six centers. The CT (grayscale) window in Hounsfield unit is $[-140, 260]$ and the PET window in SUV is $[0, 12]$, represented in a "hot" colormap.

re-annotated as described below. Details of the initial annotations of the training set can be found in [58]. In particular, 40% (80 cases) of the training radiotherapy contours were directly drawn on the CT of the PET/CT scan and thereafter used for treatment planning. The remaining 60% of the training radiotherapy contours were drawn on a different CT scan dedicated to treatment planning and were then registered to the FDG-PET/CT scan reference frame using intensity-based free-form deformable registration with the software MIM (MIM software Inc., Cleveland, OH). The initial contours of the test set were all directly drawn on the CT of the PET/CT scan.

The original contours for the CHUV center were drawn by an expert radiation oncologist for radiomics purpose [9]. The expert contoured the tumors on the PET/CT scan. The delineation from the CHUP center were obtained semi-automatically with a Fuzzy Locally Adaptive Bayesian (FLAB) segmentation [24] and corrected by an expert radiation oncologist. These contours were obtained on the PET images only.

Given the heterogeneous nature of the original contours, a re-annotation of the VOIs was performed. During the first edition of HECKTOR (HGJ, CHUS, HMR, CHUM, and CHUV), the re-annotation was supervised by an expert who is both radiologist and nuclear medicine physician. Two non-experts (organizers of the challenge) made an initial cleaning in order to facilitate the expert's work. The expert either validated or edited the VOIs. The Siemens Syngo.Via RT Image Suite was used to edit the contours in 3D with fused PET/CT images. Most of the contours were re-drawn completely, and the original segmentations were used to localize the tumor and discriminate between malignant versus benign high metabolic regions.

For the data added to the current HECKTOR edition (CHUP), the re-annotation was performed by three experts: one nuclear medicine physician, one radiation oncologist and one who is both radiologist and nuclear medicine physician. The 71 cases were divided between the three experts and each annotation was then checked by all three of them. This re-annotation was performed in a centralized fashion with the MIM software, and the verification of the contours was made possible by the MIM Cloud platform[6]. Guidelines for re-annotating the images were developed by our experts and are stated in the following.

Oropharyngeal lesions are contoured on PET/CT using information from PET and unenhanced CT acquisitions. The contouring includes the entire edges of the morphologic anomaly as depicted on unenhanced CT (mainly visualized as a mass effect) and the corresponding hypermetabolic volume, using PET acquisition, unenhanced CT and PET/CT fusion visualizations based on automatic co-registration. The contouring excludes the hypermetablic activity projecting outside the physical limits of the lesion (for example in the lumen of the airway or on the bony structures with no morphologic evidence of local invasion). The standardized nomenclature per AAPM TG-263 is "GTVp".

---

[6] https://mim-cloud.appspot.com/ as of December 2021.

For more specific situations, clinical nodal category was verified to ensure the exclusion of nearby FDG-avid and/or enlarged lymph nodes (e.g. submandibular, high level II, and retropharyngeal). In case of tonsillar fossa or base of tongue fullness/enlargement without corresponding FDG avidity, the clinical datasheet was reviewed to exclude patients with pre-radiation tonsillectomy or extensive biopsy.

Data Preprocessing Methods
No preprocessing was performed on the images to reflect the diversity of clinical data and to leave full flexibility to the participants. However, we provided various pieces of code to load, crop and resample the data, as well as to evaluate the results on our GitHub repository[7]. This code was provided as a suggestion to help the participants and to maximize transparency, but the participants were free to use other methods.

Sources of Errors
In [51], we reported an inter-observer (four observers) agreement of 0.6110 on a subset of the HECKTOR 2020 data containing 21 randomly drawn cases. Similar agreements were reported in the literature [21] with an average DSC agreement of three observers of 0.57 using only the CT images for annotation and 0.69 using both PET and CT. A source of error therefore originates from the degree of subjectivity in the annotation and correction of the expert. Another source of error is the difference in the re-annotation between the centers used in HECKTOR 2020 and the one added in HECKTOR 2021. In HECKTOR 2020, the re-annotation was checked by only one expert while for HECKTOR 2021 three experts participated in the re-annotation. Moreover, the softwares used were different.

Another source of error comes from the lack of CT images with a contrast agent for a more accurate delineation of the primary tumor.

Institutional Review Boards
Institutional Review Boards (IRB) of all participating institutions permitted the use of images and clinical data, either fully anonymized or coded, from all cases for research purposes only. Retrospective analyses were performed following the relevant guidelines and regulations as approved by the respective institutional ethical committees with protocol numbers: MM-JGH-CR15-50 (HGJ, CHUS, HMR, CHUM) and CER-VD 2018-01513 (CHUV). In the case of CHUP, ethical review and approval were waived because data were already collected for routine patient management before analysis, in which patients provided informed consent. No additional data was specifically collected for the present challenge.

---

[7] github.com/voreille/hecktor, as of December 2021.

3   Task 1: Segmentation

3.1   Methods: Reporting of Challenge Design

A summary of the information on the challenge organization is provided in Appendix 1, following the BIAS recommendations.

Assessment Aim
The assessment aim for the segmentation task is the following; evaluate the feasibility of fully automatic GTVt segmentation for H&N cancers in the oropharyngeal region via the identification of the most accurate segmentation algorithm. The performance of the latter is identified by computing the Dice Similarity Coefficient (DSC) and Hausdorff Distance (HD) at 95th percentile (HD95) between prediction and manual expert annotations.

DSC measures volumetric overlap between segmentation results and annotations. It is a good measure of segmentation for imbalanced segmentation problems, i.e. the region to segment is small as compared to the image size. DSC is commonly used in the evaluation and ranking of segmentation algorithms and particularly tumor segmentation tasks [21, 46].

In 3D, the HD computes the maximal distance between the surfaces of two segmentations. It provides an insight on how close the boundaries of the prediction and the ground truth are. The HD95 measure the 95th quantile of the distribution of surface distances instead of the maximum. This metric is more robust towards outlier segmented pixels than the HD and thus is often used to evaluate automatic algorithms [31, 39].

Assessment Method
Participants were given access to the test cases without the ground truth annotations and were asked to submit the results of their algorithms on the test cases on the AIcrowd platform.

Results were ranked using the DSC and HD95, both computed on images cropped using the provided bounding-boxes (see Sect. 2.2) in the original CT resolution. If the submitted results were in a resolution different from the CT resolution, we applied nearest-neighbor interpolation before evaluation.

The two metrics are defined for set A (ground truth volumes) and set B (predicted volumes) as follow:

$$\mathrm{DSC}(A, B) = \frac{2\,A|\cap B|}{|A| + |B|}, \tag{1}$$

where $|\cdot|$ is the set cardinality and

$$\mathrm{HD95}(A, B) = P_{95}\left\{\sup_{a\in A}\inf_{b\in B}\,\mathrm{d}(a,b), \sup_{b\in B}\inf_{a\in A}\,\mathrm{d}(a,b)\right\}, \tag{2}$$

where $\mathrm{d}(a,b)$ is the Euclidean distance between points a and b, sup and inf are the supremum and infimum, respectively. P_{95} is the 95th percentile.

The ranking was computed from the average DSC and median HD95 across all cases. Since the HD95 is unbounded, i.e. it is infinity when there is no prediction, we choose the median instead of the mean for aggregation. The two metrics are ranked separately and the final rank is obtained by Borda counting. This ranking method was used first to determine the best submission of each participating team (ranking the 1 to 5 submissions), then to obtain the final ranking (across all participants). Each participating team had the opportunity to submit up to five (valid) runs. The final ranking is reported in Sect. 4.2 and discussed in Sect. 5.

Missing values (i.e. missing predictions on one or multiple patients), did not occur in the submitted results but would have been treated as DSC of zero and a HD95 of infinity. In the case of tied rank (which was very unlikely due to the computation of the results average of 53 DSCs), we considered precision as the second ranking metric. The evaluation implementation can be found on our GitHub repository[8] and was made available to the participants to maximize transparency.

3.2   Results: Reporting of Segmentation Task Outcome

Participation. At of September 14 2021 (submission deadline), the number of registered teams was 44 for Task 1, 30 for Task 2 and 8 for Task 3. A team is made of at least one participant and not all participants that signed the End User Agreement (EUA) registered a team. Each team could submit up to five valid submissions. By the submission deadline, we had received 448 results submissions, including valid and invalid ones (i.e. not graded due to format errors). This participation was much higher than last year's challenge with 83 submissions and highlights the growing interest in the challenge.

In this section, we present the algorithms and results of participants who submitted a paper [1,8,11,15,16,20,29,33,34,38,40,44,45,47,49,52,53,55,60,61,63, 66,68]. An exhaustive list of the results can be seen on the leaderboard[9].

Segmentation: Summary of Participants' Methods. This section summarizes the approaches proposed by all teams for the automatic segmentation of the primary tumor (Task 1). Table 3 provides a synthetic comparison of the methodological choices and design. All methods are further detailed in dedicated paragraphs. The paragraphs are ordered according to the official ranking, starting with the winners of Task 1.

---

[8] github.com/voreille/hecktor, as of December 2021.
[9] https://www.aicrowd.com/challenges/miccai-2021-hecktor/leaderboards?
challenge_leaderboard_extra_id=667&challenge_round_id=879.

Table 3. Synthetic comparison of segmentation methods and results. More details are available in Sect. 3.2. The number of used models is reported in the last column when ensembling was used. "na" stands for "not available".

Team	Dice	HD95	Preprocess.				Data augmentation					Model archit.				Loss				Training/evaluation				
			iso-resampling	CT clipping	Min-max norm.	Standardization	Rotation	Scaling	Flipping	Noise addition	Other	U-Net	Attention	Res. connection	SE norm. [27]	Dice	Cross-entropy	Focal [37]	Else	Optimizer	nnU-Net [28]	LR decay	Cross-validation	Ensembling
Pengy [63]	0.7785	3.0882	✓	✓		✓	✓	✓	✓	✓	✓	✓			✓	✓			✓	SGD	✓	✓	✓	5
SJTU EIEE 2-426Lab[a] [1]	0.7733	3.0882	✓	✓	✓		✓		✓			✓		✓	✓	✓	✓	✓	✓	Adam	✓	✓	✓	9
HiLab [40]	0.7735	3.0882	✓	✓		✓						✓	✓		✓		✓		✓	Adam		✓	✓	14
BCIOQurit [67]	0.7709	3.0882	✓	✓		✓	✓	✓	✓			✓	✓					✓	✓	Adam	✓	✓		10
Aarhus Oslo [53]	0.7790	3.1549	✓				✓	✓				✓				✓	✓			Adam	✓	✓	✓	3
Fuller MDA [49]	0.7702	3.1432	✓	✓	✓	✓	✓	✓	✓		✓	✓		✓		✓				Adam		✓	✓	10
UMCG [15]	0.7621	3.1432	✓			✓						✓	✓					✓	✓	Adam		✓	✓	5
Siat [60]	0.7681	3.1549	✓	✓		✓	✓		✓			✓	✓	✓		✓		✓		na	na		✓	5
Heck Uihak [11]	0.7656	3.1549	✓	✓	✓	✓	✓		✓			✓	✓	✓		✓		✓		Adam		✓	✓	5
BMIT USYD [45]	0.7453	3.1549	✓	✓	✓	✓	✓		✓			✓				✓				Adam		✓	✓	10
DeepX [68]	0.7602	3.2700	✓	✓		✓		✓		✓	✓	✓	✓	✓		✓				Adam		✓	✓	15
Emmanuelle Bourigault [8]	0.7595	3.2700	✓	✓		✓	✓		✓			✓	✓	✓	✓	✓			✓	Adam		✓	✓	5
C235 [38]	0.7565	3.2700	✓	✓	✓	✓						✓	✓	✓		✓			✓	Adam		✓		5
Abdul Qayyum[52]	0.7487	3.2700		✓				✓		✓	✓	✓		✓		✓	✓			Adam			✓	
RedNeucon [44]	0.7400	3.2700	na	na	na	na	✓	✓	✓			✓	✓			✓	✓		✓	Adam		✓		25
DMLang [33]	0.7046	4.0265	✓	✓	✓		✓		✓	✓	✓	✓				✓				Adam				
Xuefeng [20]	0.6851	4.1932	✓	✓			✓	✓	✓	✓	✓	✓				✓	✓			SGD	✓	✓		
Qurit Tecvico [55]	0.6771	5.4208				✓						✓				✓				Adam				
Vokyj [29]	0.6331	6.1267	✓	✓		✓	✓									✓			✓	Adam		✓		
TECVICO Corp Family [16]	0.6357	6.3718	na	na	na	na						✓		✓		✓				Adam				2
BAMF health [47]	0.7795	3.0571	✓	✓		✓	✓	✓	✓	✓	✓	✓		✓		✓	✓			SGD	✓		✓	10
Wangjiao [61]	0.7628	3.2700	✓	✓		✓						✓	✓	✓	✓	✓			✓	Adam		✓	✓	6

[a] It is ranked second due to the HD95 slightly better than the third (HiLab), 3.088160269617 vs 3.088161777508, and the ranking strategy described in Sect. 3.1.

In [63], Xie and Peng (team "Pengy") used a well-tuned patch-based 3D nnU-Net [28] with standard pre-processing and training scheme, where the learning rate is adjusted dynamically using polyLR [10]. The Squeeze and Excitation (SE) normalization [27] was also one of the main ingredient of their approach. The approach is straighforward yet efficient as they ranked first for Task 1. Five models are trained in a five-fold cross-validation with random data augmentation including rotation, scaling, mirroring, Gaussian noise and Gamma correction. The five test predictions are ensembled via probability averaging for the final results.

In [1], An et al. (team "SJTU EIEE 2-426Lab") proposed a framework which is based on the subsequent application of three different U-Nets. The first U-Net is used to coarsely segment the tumor and then select a bounding-box. Then, the second network performs a finer segmentation on the smaller bounding box. Finally, the last network takes as input the concatenation of PET, CT and the previous segmentation to refine the predictions. They trained the three networks with different objectives. The first one was trained to optimize the recall rate, and the two subsequent ones were trained to optimize the Dice score. All objectives were implemented with the F-loss which includes a hyper-parameter allowing to balance between recall and Dice. The final prediction was obtained through majority voting on three different predictions: an ensemble of five nnU-Nets [28] (trained on five different folds), an ensemble of three U-Nets with SE normalization [27], and the predictions made by the proposed model.

In [40], Lu et al. (team "HiLab") employed an ensemble of various 3D U-Nets, including the eight models used in [27], winner of HECKTOR 2020, five models trained with leave-one-center-out, and one model combining a priori and a posteriori attention. In this last model, the normalized PET image was used as a priori attention map for segmentation on the CT image. Mix-up was also used, mixing PET and CT in the training set to construct a new domain to account for the domain shift in the test set. All 14 predictions were averaged and thresholded to 0.5 for the final ensembled prediction.

In [66], Yousefirizi et al. (team "BCIOqurit") used a 3D nnU-Net with SE normalization [27] trained on a leave-one-center-out with a combination of a "unified" focal and Mumford-Shah [30] losses taking the advantage of distribution, region, and boundary-based loss functions.

In [53], Ren et al. (team "Aarhus Oslo") proposed a 3D nnU-Net with various PET normalization methods, namely PET-clip and PET-sin. The former clips the Standardized Uptake Values (SUV) range in [0,5] and the latter transforms monotonic spatial SUV increase into onion rings via a sine transform of SUV. Loss functions were also combined and compared (Dice, Cross-Entropy, Focal and TopK). No strong global trend was observed on the influence of the normalization or loss.

In [49], Naser et al. (team "Fuller MDA") used an ensemble of 3D residual U-Nets trained on a 10-fold CV resulting in 10 different models. The ensemble was performed either by STAPLE or majority voting on the binarized predictions. Models with different numbers of channels were also compared. The best combination was the one with fewer feature maps and ensembled with majority voting.

In [15], De Biase et al. (team "UMCG") compared two methods: (i) Co-learning Multi-Modal PET/CT adapted from [65], which takes as input PET and CT as two separate images, outputs two masks that are averaged and (ii) Skip-scSE Multi-Scale Attention, which concatenates PET and CT in the channel dimension. The Skip-scSE models clearly outperformed the other. Ensembling (i) and (ii) provided worse results.

In [60], Wang et al. (team "Siat") used an ensemble of 3D U-Nets with multi-channel attention mechanisms. For each channel in the input data, this attention module outputs a weighted combination of filter outputs from three receptive fields over the input. A comparison with a standard 3D Vnet without attention showed the benefit of the latter.

In [11], Cho et al. team "Heck Uihak") used a backbone 3D U-Net that takes as input PET/CT images and outputs the predictions. This backbone U-Net is coupled with an attention module. The attention module was designed around a U-Net architecture and takes as input the PET images and produces attention maps. These attention maps are then multiplied with the skip connections of the backbone U-Net. The whole pipeline was trained with a sum of a Dice loss and a focal loss.

In [45], Meng et al. (team "BMIT USYD") used multi-task learning scheme to address Tasks 1 and 2. A modified 3D U-Net was used for segmentation. Its output is a voxel-wise tumor probability that is fed together with PET/CT to a 3D denseNet. Ensembling was used to produce the final output.

In [68], Yuan et al. (team "DeepX") proposed a 3D U-Net with scale attention which is referred to as Scale Attention Network (SA-Net). The skip connections were replaced by an attention block and the concatenation was replaced by a summation. The attention block takes as input the skip connections at all the scales and output an attention map which is added to the feature maps of the decoder. The attention blocks include a SE block. The encoder and decoder include ResNet-like blocks containing a SE block. An ensemble of 15 models was used for the final prediction (5 from the 5-fold CV with input size $144 \times 144 \times 144$ at $1\,\text{mm}^3$, 5 from the 5-fold CV with input size $128 \times 128 \times 128$ at $1.25 \times 1.25 \times 1.25\,\text{mm}^3$, and 5 from a leave-one-center-out CV with input size $144 \times 144 \times 144$ at $1\,\text{mm}^3$).

In [8], Bourigault et al. (team "Emmanuelle Bourigault") proposed a full scale 3D U-Net architecture with attention, residual connections and SE norm. Conditional random fields was applied as post-processing.

In [38], the authors (team "C235") proposed a model based on 3D U-Net supplemented with a simple attention module referred to as SimAM. Different from channel-wise and spatial-wise attention mechanisms, SimAM generates the corresponding weight for each pixel in each channel and spatial position. They compared their model to last year's winning algorithm based on SE Norm and report a small but consistent increase in segmentation performance when using the proposed SimAM attention module, which also resulted in models with about 20 times less parameters.

In [52], Qayyum et al. (team "Abdul Qayyum") proposed to use a 3D U-Net with 3D inception as well as squeeze and excitation modules with residual connections. They extended the 2D inception module into 3D with extra 3D depth-wise layers for semantic segmentation. The comparison with and without the inception module showed a systematic improvement associated with the latter.

In [44], Asenjo et al. (team "RedNeucon") ensembled a total of 25 models: a combination of 2D (trained on axial, sagittal and coronal planes) and 3D U-Nets, all trained on cross-validation and on the full dataset.

In [33], Lang et al. (team "DMLang") used a network based on a 3D U-Net. The main modification is that the skip connections were linked directly after the downsampling. They also optimized the kernel size and the strides of the convolutions.

In [20], Ghimire et al. (team "Xuefeng") developed a patch-based 3D U-Net with overlapping sliding window at test time. Deep supervision technique was applied to the network, where the computation of loss occurs at each decoding block. Various patch sizes, modality combination and convolution types were compared. Results suggest that larger patch size, bi-modal inputs, and conventional convolution (i.e. not dilated) was better.

In [55] Paeenafrakati et al. (team "Qurit Tecvico") proposed to use 3D U-Net or 3D U-NeTr (U-Net with transformers) to segment the GTVt. The network's input consists of a one-channel image. This image was obtained by image-level fusion techniques to combine information of both PET and CT images. They assessed ten different image fusion methods. To select the best combination of architecture and fusion method, they used a validation set of 23 images. The best combination was a U-Net architecture with the Laplacian pyramid method for fusion. This model obtained a DSC of, respectively, 0.81 and 0.68 on the validation and test set.

In [29], Muller et al. (team "Vokyj") proposed a model trained on supervoxels (obtained with Simple Linear Iterative Clustering, SLIC), motivated by the efficiency of the latter. The model is composed of an MLP encoder and graph CNN decoder. The models were trained on extracted patches of size $72 \times 72 \times 72$.

In [16], Fatan et al. (team "TECVICO Corp Family") employed a 3D U-Net with autoencoder regularization [48] trained on various fusions of PET and CT images. The best results were obtained with a Laplacian pyramid-sparse representation mixture.

Lee et al. [34] (team "Neurophet") used a dual path encoder (PET, CT) whose paths are coupled by a shared-weight cross-information module in each layer of the encoding path of the 3D U-Net architecture. The cross-attention module performs global average pooling over the feature channels resulting from convolutional blocks in both paths and feeds the resulting pooled features into a weight-shared fully connected layer. Its output, two (transformed) feature vectors are added elementwise and activated using a sigmoid function. The final output of each layer in the encoding part is obtained by multiplication of the features in each of the two paths with these cross-attention weights. The study used the generalized dice loss as training metric. Five separate models were built, using data from four centers for training and data from the 5th center for evaluation (average DSC 0.6808). Predictions on the test set (DSC 0.7367) were obtained by majority voting across the segmentation results of all 5 models.

In [47], Murugesan et al. (team "BAMF Health") proposed to ensemble the predictions of 3D nnU-Nets (with and without residual connections) using adaptive ensembling to eliminate false positives. A selective ensemble of 8 test-time augmentations and 10 folds (5 U-Nets and 5 residual U-Nets) was used for the final segmentation output.

In [61], Wang et al. (team "Wangjiao") used a combination of convolutional and transformer blocks in a U-Net model with attention (global context and channel) in the decoder. The model was trained with squeeze and excitation, and a Dice and Focal loss.

In Table 3, we summarize some of the main components of the participants' algorithms, including model architecture, preprocessing, training scheme and postprocessing.

Results. The results, including average DSC and HD95 are summarized in Table 3 with an algorithm summary. The two results at the bottom of the table without a rank were made ineligible to the ranking due to an excessive number of submissions on the HECKTOR 2020 dataset (on the online leaderboard) resulting in an overfit of the 2020 test set which represents half of the 2021 test set.

The results from the participants range from an average DSC of 0.6331 to 0.7785 and the median HD95 from 6.3718 to 3.0882. Xie and Peng. [63] (team "Pengy") obtained the best overall results with an average DSC of 0.7785 and a median HD95 of 3.0882. Examples of segmentation results (true positives on top row, and false positives on bottom row) are shown in Fig. 2.

4   Tasks 2 and 3: Outcome Prediction

In order to expand the scope of the challenge compared to the previous installment (2020) that focused on a single task dedicated to the automatic segmentation of GTVt (i.e., same as the updated Task 1 in the 2021 edition), it was decided to add a task with the aim of predicting outcome, i.e. Progression-Free Survival (PFS).

4.1   Methods: Reporting of Challenge Design

It was chosen to carry out this task on the same patients dataset used for Task 1, exploiting both the available clinical information and the multimodal FDG-PET/CT images. The available clinical factors included center, age, gender, TNM 7/8th edition staging and clinical stage, tobacco and alcohol consumption, performance status, HPV status, treatment (radiotherapy only or chemoradiotherapy). The information regarding tobacco and alcohol consumption, performance and HPV status was available only for some patients. For five patients from the training set, the weight was unknown and was set at 75kg to compute SUV values. Of note, this outcome prediction task was subdivided into two different tasks that participants could choose to tackle separately: Task 3

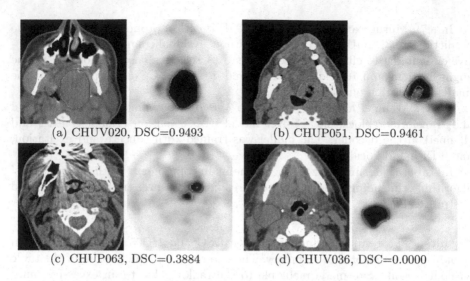

(a) CHUV020, DSC=0.9493   (b) CHUP051, DSC=0.9461

(c) CHUP063, DSC=0.3884   (d) CHUV036, DSC=0.0000

Fig. 2. Examples of results of the winning team (Pengy [63]). The automatic segmentation results (green) and ground truth annotations (red) are displayed on an overlay of 2D slices of PET (right) and CT (left) images. The reported DSC is computed on the whole image, see Eq. (1). (Color figure online)

provided the same data as Task 2, with the exception of providing, in addition, the reference expert contours (i.e., ground-truth of the GTVt). In order to avoid providing the reference contours to participants that could also participate in Task 1, we relied on a Docker-based submission procedure: participants had to encapsulate their algorithm in a Docker and submit it on the challenge platform. The organizers then ran the Dockers on the test data locally, in order to compute the performance. In such a way, the participants never had direct access to the reference contours of the test set, although they could incorporate them in their algorithms the way they saw fit.

Assessment Aim. The chosen clinical endpoint to predict was PFS. Progression was defined based on Response Evaluation Criteria In Solid Tumors (RECIST) criteria, i.e., either a size increase of known lesions (i.e., change of T and or N), or appearance of new lesions (i.e., change of N and/or M). Disease-specific death was also considered a progression event for patients previously considered stable. In the training set, participants were provided with the survival endpoint to predict, censoring and time-to-event between PET/CT scan and event (in days).

Assessment Method. For Task 2, challengers had to submit a CSV file containing the patient IDs with the outputs of the model as a predicted risk score anti-concordant with the PFS in days. For Task 3, the challengers had to submit a Docker encapsulating their method which was run by the organizers on the test set, producing the CSV file for evaluation. Thus for both tasks, the performance of the output predicted scores were evaluated using the Concordance index (C-index) [22] on the test data. The C-index quantifies the model's ability to provide an accurate ranking of the survival times based on the computed individual risk scores, generalizing the Area Under the ROC Curve (AUC). It can account for censored data and represents the global assessment of the model discrimination power. Therefore the final ranking was based on the best C-index value obtained on the test set, out of the maximum of 5 submissions per team. The C-index computation is based on the implementation found in the Lifelines library [14] and adapted to handle missing values that are counted as non-concordant.

4.2   Results: Reporting of Challenge Outcome

Participation. Thirty different teams submitted a total of 149 valid submissions to Task 2. Eighteen corresponding papers were submitted, which made the submissions eligible for final ranking and prize. Probably because of the added complexity of Task 3 requiring encapsulating the method in a Docker, only 8 teams submitted a total of 27 valid submissions. All these 8 teams also participated in Task 2, with 7 corresponding papers.

Outcome Prediction: Summary of Participants' Methods. The following describes the approach of each team participating in Task 2 (and 3 for some), in the order of the Task 2 ranking. Table 4 provides a synthetic comparison of the methodological choices and designs for these tasks.

In [54], Saeed et al. (team "BiomedIA") first experimented with the clinical variables and determined that better prediction was achieved using only variables with complete values, compared to using all variables with imputing missing values. They elected to first implement a fusion of PET and CT images by averaging them into a new single PET/CT image that would be further cropped (2 different sizes of $50 \times 50 \times 50$ and $80 \times 80 \times 80$ were tested) to form the main input to their solution based on a 3D CNN (Deep-CR) trained to extract features which were then fed into Multi-Task Logistic Regression (MTLR, a sequence of logistic regression models created at various timelines to evaluate the probability of the event happening) improved by integrating neural networks to achieve nonlinearity, along with the clinical variables. Two different models were compared as the input to MTLR: either a CNN with 3 paths (for PET, CT and fused PET/CT) or only one using only fused PET/CT. The batch size, learning rate, and dropout were experimentally set to 16, 0.016, and 0.2 respectively for the training. The model was trained for 100 epochs using Adam optimizer. No cross-validation or data augmentation was used. Of note, the results of CNN and MTLR (i.e., exploiting both images and clinical variables) were averaged with the prediction of a Cox model using only clinical variables to obtain the best result. This team won Task 2 with a C-index of 0.72 but did not participate in Task 3.

Table 4. Synthetic comparison of outcome prediction methods. More details are available in Sect. 4.2. All participants of task 3 also participated in task 2.

Team	C-index Task 2	Iso-resampling	CT clipping	Min-max norm.	Standardization	PET/CT fusion	Further cropping	Relies on Task 1	Additional segm.	No segmentation	Deep features	Large radiomics set	Volume, shape	IBSI compliant	Ensembling	Deep model	Algo. RF, SVM...	Feature selection	PET as input	CT as input	PET/CT fusion	Use clinical var.	Imputed missing	Cross-val.	Augmentation	C-index Task 3	GT masks	Task 1 masks	PET thresh. masks
BioMedIA [54]	0.7196				✓		✓	✓		✓	✓					✓								✓	✓	na			
Fuller MDA [50]	0.6938		✓	✓	✓			✓	✓						✓	✓			✓	✓		✓			✓	0.6978	✓	✓	
Qurit Tecvico [55]	0.6828			✓			✓					✓	✓		✓		✓	✓			✓	✓		✓		na			
BMIT_USYD [45]	0.6710		✓	✓	✓		✓	✓			✓				✓		✓			✓	✓		✓		✓	na			
DMLang [33]	0.6681		✓	✓			✓				✓				✓				✓	✓		✓		✓	na			
TECVICO_C. [16]	0.6608				✓	✓	✓				✓		✓		✓	✓				✓	✓		✓		na			
BAMF Health [47]	0.6602		✓	✓	✓			✓	✓		✓		✓		✓	✓	✓	✓			✓	✓	✓		0.6602			✓
ia-h-ai [57]	0.6592							✓		✓		✓		✓	✓	✓	✓	✓				✓			0.6592			✓
Neurophet [34]	0.6495				✓					✓					✓					✓	✓			na				
UMCG [41]	0.6445	✓	✓	✓	✓			✓	✓					✓	✓			✓	✓	✓	✓			✓	0.6373	✓	✓	
Aarhus Oslo [26]	0.6391																✓					✓		na				
RedNeucon [44]	0.6280				✓	✓				✓				✓	✓					✓	✓			na				
Emmanuelle B. [8]	0.6223	✓	✓		✓		✓			✓	✓		✓	✓	✓	✓	✓	✓			✓	✓	✓	✓	na			
BCIOQurit	0.6116 [66]	✓	✓		✓		✓			✓	✓		✓	✓	✓	✓	✓	✓				✓	0.4903	✓				
Vokyj [29]	0.5937	✓	✓		✓		✓						✓		✓							✓		✓	na			
Xuefeng [20]	0.5510	✓	✓		✓		✓						✓			✓								0.5089	✓			
DeepX [68]	0.5290	✓	✓		✓		✓				✓		✓			✓				✓	✓		✓		na			

In [50], Naser et al. (team "Fuller MDA") also adopted an approach based on deep learning. Clinical variables without missing values were transformed into an image matrix in order to be fed along with PET and CT images (rescaled and z-score normalized) as separate channels to a DenseNet121 CNN. Adopting a 10-fold cross-validation scheme, the model was trained either only with 2 channels (PET and CT) or 3 (adding the clinical), with data augmentation, for 800 iterations with a decreasing learning rate, the Adam optimizer and a negative log-likelihood loss. Of note, the PFS was discretized into 20 discrete intervals for the output of the network. Two different approaches of ensembling the various models obtained over the 10 folds (consensus or average) were implemented. The best result (0.694, rank 2) was obtained with the Image+Clinical consensus model. The team also participated in Task 3 where they used ground-truth masks as an additional input channel to the same network [59], achieving the first rank with a C-index of 0.70.

In [55], Paeenafrakati et al. (team "Qurit Tecvico") implemented a classical radiomics approach, where a large set of IBSI-compliant features were extracted with the SERA package [6] from the delineated tumor (based on the output of their solution for Task 1) in PET, CT as well as a fusion of PET/CT (of note, 10 different fusion techniques were explored). The features were then selected through 13 different dimensionality reduction techniques and 15 different selection methods and combined along with clinical variables, into several models with 5-fold cross-validation (the entire training set was used for each approach) through the use of 8 different survival prediction algorithms. The best performance (0.68) in the test set was obtained with an ensemble voting of these various algorithms, obtaining third rank in Task 2 (the team did not participate in Task 3).

In [45], Meng et al. (team "BMIT USYD") proposed a single unified framework to achieve both segmentation (Task 1) and outcome prediction (Task 2, no participation in Task 3). They first selected a few relevant clinical variables to take into account by performing a univariate/multivariate analysis, retaining only HPV status, performance status, and M stage. Their proposed model is composed of two main components: a U-Net based network for segmentation and a DenseNet based cascaded survival network. Both extract deep features that are fed into fully connected layers for outcome prediction and are trained in an end-to-end manner to minimize the combined loss of segmentation and survival prediction losses, with Adam optimizer, a batch size of 8 for 10000 iterations, with a decreasing learning rate. Clinical factors were concatenated in the non-activated fully connected layer. Of note, both the segmentation output and the cropped, normalized PET and CT images are fed to the DenseNet cascaded survival network. Data augmentation (random translations, rotations and flipping) was applied. Ten different models were trained, the first 5 through leave-one-center-out cross-validation and the next five with 5-fold cross-validation. The ensemble of these achieved a C-index of 0.671 in the test set.

In [33], Lang et al. (team "DMlang") relied on the segmentation output of Task 1 (or on the reference contours in training) to generate cropped bounding-boxes as inputs to their approach for predicting outcome, which relied on extracting deep features from PET and CT images thanks to a pre-trained C3D network designed to classify video clips. In order to feed PET and CT images to this C3D model, each 3 consecutive slices were fed to the color channels. The obtained PET and CT features were then concatenated and fed to a dense layer, which was then concatenated with clinical variables. Each output neuron represented the conditional probability of surviving a discrete time interval (the best model involved layers of size 512 and 256 and an output size of 15 corresponding to time intervals covering a maximum of 10 years of survival with the first 5 years split into intervals of half a year and all subsequent intervals with a width of one year). The same data augmentation as for the segmentation task was used. For training this network, a batch size of 16 was applied and 75 epochs were used with the Adam optimizer to minimize the negative log-likelihood. For model optimization, hyper-parameters were tuned manually. Of note, the team did not rely on ensemble of models nor on cross-validation, but generated a single stratified split of the training data. The trained model achieved a C-index of 0.668. The team did not participate in Task 3.

In [16], Fatan et al. (team "TECVICO Corp Family") used a similar PET/CT fusion approach (5 different techniques) and cropping as the team "Qurit_Tecvivo", extracted 215 IBSI-compliant radiomics features with the same package (SERA), that were fed into a number of feature selection techniques (7) and classifiers (5). They did not perform an ensemble of these but selected the best model in cross-validation during training. The best combination (LP-SR fusion and the classifier GlmBoost) obtained 0.66 in the test set. They did not participate in Task 3.

In [47], Murugesan et al. (team "BAMF Health") participated in both Tasks 2 and 3. Interestingly, their best results were obtained using the tumor masks by their segmentation method of Task 1, instead of the reference contours. Their solution was based on standard IBSI-compliant radiomics features extracted with Pyradiomics from PET and CT images after z-score normalization of intensities. In addition, in-house features calculating the number of uptakes and their volumes in each PET/CT were calculated through thresholding of PET SUVs. All clinical variables were exploited, missing values were imputed using the mean value of provided variables. Before further exploitation of the radiomics features, they were standardized using their mean and standard deviation. Then principal component analysis was applied to the features, capturing 95 of information. Variable importance combined with fast unified random forests for survival, regression, and classification was used for modeling through repeated random sub-sampling validation over 100 multiple random splits, in order to look for an optimal combination of features and to optimize hyper-parameters. The best result in the test set was obtained with a model relying on PCA components, with a 0.66 C-index (for both Tasks 2 and 3).

In [57], Starke et al. (team "ia-h-ai") built a strategy based on standard radiomics modeling, addressing both Tasks 2 and 3. They first strategically split the training data into 3 folds, ensuring that for each split, one of the centers is present in the validation set but not the training. Clinical factors were all considered, by imputing missing values through k-nearest neighbor ($k = 20$). They used either the provided reference volumes or alternative ones obtained through thresholding the PET intensities with SUV > 2.5. 172 IBSI-compliant hand-crafted features were then extracted from both PET and CT images volumes of interest using Pyradiomics. They established some baseline models through Cox proportional hazards models exploiting only the clinical variables, then moved to more advanced modeling relying on random survival forest, still using only clinical variables. In order to add image features to the models, they performed feature selection through three different processes: stability (L1-regularized Cox regression applied to multiple bootstrapped datasets for a range of regularization strength parameters), permutation-based feature importance and finally sequential feature selection. This allowed them to retain only a small number of features for the actual modeling step, where they compared different approaches using random forest survival (300 trees): fully automated feature selection and combination or different ways of manually selecting features, including a priori selection based on literature. They consistently obtained better performance on the test set by relying on the alternative volumes of interest (thresholded at SUV > 2.5, leading to volumes larger than the reference ground-truth contours), and models with hand-picked features, contrary to fully automatic selection that demonstrated overfitting.

In [34], Lee et al. (team "Neurophet") exploited only clinical variables (missing values were coded as 0 or −1 depending on the variable) and segmented volumes from Task 1 (i.e. only 1 feature, the tumor volume) to train a random forest survival model through 5-fold randomized cross-validation with 100 iterations. Of note, the center ID was added as a clinical factor. The proposed model achieved a C-index of 0.65 on the test set, with a higher performance than the same model without tumor volume (0.64). The team did not participate in Task 3.

In [41], Ma et al. (team "UMCG") proposed a pipeline based on deep learning as well, consisting of three parts: 1) the pyramid autoencoder of a 3D Resnet extracting image features from both CT and PET, 2) a feed-forward feature selection to remove the redundant image and clinical features, and 3) a Deep-Surv (a Cox deep network) for survival prediction. Clinical variables were used but missing values were not imputed, rather described as an additional class (i.e., unknown). PET and CT images were pre-processed and a new PET/CT image obtained by summation of PET and CT was used as a third input to the autoencoder. The segmentation masks were not used for Task 2, but were used for Task 3 in order to extract the tumor region in two different ways, both being used as inputs to the network. This pipeline was trained on using different splits of the training set (leave-one-center out and random selection of 179 patients for training and 45 for validations), resulting in 6-fold cross-validation. The Autoencoders were trained using the Adam optimizer with the initial learning rate 0.001 and data augmentation for 80 epochs. The official DeepSurv was trained for 5000 steps with the default settings. A total of 30 DeepSurv models were trained in each fold and the 3 models with the highest validation set C-index were selected. In total 18 models were obtained and their predicted risk scores are averaged to obtain the final result: 0.6445 and 0.6373 C-index in the test set for Task 2 and 3 respectively.

In [26], Ren et al. (team "Aarhus Oslo") team compared a conventional radiomics approach (although without tumor delineation, i.e., features were extracted from the whole bounding-box) and a deep learning approach in Task 2 only. Both used the provided bounding-box of PET and CT images as inputs, and in the case of the deep learning approach, an additional pre-processing step was applied to PET images in order to reduce the variability of images due to various centers based on a sin transform. For the standard radiomics approach, only clinical variables without missing values were exploited, whereas they were not used in the case of the deep learning approach. In the standard radiomics modeling, over 100 IBSI-compliant features were calculated but only a handful were manually selected based on literature and further used: one from CT and 4 from PET. These features (and clinical variables) were then fed to 2 ensemble models: random forest and gradient boosting. Hyper-parameters (number of trees, maximum depth for each tree, and learning rate, loss function tuning) were tuned using grid-search, and models were trained and evaluated using 5-fold cross-validation. In the case of deep learning, only CT and PET-sin images were used as input of a CNNs built with the encoder part of the SE Norm U-Net model [27] with three fully connected layers (4096, 512, and 1 units) added to the top. Five-fold cross-validation was also used. Each model was trained for 150 epochs using the Adam optimizer with a batch size of 4. The initial learning rate was set to 3e-6 and the loss was defined as a fusion of the Canberra distance loss and Huber loss ($\delta = 1$). Based on the results of cross-validation in training, the four following models were evaluated on the test set: Gradient boosting trained on either clinical factors (either all or only uncensored data) or both clinical factors and selected radiomics features and ensemble based on mean predicted values of five-fold deep learning models trained on FDG-PET/CT. All models had near-random performance in the test set, except the clinical-only model built with gradient boosting (0.66).

In [44], Asenjo et al. (team "RedNeucon") implemented a conventional radiomics approach based on the extraction of handcrafted features from PET and CT with a Matlab toolbox, from the reference contour volumes and the segmentation output of Task 1, as well as an additional volume of interest generated by determining a two pixel inward and outward the contours to get a tumor "boundary region". Only clinical variables without missing values were used. Features were then selected after ranking according to 2 methods, ranking for classification using a Fisher F-Test and an algorithm based on K-nearest neighbors. When two features showed a correlation above 0.5, the best one was kept. Three different modeling algorithms were compared in 5-fold cross-validation: Gaussian Process Regression (GPR), an Ensembled Bagged of trees and a Support Vector Machine. The best result on the test set (0.628) was obtained with the GPR with 35 features.

In [8], Bourigault et al. (team "Emmanuelle Bourigault") proposed a Cox proportional hazard regression model using a combination of clinical, radiomic, and deep learning features from PET/CT images. All clinical variables were exploited, after imputing missing values using a function of available ones. IBSI-compliant handcrafted radiomics features including wavelet-filtered ones were calculated using Pyradiomics and were combined with deep features from the 3D U-Net used in the segmentation Task 1, in addition to clinical variables. Spearman rank correlation above 0.8 was used to eliminate intercorrelated features. Feature selection was performed using Lasso regression with 5-fold cross-validation, reducing the set of 270 variables to 70 (7 clinical, 14 radiomics and 49 deep). Three different models were implemented for modeling: Cox proportional hazard regression model, random survival forest and Deepsurv (a Cox proportional hazards deep neural network). All three models were trained with different combinations of the selected clinical, radiomics (PET, CT or PET/CT) and deep features. The best performance in validation was obtained with the Cox model using clinical + CT radiomics + deep learning features, although in the test set its final performance was 0.62.

In [66], Yousefirizi et al. (team "BCIOqurit") proposed training a proportional hazard Cox model with a multilayer perceptron neural net backbone to predict the score for each patient. This Cox model was trained on a number of PET and CT radiomics features extracted from the segmented lesions, patient demographics, and encoder features provided from the penultimate layer of a multi-input 2D PET/CT convolutional neural network tasked with predicting time-to-event for each lesion. A grid search over several feature selection and classifiers methods identified 192 unique combinations of radiomics features that were used to train the overall Cox model with the Adam optimizer, a learning rate of 0.0024, a batch size of 32, and an early stopping method that monitored the validation loss. A 10-fold cross-validation scheme was used and an ensemble model of these achieved a C-index score of 0.612 in the test set.

In [29], Muller et al. (team "Vokyj") proposed to fit a Weibull accelerated failure time model with clinical factors and the shape descriptors of the segmented tumor (output of Task 1). M-stage and two shape features (Euler number and Surface Area) were the most predictive of PFS, the model achieving a performance of 0.59 in the test set. The team did not participate in Task 3.

In [20], Ghimire et al. (team "Xuefeng") implemented a straightforward approach that consisted in calculating the tumor volume and tumor surface area of the Task 1 segmentation outputs, as well as the classification output from the segmentation network trained to classify the input images into 6 different classes of PFS (which was first discretized into 6 bins). These imaging features were then combined with all available clinical factors, for which missing values were imputed with the median value for numerical variables and mode value for categorical ones. All features were then normalized to zero mean and 1 standard deviation for a linear model to be fitted to the training data. The model was applied to both Tasks 2 and 3, using the reference contours instead of the Task 1 segmentation results, leading to C-index values of 0.43 and 0.51 respectively.

In [68], Yuan et al. (team "DeepX") implemented a standard radiomics approach, extracting more than 200 IBSI-compliant handcrafted features with Pyradiomics, from both PET and CT images using the segmentation output of Task 1, which were then manually ranked and selected according to their concordance index. Regarding clinical variables, only age was used. The 7 selected features were evaluated independently or combined through averaging concordance ranking, obtaining their best C-index of 0.53 in the test set.

5   Discussion: Putting the Results into Context

Outcomes and findings of participating methods are summarized in Sect. 5.1 for all three tasks. In general, we observed that simplicity was beneficial for generalization and that sophisticated methods tend to overfit the training/validation Despite the diversity in terms of centers and image acquisition, no specific feature or image harmonization method was employed, which could be one avenue for improving generalization abilities of the methods in all tasks [7].

The combined scope of the three proposed tasks also allowed the emergence of very interesting findings concerning the relationship of the GTVt contouring task and PFS prediction. In a nutshell, ground truth ROIs were not providing top results, even though they were re-annotated in a centralized fashion to be dedicated for radiomics [17]. Simple PET thresholded and bounding-boxes for deep learning outperformed the use of ground truth ROI. This suggests that algorithms looking elsewhere than the GTVt is beneficial (e.g. tumoral environment, nodal metastases). Fully automatic algorithms are expected to provide optimal results, which was already highlighted by several papers in the context of the HECKTOR challenge [2,18,47,50,57]. This potentially obviates the need for GTVt contouring, opening avenues for reproducible and large scale radiomics studies including thousands potential subjects.

5.1   Outcomes and Findings

A major benefit of this challenge is to compare various algorithms developed by teams from all around the world on the same dataset and task, with held-out test data.

We distinguish here between the technical and biomedical impact. The main technical impact of the challenge is the comparison of state-of-the-art algorithms on the provided data. We identified key elements for addressing the task: 3D U-Net, preprocessing, normalization, data augmentation and ensembling, as summarized in Tables 3 and 4. The main biomedical impact of the results is the opportunity to generate large cohorts with automatic tumor segmentation for comprehensive radiomics studies, as well as to define and further push state of the art performance.

Task 1: Automatic Segmentation of the GTVt
The best methods obtain excellent results with DSCs above 0.75, better than inter-observer variability (DSC 0.61) performed on a subset of our data and similar variability reported in the literature (DSCs of 0.57 and 0.69 on CT and PET/CT respectively) [21]. Note that without injected contrast CT, delineating the exact contour of the tumor is very difficult. Thus, the inter-observer DSC could be low only due to disagreements at the border of the tumor, without taking into account the error rate due to the segmentation of non-malignant structures (if any). For that reason, defining the task as solved solely based on the DSC is not sufficient. In the context of this challenge, we can therefore define the task as solved if the algorithms follow these three criteria:

1. Higher or similar DSC than inter-observers agreement.
2. Detect all the primary tumors in the oropharynx region (i.e. segmentation not evaluated at the pixel level, rather at the occurrence level).
3. Similarly, detect only the primary tumors in the oropharynx region (discarding lymph nodes and other potentially false positives).

According to these criteria, the task is partially solved. The first criterion, evaluating the segmentation at the pixel level, is fulfilled. At the occurrence level (criteria 2 and 3), however, even the algorithms with the highest DSC output FP and FN regions. Besides, there is still a lot of work to do on highly related tasks, including the segmentation of lymph nodes, the development of super-annotator ground truth as well as the agreement of multiple annotators, and, finally, the prediction of patient outcome following the tumor segmentation.

Similarly to last year's challenge, we identified the same key elements that cause the algorithms to fail in poorly segmented cases. These elements are as follows; low FDG uptake on PET, primary tumor that looks like a lymph node, abnormal uptake in the tongue and tumor present at the border of the oropharynx region. Some examples are illustrated in Fig. 1.

Tasks 2 and 3: Predicting PFS

The challengers relied on a variety of approaches and tackled the task quite differently (Table 4). A few teams relied on deep learning exclusively, whereas others exploited more classical radiomics pipelines. Some teams also implemented various combinations of both. PET and CT images were also exploited in several different ways. Either as separate inputs or through various fusion techniques, for either deep learning or classical radiomics analysis. Interestingly, despite the recent rise of interest in the development of methods dedicated to the harmonization of multicentric data, either in the image domain through image processing or deep learning based image synthesis [12] or in the features domain through batch-harmonization techniques such as ComBat [13], none of the teams implemented specific multicentric harmonization techniques, beyond usual approaches to take into account the diversity of the images in the training and testing sets by relying on, for example, leave-one-center-out cross-validation and image intensities rescaling or z-score normalization. The use of clinical variables was also the opportunity for challengers to deploy different approaches. Among the methods using deep networks, some encoded the clinical information into images to feed them as input to the deep networks, whereas others integrated them as vectors concatenated in other layers. Some teams elected to rely only on clinical factors without missing values, whereas others implemented some way of imputing missing values in order to exploit all available variables. In addition, some teams pre-selected only a subset of the clinical variables with prior knowledge. Interestingly, some challengers obtained their best performance by building models relying only on clinical variables. Finally, most teams who participated in Task 1 relied on their segmentation output in Tasks 2 and 3, however, a few explored additional or alternative volumes of interest. Interestingly for Task 3, some challengers obtained better results using alternative segmentation or Task 1 outputs instead of the provided reference contours.

Predicting PFS was the Objective of Both Tasks 2 and 3

The only difference was that the GTVt ROI was provided for Task 3, but not for Task 2. One surprising trend showed that the predictive performance was found to be slightly higher when the GTVt ROI was not used (Task 2), which could be the result of the following. First, fewer teams participated in Task 3, which can be partially explained by the requirement to submit a Docker container instead of direct prediction of hazard scores. Second, for deep learning-based radiomics, using input ROIs is less straightforward than handcrafted radiomics, which makes input contour less relevant. Nevertheless, Starke et al. [57] used a classical radiomics pipeline and observed that ROIs based on a simple PET-based thresholding approach systematically outperformed a model based on features extracted from the provided GTVt. This suggests that prognostically relevant information is contained not only in the primary tumor area, but also in other (metabolically active) parts such as the lymph nodes. Similar results have been obtained recently in different tumor localizations. For instance, it was shown in uterine cancer that radiomics features extracted from the entire uterus organ in MRI rather than the tumor only led to more accurate models [62]. In cervical cancer patients, specific SUV thresholds in PET images led to more accurate metrics [35], even though this threshold might not be the more accurate to delineate the metabolic uptake tumor volume. Finally, a study in non-small cell lung cancer recently showed that radiomics features extracted from a large volume of interest containing the primary tumor and the surrounding healthy tissues in PET/CT images could be used to train models as accurate as those trained on features extracted from the delineate tumor, provided a consensus of several machine learning algorithms is used for the prediction [56].

5.2   Limitations of the Challenge

The dataset provided in this challenge suffers from several limitations. First, the contours were mainly drawn based on the PET/CT fusion which is not sufficient to clearly delineate the tumor. Other methods such as MRI with gadolinium or contrast CT are the gold standard to obtain the true contours for radiation oncology. Since the target clinical application is radiomics, however, the precision of the contours is not as important as for radiotherapy planning.

Another limitation comes from the definition of the task, given that only one segmentation was drawn on the fusion of PET and CT. For radiomics analysis, it could be beneficial to consider one segmentation per modality since the PET signal is often not contained in the fusion-based segmentation due to the poor spatial resolution of this modality.

6   Conclusions

This paper presented a general overview of the HECKTOR challenge including the data, the participation, main results and discussions. The proposed tasks were the segmentation of the primary tumor in oropharyngeal cancer as well as the PFS prediction. The participation was high, with 20, 17, and 6 eligible teams for tasks 1, 2, and 3, respectively. The participation doubled compared to the previous edition, which shows the growing interest in automatic lesion segmentation for H&N cancer.

The task proposed this year was to segment the primary tumor in PET/CT images. This task is not as simple as thresholding the PET image since we target only the primary tumor and the region covered by high PET activation is often too large, going beyond the limits of the tumor tissues. Deep learning methods based on U-Net models were mostly used in the challenge. Interesting ideas were implemented to combine PET and CT complementary information. Model ensembling, as well as data preprocessing and augmentation, seem to have played an important role in achieving top-ranking results.

Acknowledgments. The organizers thank all the teams for their participation and valuable work. This challenge and the winner prizes were sponsored by Siemens Healthineers Switzerland, Bioemtech Greece and Aquilab France (500€ each, for Task 1, 2 and 3 respectively). The software used to centralise the quality control of the GTVt regions was MIM (MIM software Inc., Cleveland, OH), which kindly supported the challenge via free licences. This work was also partially supported by the Swiss National Science Foundation (SNSF, grant 205320_179069) and the Swiss Personalized Health Network (SPHN, via the IMAGINE and QA4IQI projects).

Appendix 1: Challenge Information

In this appendix, we list important information about the challenge as suggested in the BIAS guidelines [43].

Challenge Name
HEad and neCK TumOR segmentation and outcome prediction challenge (HECKTOR) 2021

Organizing Team
(Authors of this paper) Vincent Andrearczyk, Valentin Oreiller, Sarah Boughdad, Catherine Cheze Le Rest, Hesham Elhalawani, Mario Jreige, John O. Prior, Martin Vallières, Dimitris Visvikis, Mathieu Hatt and Adrien Depeursinge

Life Cycle Type
A fixed submission deadline was set for the challenge results.

Challenge Venue and Platform
The challenge is associated with MICCAI 2021. Information on the challenge is available on the website, together with the link to download the data, the submission platform and the leaderboard[10].

Participation Policies

(a) Task 1: Algorithms producing fully-automatic segmentation of the test cases were allowed. Task 2 and 3: Algorithms producing fully-automatic PFS risk score prediction of the test cases were allowed.

(b) The data used to train algorithms was not restricted. If using external data (private or public), participants were asked to also report results using only the HECKTOR data.

(c) Members of the organizers' institutes could participate in the challenge but were not eligible for awards.

(d) Task 1: The award was 500 euros, sponsored by Siemens Healthineers Switzerland. Task 2: The award was 500 euros, sponsored by Aquilab. Task 3: The award was 500 euros, sponsored by Bioemtech.

(e) Policy for results announcement: The results were made available on the AIcrowd leaderboard and the best three results of each task were announced publicly. Once participants submitted their results on the test set to the challenge organizers via the challenge website, they were considered fully vested

---

[10] www.aicrowd.com/challenges/hecktor.

in the challenge, so that their performance results (without identifying the participant unless permission was granted) became part of any presentations, publications, or subsequent analyses derived from the challenge at the discretion of the organizers.

(f) Publication policy: This overview paper was written by the organizing team's members. The participating teams were encouraged to submit a paper describing their method. The participants can publish their results separately elsewhere when citing the overview paper, and (if so) no embargo will be applied.

Submission Method

Submission instructions are available on the website[11] and are reported in the following. Task 1: Results should be provided as a single binary mask per patient (1 in the predicted GTVt) in .nii.gz format. The resolution of this mask should be the same as the original CT resolution and the volume cropped using the provided bounding-boxes. The participants should pay attention to saving NIfTI volumes with the correct pixel spacing and origin with respect to the original reference frame. The NIfTI files should be named [PatientID].nii.gz, matching the patient names, e.g. CHUV001.nii.gz and placed in a folder. This folder should be zipped before submission. If results are submitted without cropping and/or resampling, we will employ nearest neighbor interpolation given that the coordinate system is provided.

Task 2: Results should be submitted as a CSV file containing the patient ID as "PatientID" and the output of the model (continuous) as "Prediction". An individual output should be anti-concordant with the PFS in days (i.e., the model should output a predicted risk score).

Task 3: For this task, the developed methods will be evaluated on the testing set by the organizers by running them within a docker provided by the challengers. Practically, your method should process one patient at a time. It should take 3 nifty files as inputs (file 1: the PET image, file 2: the CT image, file 3: the provided ground-trugh segmentation mask, all 3 files have the same dimensions, the ground-truth mask contains only 2 values: 0 for the background, 1 for the tumor), and should output the predicted risk score produced by your model.

Participants were allowed five valid submissions per task. The best result was reported for each task/team. For a team submitting multiple runs to task one, the best result was determined as the highest ranking result within these runs (see ranking description in Sect. 3.1).

Challenge Schedule

The schedule of the challenge, including modifications, is reported in the following.

– the release date of the training cases: ~~June 01~~ June 04 2021
– the release date of the test cases: ~~Aug. 01~~ Aug. 06 2021

---

[11] https://www.aicrowd.com/challenges/miccai-2021-hecktor#results-submission-format.

- the submission date(s): opens Sept. 01 2021 closes Sept. 10 Sept. 14 2021 (23:59 UTC-10)
- paper abstract submission deadline: Sept. 15 2021 (23:59 UTC-10)
- full paper submission deadline: Sept. 17 2021 (23:59 UTC-10)
- the release date of the ranking: ~~Sept. 17 2021~~ Sept. 27 2021
- associated workshop days: Sept. 27 2021

Ethics Approval

Montreal: CHUM, CHUS, HGJ, HMR data (training): The ethics approval was granted by the Research Ethics Committee of McGill University Health Center (Protocol Number: MM-JGH-CR15-50).

Lausanne: CHUV data (testing): The ethics approval was obtained from the Commission cantonale (VD) d'éthique de la recherche sur l'être humain (CER-VD) with protocol number: 2018-01513.

Poitiers: CHUP data (partly training and testing): The fully anonymized data originates from patients who consent to the use of their data for research purposes.

Data Usage Agreement

The participants had to fill out and sign an end-user-agreement in order to be granted access to the data. The form can be found under the Resources tab of the HECKTOR website.

Code Availability

The evaluation software was made available on our github page[12]. The participating teams decided whether they wanted to disclose their code (they were encouraged to do so).

Conflict of Interest

No conflict of interest applies. Fundings are specified in the acknowledgments. Only the organizers had access to the test cases' ground truth contours.

Author contributions

Vincent Andrearczyk:

Design of the tasks and of the challenge, writing of the proposal, development of baseline algorithms, development of the AIcrowd website, writing of the overview paper, organization of the challenge event, organization of the submission and reviewing process of the participants' papers.

Valentin Oreiller:

Design of the tasks and of the challenge, writing of the proposal, development of the AIcrowd website, development of the evaluation code, writing of the overview paper, organization of the challenge event, organization of the submission and reviewing process of the papers.

---

[12] github.com/voreille/hecktor.

Sarah Boughdad:
Design of the tasks and of the challenge, annotations.

Catherine Cheze Le Rest:
Design of the tasks and of the challenge, annotations.

Hesham Elhalawani:
Design of the tasks and of the challenge, annotations.

Mario Jreige:
Design of the tasks and of the challenge, quality control/annotations, annotations, revision of the paper and accepted the last version of the submitted paper.

John O. Prior:
Design of the tasks and of the challenge, revision of the paper and accepted the last version of the submitted paper.

Martin Vallières:
Design of the tasks and of the challenge, provided the initial data and annotations for the training set [58], revision of the paper and accepted the last version of the submitted paper.

Dimitris Visvikis:
Design of the task and challenge.

Mathieu Hatt:
Design of the tasks and of the challenge, writing of the proposal, writing of the overview paper, organization of the challenge event.

Adrien Depeursinge:
Design of the tasks and of the challenge, writing of the proposal, writing of the overview paper, organization of the challenge event.

Appendix 2:   Image Acquisition Details

HGJ: For the PET portion of the FDG-PET/CT scan, a median of 584 MBq (range: 368–715) was injected intravenously. After a 90-min uptake period of rest, patients were imaged with the PET/CT imaging system. Imaging acquisition of the head and neck was performed using multiple bed positions with a median of 300 s (range: 180–420) per bed position. Attenuation corrected images were reconstructed using an ordered subset expectation maximization (OSEM) iterative algorithm and a span (axial mash) of 5. The FDG-PET slice thickness resolution was 3.27 mm for all patients and the median in-plane resolution was 3.52×3.52 mm^2 (range: 3.52–4.69). For the CT portion of the FDG-PET/CT scan, an energy of 140 kVp with an exposure of 12 mAs was used. The CT slice thickness resolution was 3.75 mm and the median in-plane resolution was 0.98 $\times$ 0.98 mm^2 for all patients.

CHUS: For the PET portion of the FDG-PET/CT scan, a median of 325 MBq (range: 165–517) was injected intravenously. After a 90-min uptake period of rest, patients were imaged with the PET/CT imaging system. Imaging acquisition of the head and neck was performed using multiple bed positions with a median of 150 s (range: 120–151) per bed position. Attenuation corrected images were reconstructed using a LOR-RAMLA iterative algorithm. The FDG-PET slice thickness resolution was 4 mm and the median in-plane resolution was 4×4 mm^2 for all patients. For the CT portion of the FDG-PET/CT scan, a median energy of 140 kVp (range: 12–140) with a median exposure of 210 mAs (range: 43–250) was used. The median CT slice thickness resolution was 3 mm (range: 2–5) and the median in-plane resolution was 1.17×1.17 mm^2 (range: 0.68–1.17).

HMR: For the PET portion of the FDG-PET/CT scan, a median of 475 MBq (range: 227–859) was injected intravenously. After a 90-min uptake period of rest, patients were imaged with the PET/CT imaging system. Imaging acquisition of the head and neck was performed using multiple bed positions with a median of 360 s (range: 120–360) per bed position. Attenuation corrected images were reconstructed using an ordered subset expectation maximization (OSEM) iterative algorithm and a median span (axial mash) of 5 (range: 3–5). The FDG-PET slice thickness resolution was 3.27 mm for all patients and the median in-plane resolution was 3.52×3.52 mm^2 (range: 3.52–5.47). For the CT portion of the FDG-PET/CT scan, a median energy of 140 kVp (range: 120–140) with a median exposure of 11 mAs (range: 5–16) was used. The CT slice thickness resolution was 3.75 mm for all patients and the median in-plane resolution was 0.98×0.98 mm^2 (range: 0.98–1.37).

CHUM: For the PET portion of the FDG-PET/CT scan, a median of 315 MBq (range: 199–3182) was injected intravenously. After a 90-min uptake period of rest, patients were imaged with the PET/CT imaging system. Imaging acquisition of the head and neck was performed using multiple bed positions with a median of 300 s (range: 120–420) per bed position. Attenuation corrected images were reconstructed using an ordered subset expectation maximization (OSEM) iterative algorithm and a median span (axial mash) of 3 (range: 3–5). The median FDG-PET slice thickness resolution was 4 mm (range: 3.27–4) and the median in-plane resolution was 4×4 mm^2 (range: 3.52–5.47). For the CT portion of the FDG-PET/CT scan, a median energy of 120 kVp (range: 120–140) with a median exposure of 350 mAs (range: 5–350) was used. The median CT slice thickness resolution was 1.5 mm (range: 1.5–3.75) and the median in-plane resolution was 0.98×0.98 mm^2 (range: 0.98–1.37).

CHUV: The patients fasted at least 4 h before the injection of 4 Mbq/kg of (18F)-FDG (Flucis). Blood glucose levels were checked before the injection of (18F)-FDG. If not contra-indicated, intravenous contrast agents were administered before CT scanning. After a 60-min uptake period of rest, patients were imaged with the PET/CT imaging system. First, a CT (120 kV, 80 mA, 0.8-s rotation time, slice thickness 3.75 mm) was performed from the base of the skull to the mid-thigh. PET scanning was performed immediately after acquisition of the CT. Images were acquired from the base of the skull to the mid-thigh (3 min/bed position). PET images were reconstructed by using an ordered-subset expectation maximization iterative reconstruction (OSEM) (two iterations, 28 subsets) and an iterative fully 3D (DiscoveryST). CT data were used for attenuation calculation.

CHUP: PET/CT acquisition began after 6 h of fasting and 60 ± 5 min after injection of 3 MBq/kg of 18F-FDG (421 ± 98 MBq, range 220–695 MBq). Non-contrast-enhanced, non-respiratory gated (free breathing) CT images were acquired for attenuation correction (120 kVp, Care Dose® current modulation system) with an in-plane resolution of $0.853 \times 0.853 \, mm^2$ and a 5 mm slice thickness. PET data were acquired using 2.5 min per bed position routine protocol and images were reconstructed using a CT-based attenuation correction and the OSEM-TrueX-TOF algorithm (with time-of-flight and spatial resolution modeling, 3 iterations and 21 subsets, 5 mm 3D Gaussian post-filtering, voxel size $4 \times 4 \times 4 \, mm^3$).

References

1. An, C., Chen, H., Wang, L.: A coarse-to-fine framework for head and neck tumor segmentation in CT and PET images. In: Andrearczyk, V., Oreiller, V., Hatt, M., Depeursinge, A. (eds.) HECKTOR 2021. LNCS, vol. 13209, pp. 50–57. Springer, Cham (2022)
2. Andrearczyk, V., et al.: Multi-task deep segmentation and radiomics for automatic prognosis in head and neck cancer. In: Rekik, I., Adeli, E., Park, S.H., Schnabel, J. (eds.) PRIME 2021. LNCS, vol. 12928, pp. 147–156. Springer, Cham (2021). https://doi.org/10.1007/978-3-030-87602-9_14
3. Andrearczyk, V., Oreiller, V., Depeursinge, A.: Oropharynx detection in PET-CT for tumor segmentation. In: Irish Machine Vision and Image Processing (2020)
4. Andrearczyk, V., et al.: Overview of the HECKTOR challenge at MICCAI 2020: automatic head and neck tumor segmentation in PET/CT. In: Andrearczyk, V., Oreiller, V., Depeursinge, A. (eds.) HECKTOR 2020. LNCS, vol. 12603, pp. 1–21. Springer, Cham (2021). https://doi.org/10.1007/978-3-030-67194-5_1
5. Andrearczyk, V., et al.: Automatic segmentation of head and neck tumors and nodal metastases in PET-CT scans. In: International Conference on Medical Imaging with Deep Learning (MIDL) (2020)
6. Ashrafinia, S.: Quantitative nuclear medicine imaging using advanced image reconstruction and radiomics. Ph.D. thesis, The Johns Hopkins University (2019)
7. Atul Mali, S., et al.: Making radiomics more reproducible across scanner and imaging protocol variations: a review of harmonization methods. J. Pers. Med. 11(9), 842 (2021)
8. Bourigault, E., McGowan, D.R., Mehranian, A., Papiez, B.W.: Multimodal PET/CT tumour segmentation and prediction of progression-free survival using a full-scale UNet with attention. In: Andrearczyk, V., Oreiller, V., Hatt, M., Depeursinge, A. (eds.) HECKTOR 2021. LNCS, vol. 13209, pp. 189–201. Springer, Cham (2022)
9. Castelli, J., et al.: PET-based prognostic survival model after radiotherapy for head and neck cancer. Eur. J. Nucl. Med. Mol. Imaging 46(3), 638–649 (2018). https://doi.org/10.1007/s00259-018-4134-9

10. Chen, L., Papandreou, G., Kokkinos, I., Murphy, K., Yuille, A.L.: DeepLab: semantic image segmentation with deep convolutional nets, atrous convolution, and fully connected CRFs. CoRR abs/1606.00915 (2016)
11. Cho, M., Choi, Y., Hwang, D., Yie, S.Y., Kim, H., Lee, J.S.: Multimodal spatial attention network for automatic head and neck tumor segmentation in FDG-PET and CT images. In: Andrearczyk, V., Oreiller, V., Hatt, M., Depeursinge, A. (eds.) HECKTOR 2021. LNCS, vol. 13209, pp. 75–82. Springer, Cham (2022)
12. Choe, J., et al.: Deep learning-based image conversion of CT reconstruction kernels improves radiomics reproducibility for pulmonary nodule. Radiology **292**(2), 365–373 (2019)
13. Da-ano, R., et al.: Performance comparison of modified ComBat for harmonization of radiomic features for multicentric studies. Sci. Rep. **10**(1), 102488 (2020)
14. Davidson-Pilon, C.: lifelines: survival analysis in Python. J. Open Source Softw. **4**(40), 1317 (2019)
15. De Biase, A., et al.: Skip-SCSE multi-scale attention and co-learning method for oropharyngeal tumor segmentation on multi-modal PET-CT images. In: Andrearczyk, V., Oreiller, V., Hatt, M., Depeursinge, A. (eds.) HECKTOR 2021. LNCS, vol. 13209, pp. 109–120. Springer, Cham (2022)
16. Fatan, M., Hosseinzadeh, M., Askari, D., Sheykhi, H., Rezaeijo, S.M., Salmanpoor, M.R.: Fusion-based head and neck tumor segmentation and survival prediction using robust deep learning techniques and advanced hybrid machine learning systems. In: Andrearczyk, V., Oreiller, V., Hatt, M., Depeursinge, A. (eds.) HECKTOR 2021. LNCS, vol. 13209, pp. 211–223. Springer, Cham (2022)
17. Fontaine, P., et al.: Cleaning radiotherapy contours for radiomics studies, is it worth it? A head and neck cancer study. Clin. Transl. Radiat. Oncol. **33**, 153–158 (2022)
18. Fontaine, P., et al.: Fully automatic head and neck cancer prognosis prediction in PET/CT. In: Syeda-Mahmood, T., et al. (eds.) ML-CDS 2021. LNCS, vol. 13050, pp. 59–68. Springer, Cham (2021). https://doi.org/10.1007/978-3-030-89847-2_6
19. Foster, B., Bagci, U., Mansoor, A., Xu, Z., Mollura, D.J.: A review on segmentation of positron emission tomography images. Comput. Biol. Med. **50**, 76–96 (2014)
20. Ghimire, K., Chen, Q., Feng, X.: Head and neck tumor segmentation with deeply-supervised 3D UNet and progression-free survival prediction with linear model. In: Andrearczyk, V., Oreiller, V., Hatt, M., Depeursinge, A. (eds.) HECKTOR 2021. LNCS, vol. 13209, pp. 141–149. Springer, Cham (2022)
21. Gudi, S., et al.: Interobserver variability in the delineation of gross tumour volume and specified organs-at-risk during IMRT for head and neck cancers and the impact of FDG-PET/CT on such variability at the primary site. J. Med. Imaging Radiat. Sci. **48**(2), 184–192 (2017)
22. Harrell, F.E., Califf, R.M., Pryor, D.B., Lee, K.L., Rosati, R.A.: Evaluating the yield of medical tests. JAMA **247**(18), 2543–2546 (1982)
23. Hatt, M., et al.: The first MICCAI challenge on PET tumor segmentation. Med. Image Anal. **44**, 177–195 (2018)
24. Hatt, M., Le Rest, C.C., Turzo, A., Roux, C., Visvikis, D.: A fuzzy locally adaptive Bayesian segmentation approach for volume determination in PET. IEEE Trans. Med. Imaging **28**(6), 881–893 (2009)
25. Hatt, M., et al.: Classification and evaluation strategies of auto-segmentation approaches for PET: report of AAPM task group No. 211. Med. Phys. **44**(6), e1–e42 (2017)

26. Huynh, B.N., Ren, J., Groendahl, A.R., Tomic, O., Korreman, S.S., Futsaether, C.M.: Comparing deep learning and conventional machine learning for outcome prediction of head and neck cancer in PET/CT. In: Andrearczyk, V., Oreiller, V., Hatt, M., Depeursinge, A. (eds.) HECKTOR 2021. LNCS, vol. 13209, pp. 318–326. Springer, Cham (2022)

27. Iantsen, A., Visvikis, D., Hatt, M.: Squeeze-and-excitation normalization for automated delineation of head and neck primary tumors in combined PET and CT images. In: Andrearczyk, V., Oreiller, V., Depeursinge, A. (eds.) HECKTOR 2020. LNCS, vol. 12603, pp. 37–43. Springer, Cham (2021). https://doi.org/10.1007/978-3-030-67194-5_4

28. Isensee, F., Jaeger, P.F., Kohl, S.A., Petersen, J., Maier-Hein, K.H.: nnU-Net: a self-configuring method for deep learning-based biomedical image segmentation. Nat. Methods **18**(2), 203–211 (2021)

29. Juanco-Müller, Á.V., Mota, J.F.C., Goatman, K., Hoogendoorn, C.: Deep supervoxel segmentation for survival analysis in head and neck cancer patients. In: Andrearczyk, V., Oreiller, V., Hatt, M., Depeursinge, A. (eds.) HECKTOR 2021. LNCS, vol. 13209, pp. 257–265. Springer, Cham (2022)

30. Kim, B., Ye, J.C.: Mumford-Shah loss functional for image segmentation with deep learning. IEEE Trans. Image Process. **29**, 1856–1866 (2019)

31. Kuijf, H.J., et al.: Standardized assessment of automatic segmentation of white matter hyperintensities and results of the WMH segmentation challenge. IEEE Trans. Med. Imaging **38**(11), 2556–2568 (2019)

32. Kumar, A., Fulham, M., Feng, D., Kim, J.: Co-learning feature fusion maps from PET-CT images of lung cancer. IEEE Trans. Med. Imaging **39**, 204–217 (2019)

33. Lang, D.M., Peeken, J.C., Combs, S.E., Wilkens, J.J., Bartzsch, S.: Deep learning based GTV delineation and progression free survival risk score prediction for head and neck cancer patients. In: Andrearczyk, V., Oreiller, V., Hatt, M., Depeursinge, A. (eds.) HECKTOR 2021. LNCS, vol. 13209, pp. 150–159. Springer, Cham (2022)

34. Lee, J., Kang, J., Shin, E.Y., Kim, R.E.Y., Lee, M.: Dual-path connected CNN for tumor segmentation of combined PET-CT images and application to survival risk prediction. In: Andrearczyk, V., Oreiller, V., Hatt, M., Depeursinge, A. (eds.) HECKTOR 2021. LNCS, vol. 13209, pp. 248–256. Springer, Cham (2022)

35. Leseur, J., et al.: Pre- and per-treatment 18F-FDG PET/CT parameters to predict recurrence and survival in cervical cancer. Radiother. Oncol. J. Eur. Soc. Ther. Radiol. Oncol. **120**(3), 512–518 (2016)

36. Li, L., Zhao, X., Lu, W., Tan, S.: Deep learning for variational multimodality tumor segmentation in PET/CT. Neurocomputing **392**, 277–295 (2019)

37. Lin, T.Y., Goyal, P., Girshick, R., He, K., Dollár, P.: Focal loss for dense object detection. In: Proceedings of the IEEE International Conference on Computer Vision, pp. 2980–2988 (2017)

38. Liu, T., Su, Y., Zhang, J., Wei, T., Xiao, Z.: 3D U-net applied to simple attention module for head and neck tumor segmentation in PET and CT images. In: Andrearczyk, V., Oreiller, V., Hatt, M., Depeursinge, A. (eds.) HECKTOR 2021. LNCS, vol. 13209, pp. 50–57. Springer, Cham (2022)

39. Liu, Z., et al.: Automatic segmentation of clinical target volume used for post-modified radical mastectomy radiotherapy with a convolutional neural network. Front. Oncol. **10**, 3268 (2020)

40. Lu, J., Lei, W., Gu, R., Wang, G.: Priori and posteriori attention for generalizing head and neck tumors segmentation. In: Andrearczyk, V., Oreiller, V., Hatt, M., Depeursinge, A. (eds.) HECKTOR 2021. LNCS, vol. 13209, pp. 134–140. Springer, Cham (2022)

41. Ma, B., et al.: Self-supervised multi-modality image feature extraction for the progression free survival prediction in head and neck cancer. In: Andrearczyk, V., Oreiller, V., Hatt, M., Depeursinge, A. (eds.) HECKTOR 2021. LNCS, vol. 13209, pp. 202–210. Springer, Cham (2022)
42. Maier-Hein, L., et al.: Why rankings of biomedical image analysis competitions should be interpreted with care. Nat. Commun. **9**(1), 1–13 (2018)
43. Maier-Hein, L., et al.: BIAS: transparent reporting of biomedical image analysis challenges. Med. Image Anal. **66**, 101796 (2020)
44. Martinez-Larraz, A., Asenjo, J.M., Rodríguez, B.A.: PET/CT head and neck tumor segmentation and progression free survival prediction using deep and machine learning techniques. In: Andrearczyk, V., Oreiller, V., Hatt, M., Depeursinge, A. (eds.) HECKTOR 2021. LNCS, vol. 13209, pp. 168–178. Springer, Cham (2022)
45. Meng, M., Peng, Y., Bi, L., Kim, J.: Multi-task deep learning for joint tumor segmentation and outcome prediction in head and neck cancer. In: Andrearczyk, V., Oreiller, V., Hatt, M., Depeursinge, A. (eds.) HECKTOR 2021. LNCS, vol. 13209, pp. 160–167. Springer, Cham (2022)
46. Moe, Y.M., et al.: Deep learning for automatic tumour segmentation in PET/CT images of patients with head and neck cancers. In: Medical Imaging with Deep Learning (2019)
47. Murugesan, G.K., et al.: Head and neck primary tumor segmentation using deep neural networks and adaptive ensembling. In: Andrearczyk, V., Oreiller, V., Hatt, M., Depeursinge, A. (eds.) HECKTOR 2021. LNCS, vol. 13209, pp. 224–235. Springer, Cham (2022)
48. Myronenko, A.: 3D MRI brain tumor segmentation using autoencoder regularization. In: Crimi, A., Bakas, S., Kuijf, H., Keyvan, F., Reyes, M., van Walsum, T. (eds.) BrainLes 2018. LNCS, vol. 11384, pp. 311–320. Springer, Cham (2019). https://doi.org/10.1007/978-3-030-11726-9_28
49. Naser, M.A., et al.: Head and neck cancer primary tumor auto segmentation using model ensembling of deep learning in PET-CT images. In: Andrearczyk, V., Oreiller, V., Hatt, M., Depeursinge, A. (eds.) HECKTOR 2021. LNCS, vol. 13209, pp. 121–133. Springer, Cham (2022)
50. Naser, M.A., et al.: Progression free survival prediction for head and neck cancer using deep learning based on clinical and PET-CT imaging data. In: Andrearczyk, V., Oreiller, V., Hatt, M., Depeursinge, A. (eds.) HECKTOR 2021. LNCS, vol. 13209, pp. 287–299. Springer, Cham (2022)
51. Oreiller, V., et al.: Head and neck tumor segmentation in PET/CT: the HECKTOR challenge. Med. Image Anal. **77**, 102336 (2021)
52. Qayyum, A., Benzinou, A., Mazher, M., Abdel-Nasser, M., Puig, D.: Automatic segmentation of head and neck (H&N) primary tumors in PET and CT images using 3D-Inception-ResNet model. In: Andrearczyk, V., Oreiller, V., Hatt, M., Depeursinge, A. (eds.) HECKTOR 2021. LNCS, vol. 13209, pp. 58–67. Springer, Cham (2022)
53. Ren, J., Huynh, B.N., Groendahl, A.R., Tomic, O., Futsaether, C.M., Korreman, S.S.: PET normalizations to improve deep learning auto-segmentation of head and neck in 3D PET/CT. In: Andrearczyk, V., Oreiller, V., Hatt, M., Depeursinge, A. (eds.) HECKTOR 2021. LNCS, vol. 13209, pp. 83–91. Springer, Cham (2022)
54. Saeed, N., Al Majzoub, R., Sobirov, I., Yaqub, M.: An ensemble approach for patient prognosis of head and neck tumor using multimodal data. In: Andrearczyk, V., Oreiller, V., Hatt, M., Depeursinge, A. (eds.) HECKTOR 2021. LNCS, vol. 13209, pp. 278–286. Springer, Cham (2022)

55. Salmanpour, M.R., Hajianfar, G., Rezaeijo, S.M., Ghaemi, M., Rahmim, A.: Advanced automatic segmentation of tumors and survival prediction in head and neck cancer. In: Andrearczyk, V., Oreiller, V., Hatt, M., Depeursinge, A. (eds.) HECKTOR 2021. LNCS, vol. 13209, pp. 202–210. Springer, Cham (2022)

56. Sepehri, S., Tankyevych, O., Iantsen, A., Visvikis, D., Cheze Le Rest, C., Hatt, M.: Accurate tumor delineation vs. rough volume of interest analysis for 18F-FDG PET/CT radiomic-based prognostic modeling in non-small cell lung cancer. Front. Oncol. 292(2), 365–373 (2021)

57. Starke, S., Thalmeier, D., Steinbach, P., Piraud, M.: A hybrid radiomics approach to modeling progression-free survival in head and neck cancers. In: Andrearczyk, V., Oreiller, V., Hatt, M., Depeursinge, A. (eds.) HECKTOR 2021. LNCS, vol. 13209, pp. 266–277. Springer, Cham (2022)

58. Vallières, M., et al.: Radiomics strategies for risk assessment of tumour failure in head-and-neck cancer. Sci. Rep. 7(1), 1–14 (2017)

59. Wahid, K.A., et al.: Combining tumor segmentation masks with PET/CT images and clinical data in a deep learning framework for improved prognostic prediction in head and neck squamous cell carcinoma. In: Andrearczyk, V., Oreiller, V., Hatt, M., Depeursinge, A. (eds.) HECKTOR 2021. LNCS, vol. 13209, pp. 300–307. Springer, Cham (2022)

60. Wang, G., Huang, Z., Shen, H., Hu, Z.: The head and neck tumor segmentation in PET/CT based on multi-channel attention network. In: Andrearczyk, V., Oreiller, V., Hatt, M., Depeursinge, A. (eds.) HECKTOR 2021. LNCS, vol. 13209, pp. 38–49. Springer, Cham (2022)

61. Wang, J., Peng, Y., Guo, Y., Li, D., Sun, J.: CCUT-Net: pixel-wise global context channel attention UT-Net for head and neck tumor segmentation. In: Andrearczyk, V., Oreiller, V., Hatt, M., Depeursinge, A. (eds.) HECKTOR 2021. LNCS, vol. 13209, pp. 318–326. Springer, Cham (2022)

62. Xie, H., Zhang, X., Ma, S., Liu, Y., Wang, X.: Preoperative differentiation of uterine sarcoma from leiomyoma: comparison of three models based on different segmentation volumes using radiomics. Mol. Imaging Biol. 21(6), 1157–64 (2019)

63. Xie, J., Peng, Y.: The head and neck tumor segmentation based on 3D U-Net. In: Andrearczyk, V., Oreiller, V., Hatt, M., Depeursinge, A. (eds.) HECKTOR 2021. LNCS, vol. 13209, pp. 92–98. Springer, Cham (2022)

64. Xu, L., et al.: Automated whole-body bone lesion detection for multiple myeloma on 68Ga-pentixafor PET/CT imaging using deep learning methods. Contrast Media Mol. Imaging (2018)

65. Xue, Z., et al.: Multi-modal co-learning for liver lesion segmentation on PET-CT images. IEEE Trans. Med. Imaging 40, 3531–3542 (2021)

66. Yousefirizi, F., et al.: Segmentation and risk score prediction of head and neck cancers in PET/CT volumes with 3D U-Net and Cox proportional hazard neural networks. In: Andrearczyk, V., Oreiller, V., Hatt, M., Depeursinge, A. (eds.) HECKTOR 2021. LNCS, vol. 13209, pp. 236–247. Springer, Cham (2022)

67. Yousefirizi, F., Rahmim, A.: GAN-based bi-modal segmentation using Mumford-shah loss: application to head and neck tumors in PET-CT images. In: Andrearczyk, V., Oreiller, V., Depeursinge, A. (eds.) HECKTOR 2020. LNCS, vol. 12603, pp. 99–108. Springer, Cham (2021). https://doi.org/10.1007/978-3-030-67194-5_11

68. Yuan, Y., Adabi, S., Wang, X.: Automatic head and neck tumor segmentation and progression free survival analysis on PET/CT images. In: Andrearczyk, V., Oreiller, V., Hatt, M., Depeursinge, A. (eds.) HECKTOR 2021. LNCS, vol. 13209, pp. 179–188. Springer, Cham (2022)

69. Zhao, X., Li, L., Lu, W., Tan, S.: Tumor co-segmentation in PET/CT using multi-modality fully convolutional neural network. Phys. Med. Biol. **64**(1), 015011 (2018)
70. Zhong, Z., et al.: 3D fully convolutional networks for co-segmentation of tumors on PET-CT images. In: 2018 IEEE 15th International Symposium on Biomedical Imaging (ISBI 2018), pp. 228–231. IEEE (2018)

CCUT-Net: Pixel-Wise Global Context Channel Attention UT-Net for Head and Neck Tumor Segmentation

Jiao Wang, Yanjun Peng$^{(\boxtimes)}$, Yanfei Guo, Dapeng Li, and Jindong Sun

College of Computer Science and Engineering, Shandong University of Science and Technology, Qingdao 266590, Shandong, China

Abstract. Automatic segmentation of head and neck (H&N) primary tumors in FDG-PET/CT images is significant for the treatment of cancer. In this paper, the pixel-wise global context channel attention U-shaped transformer net (CCUT-Net) was proposed by using the long-range relational information, global context information, and channel information to improve the robustness and effectiveness of tumor segmentation. First, we used the convolutional neural network (CNN) and transformer fusion encoder to obtain the feature image, which not only captured the remote dependency of the image but also reduced the impact of small datasets on the performance of the transformer. Meanwhile, this was the first time to apply the fusion of CNN and transformer to the segmentation of H&N tumors in FDG-PET/CT images. Furthermore, we proposed the pixel-wise global context channel attention module in the decoder that combined the global context information and channel information of the image. It not only considered the overall information of the image but also paid attention to the FDG-PET and CT channel information, using the advantages of the two modes to accurately localize the position and segment the boundary of the tumor. Finally, in the encoder and decoder, we applied squeeze and excitation (SE) normalization to speed up the model training and promote model convergence. We evaluated our model on the test dataset of the head and neck tumor challenge with a final dice similarity coefficient (DSC) of 0.763 and a hausdorff distance-95% (HD95) of 3.270, which showed that our method was robust in tumor segmentation. (Team name:wangjiao)

Keywords: Head and neck · Segmentation · Transformer · Context channel attention · Squeeze and excitation normalization

1 Introduction

Head and Neck (H&N) cancers are among the most common cancers worldwide with high incidence. Oropharyngeal cancer is a kind of malignant tumor of H&N, accounting for 1.3% of malignant tumors of the whole body and 4.2% of H&N. Radiotherapy is the first choice for H&N cancers. Target delineation is a basic

© Springer Nature Switzerland AG 2022
V. Andrearczyk et al. (Eds.): HECKTOR 2021, LNCS 13209, pp. 38–49, 2022.
https://doi.org/10.1007/978-3-030-98253-9_2

and important step in the process of radiotherapy planning. At present, it is mainly done by radiologists, which is very time-consuming and tedious. It is also easily affected by the subjective experience of radiologists. Therefore, the automatic segmentation of H&N primary tumors is important for practical clinical work.

Computed tomography (CT) imaging is necessary for radiotherapy planning. However, there is no clear feature of the H&N tumor in CT images. H&N tumors are highlighted in the fluorodeoxyglucose positron emission tomography (FDG-PET) images, but the spatial resolution of the FDG-PET images is low [1]. Some normal tissues are also highlighted in the FDG-PET images, such as the brain. Considering both FDG-PET and CT images, the FDG-PET images provide the position information of tumors, and the CT images indicate the position and boundary of normal tissue. The fusion of the two image modes will provide more accurate information for target delineation. Aiming to study the automatic segmentation of H&N tumors, the MICCAI 2021 Head and Neck Tumor Segmentation challenge (HECKTOR) [2] offered the opportunity to participants to develop automatic bi-modal approaches for the 3D segmentation of H&N tumors in FDG-PET/CT scans, focusing on oropharyngeal cancers.

In recent years, medical image segmentation based on deep learning has achieved similar results as human experts. Especially the application of convolutional neural network (CNN) makes deep learning very popular. As a CNN, U-Net is widely used in medical image segmentation. Influenced by natural language processing (NLP), transformers have been applied to image processing recently [3]. In this paper, we used the U-Net as the model structure and fused CNN and transformer block as the encoder to extract features and obtain the long-distance relations in the images [4]. The pixel-wise global context channel attention (CCA) module was used in the decoder to analyze the global information and channel information. It generated a high-resolution feature map to improve the automatic segmentation of H&N tumors. In the whole network model, we also used squeeze and excitation (SE) normalization [5] to optimize the model training.

2 Method

2.1 Data Preprocessing

The original data were preprocessed for network training. According to the bounding box location file, the bounding box patches which contained the oropharyngeal region were extracted from the original images. The FDG-PET and CT images were interpolated to the isotropic resolution of $1 \times 1 \times 1 \, \text{mm}^3$ by the trilinear interpolation. The masks of the primary gross tumor volume (GTVt) were interpolated to the isotropic resolution of $1 \times 1 \times 1 \, \text{mm}^3$ by the nearest interpolation. To ensure that the tumors in the images were complete, we did not crop the bounding box patches, so the sizes of the input images were $144 \times 144 \times 144$.

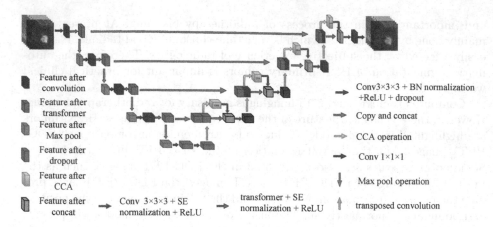

Fig. 1. Architecture of our network. Input is the fused FDG-PET/CT image of the size of $2 \times 144 \times 144 \times 144$. The encoder consists of convolutional neural network (CNN) and transformer operation. The max pool operation is used for down sampling. The decoder consists of the proposed pixel-wise global context channel attention (CCA) block and convolution blocks.

After resampling, the FDG-PET images and CT images were intensity normalized. The intensity value of the CT images was clipped to the range of $[-1024, 1024]$ Hounsfield Units (HU), and then divided by 1024 to map to the range of $[-1, 1]$. The FDG-PET image was normalized by the Z-Score method.

2.2 Network Architecture

Our model took U-Net as the overall network architecture and adopted SE normalization. The encoder combined the traditional CNN and transformer. The CNN extracted features from the input image and transmitted them to the transformer block, which not only reduced the impact of the lack of large-scale pre-training on the transformer but also made use of the ability of the transformer to capture remote relationships. In the decoder, we used the CCA block. On the one hand, using the global context information and channel information based on pixel-level grasped the overall information of the image and made use of the remote relationship captured by the transformer block. On the other hand, it also took full account of the fusion information of the FDG-PET and CT images and effectively combined their advantages.

Figure 1 showed the architecture of our network. The encoder contained a total of 4 down sampling blocks and a convolution block. Each down sampling block consisted of two parts, a convolution operation, and a transformer block. The convolution operation's kernel size was 3, stride was 1, and padding was 1. From top to bottom, the patch sizes of the transformer blocks were 16, 8, 4, and 2. The max pool operation with kernel size and stride of 2 was used for down sampling to reduce the size of the feature map. The decoder first used the transposed convolution operation with the convolution kernel size of

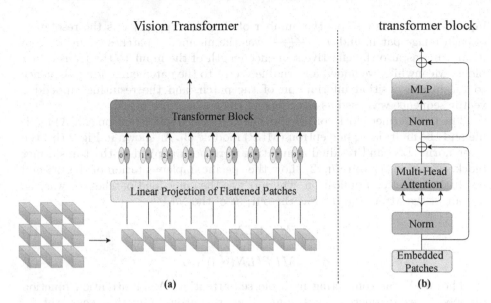

Fig. 2. Module overview. (a) The feature map extracted by the CNN is divided into multiple patches of the same size. Position embeddings are added to the linear embedded sequence of these patches as the input of the transformer block. (b) The transformer block consists of the MSA and MLP module.

4, the stride of 2, and the padding of 1 to increase the feature map. Then it was concatenated with the encoder feature map of the same size by the skip connection. The obtained feature map was sequentially input into the CCA block and a convolution operation. Then it was fused with the results. At last, a convolution operation was applied to complete the decoder. At the end of the network, we used the convolution operation with the kernel of $1 \times 1 \times 1$ to map the channel number to 1 and get our predicted feature map.

2.3   Transformer Block

The transformer is first used in the NLP and has become its preferred model. Inspired by the NLP, some papers on computer vision have also studied the application of transformers in recent years.

In this paper, we used the vision transformer [3]. Because the transformer has an ideal effect on large-scale datasets or pre-trained models, we fused it with CNN to reduce the influence of small datasets on transformer performance. As shown in Fig. 2, the feature map extracted by the CNN was divided into multiple patches of the same size. Then the patches were mapped to a linear embedded sequence. Position embeddings were added to the linear embedded sequence of these patches as the input of the transformer.

To process a 3D image, it was necessary to reshape the image $x \in R^{c \times d \times h \times w}$ into a flattened patch $x_p \in R^{n \times (p^3 \cdot c)}$, where (d, h, w) was the resolution of the input feature image, c was the number of channels, (p, p, p) was the resolution of each image patch, and $n = \frac{d \times h \times w}{p^3}$ was the number of patches resulting from it, and it was also the effective sequence length of the input to the transformer block. Meanwhile, we added a learnable vector to the patch embedding sequence to retain the position information of the patch, and the resulting embedded vector sequence was used as the input of the transformer block.

The transformer block consisted of the multi-head self-attention (MSA) module and the multi-layer perceptron (MLP) module [3]. As shown in Fig. 2 (b), the layer norm (LN) and residual connection were both applied in the transformer block. Equation (1) and Eq. (2) show the specific implementation of the LN and residual connection application, where α was the embedded patches, α' was the output of the MSA, and α'' was the output of the MLP.

$$\alpha' = MSA(LN(\alpha)) + \alpha \tag{1}$$

$$\alpha'' = MLP(LN(\alpha')) + \alpha' \tag{2}$$

The MSA was connecting multiple self-attention. A self-attention function mapped a set of queries, keys, and values to outputs. Queries, keys, values, outputs were all vectors. The outputs were the weighted sum of the values, where the weights, which represented the correlation between the queries and the current keys, assigned to values were calculated through a correlation function [4]. The calculation process was shown in Eq. (3),

$$Self - attention(Q, K, V) = softmax(\frac{QK^T}{\sqrt{d_k}})V \tag{3}$$

where Q, K, V were query, key, value vectors, d_k was the dimension of the query vectors. The number of heads of the MSA module was set to 8 in the paper.

The MLP module contained two linear layers and a GELU activation function. The GELU activation function was calculated according to Eq. (4), where x was the input value, X was a gaussian random variable with zero mean and unit variance, $P(X \leq x)$ was the probability that X was less than or equal to a given value x.

$$GELU(X) = x \times P(X \leq x) = x \times \phi(x), x \sim N(0, 1) \tag{4}$$

2.4   Pixel-Wise Global Context Channel Attention

Inspired by the GCNet [6] and the ECA-Net [7], the paper proposed the CCA block, which used the attention mechanism to give more weight to important features in the process of learning, making it play a greater role in the process of network training. Figure 3 showed its specific structure.

As shown in Fig. 3, first, the context modeling and transform parts made up the global context (GC) block. x was input to the context modeling part, which aggregated the features of all locations to form the global context features [6]. It would further analyze the features extracted by the transformer block of the

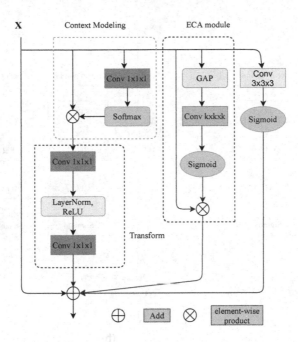

Fig. 3. The structure of the pixel-wise global context channel attention (CCA) block.

encoder. Then the obtained global context features were input into the transform module in the CCA, which strengthened the interdependence between different position features. Second, an efficient lightweight channel attention module [7] was used to capture the dependence between channels and extract effective channel information to make full use of the advantages of the FDG-PET and CT images. This part made use of the efficient channel attention (ECA) module in ECA-Net [7]. The local cross-channel interaction strategy was realized by one-dimensional convolution. The kernel size k of one-dimensional convolution determined the interaction range of local cross-channels. The value of k can be determined according to the number of channels in the image. In this paper, k was 3. Finally, the convolution operation and sigmoid activation function were used to analyze the position information based on the pixel level. It improved the extraction of tumor position information and edge features. At the end of the CCA module, add operation was used to fuse the above four parts to realize the fusion of different attention mechanisms.

Compared with the original GC block and ECA module, we combined them and added the pixel-level feature extraction and analysis, which considered both context information and channel information, and made full use of global features and local tiny features. In addition, we also added residual connections to the model to help avoid the disappearance of gradients in the training process and speed up the convergence of the model.

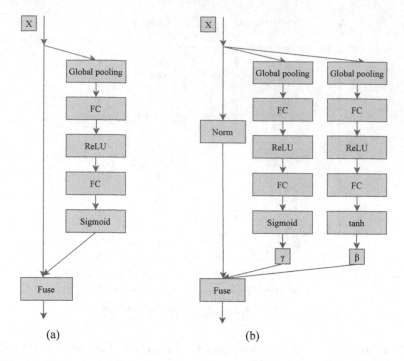

(a)   (b)

Fig. 4. SE block and SE Normalization. (a) SE block (b) SE Normalization

2.5 Squeeze and Excitation Normalization

In the encoder and decoder, we used SE normalization [5]. SE normalization normalizes all channels of each sample using the mean $\mu = E[x]$ and standard deviation $\sigma = \sqrt{Var[x] + \epsilon}$, where ϵ was to prevent division by 0, i.e.

$$x' = \frac{1}{\sigma}(x - \mu) \tag{5}$$

Then two parameters γ, β were applied to scale and shift the normalized value x'. γ and β could be learned according to the input x in SE Normalization, and the mapping relationship between them was modeled by SE block, i.e.

$$\gamma = f_\gamma(x) \tag{6}$$

$$\beta = f_\beta(x) \tag{7}$$

where f_γ was the original SE block with the sigmoid activation function and f_β was the SE block with the tanh activation function.

As shown in Fig. 4, (a) was the original SE block and (b) was SE normalization. In the SE block, the global average pooling (GAP) was first applied to compress each channel into a single descriptor. Then, two fully connected (FC) layers were designed to capture nonlinear cross-channel dependencies. The first

FC layer was realized with the compression rate r to form the bottleneck to control the complexity of the model. In this paper, we set r to a fixed value of 2. In SE normalization, the parameters γ, β after SE block modeling acted on the original normalized value to obtain a new one.

2.6   Loss Function

Our loss function used the sum of the Dice Loss [8] and the Focal Loss [9], as shown in Eq. (8):

$$L = L_{Dice} + L_{Focal} \tag{8}$$

where the Dice Loss and the Focal Loss were defined as follows:

$$L_{Dice}(y, \hat{y}) = 1 - \frac{2\sum_i^N y_i \hat{y}_i + 1}{\sum_i^N y_i + \sum_i^N \hat{y}_i + 1} \tag{9}$$

$$L_{Focal}(y, \hat{y}) = -\frac{1}{N} \sum_i^N y_i (1 - \hat{y}_i)^\lambda ln(\hat{y}_i) \tag{10}$$

where y_i was the label of the $i-th$ voxel, $\hat{y}_i$ was the prediction probability of the $i-th$ voxel, N was the total number of all voxels, and the λ in Focal Loss was set to 2.

3   Experiments

3.1   Dataset

HECKTOR provided a dataset of 325 patients from six centers. The training dataset included a total of 224 patients, which was comprised of the FDG-PET images, the CT images, and the GTVt masks. They were all in NIFTI format. There were 101 patients in the test dataset, and each case contained only the FDG-PET images and CT images in NIFTI format. In addition, both the training dataset and the test dataset contained the bounding box location files and patient information files. All images have been re-annotated by experts for the challenge.

3.2   Experiment Settings

Our network used the artificial intelligence development platform AI station for management and training. GeForce RTX2080Ti provided powerful computing resources and achieved efficient computing power support. In the process of training, our batch size was set to 2, the initial learning rate was 10^{-3}. The cosine annealing method was used to adjust the learning rate to speed up the convergence of the network. Adam was used as the optimizer, whose parameter β_1 was set to 0.9 to calculate the gradient, and β_2 was set to 0.99 to calculate the running average of the square of the gradient. A total of three hundred epochs of training were held.

Table 1. Different models can be trained by different datasets. The model results corresponding to the five centers and the randomly divided validation dataset are provided.

Center	DSC	HD95	Precision	Sensitivity
CHGJ	0.7481	3.9062	0.6831	0.8845
CHMR	0.6725	6.3175	0.7448	0.6711
CHUM	0.6746	3.2504	0.7318	0.7059
CHUS	0.6867	3.6613	0.7409	0.6983
CHUP	0.7402	4.1932	0.8279	0.7167
Random	0.7508	2.7057	0.7764	0.8048

Evaluation metrics included dice similarity coefficient (DSC), precision, sensitivity, and hausdorf distance at 95% (HD95), i.e.

$$DSC : DSC = \frac{2TP}{2TP + FP + FN} \tag{11}$$

$$Precision : Prec = \frac{TP}{TP + FP} \tag{12}$$

$$Sensitivity : Sens = \frac{TP}{TP + FN} \tag{13}$$

$$Hausdorff Distance : Haus = max\{sup_{p \in P} inf_{t \in T} d(p, t), sup_{t \in T} inf_{p \in P} d(t, p)\} \tag{14}$$

where TP, FP, FN were true positive, false positive, and false negative. P represented that the model prediction was positive, and T represented that the ground truth annotation was positive. sup was the supremum and inf was the infimum.

3.3   Results

In the experiments, we trained the five centers of the training set as validation sets to get five model results to prove the robustness of the model. At the same time, 24 patients were randomly divided from the whole training set as a validation set to prove the effectiveness of the model. The experimental results were shown in Table 1. Due to the different data sizes of the validation sets, there were some differences in the evaluation metric scores of the five centers. Among them, the CHGJ center's DSC and the CHUP center's DSC reached 0.74, indicating that the model had a certain degree of robustness. The HD95 of the CHMR center was different from the other centers, which indicated that this model had a relatively poor segmentation effect on the tumor boundary of the CHMR dataset. The model of randomly dividing the training set was the best because the training set contained all the center samples, which showed that the model was effective for tumor segmentation.

Table 2. The metric scores of the CCUT-Net and the pixel-wise global context channel attention U-shaped Net (CCU-Net) on the randomly divided validation dataset.

Model	DSC	$HD95$	$Precision$	$Sensitivity$
CCUT-Net	0.7508	2.7057	0.7764	0.8048
CCU-Net	0.6974	3.7430	0.7471	0.7215

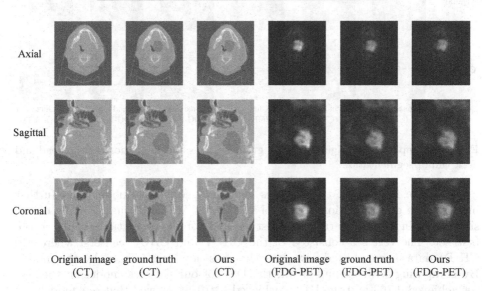

| | Original image (CT) | ground truth (CT) | Ours (CT) | Original image (FDG-PET) | ground truth (FDG-PET) | Ours (FDG-PET) |

Fig. 5. Example of the original images, ground truths and predictions of the head and neck tumor segmentation.

To further demonstrate the effectiveness of the transformer module in our proposed model, we replaced the transformer module in the model with the traditional convolution module and did experiments on the randomly divided training set and validation set. As shown in Table 2, the evaluation metric scores with the transformer module were better than that only with the CNN module, which showed that the model with transformer module was easier to learn effective information and had better segmentation performance than the model with the CNN module only.

To observe the segmentation effect of our proposed model more intuitively, we visualized the images. Figure 5 showed the original images, ground truths, and predicted segmentation results of a patient. We gave the original FDG-PET and CT images respectively to compare the differences between FDG-PET and CT images. In Fig. 5, the normal tissue structures in the CT images were clearer, while in the FDG-PET images, they were fuzzy. But areas with high metabolic activity were shown as highlighted in the FDG-PET images. From the visualization results, we find that our model could accurately locate and segment the tumor.

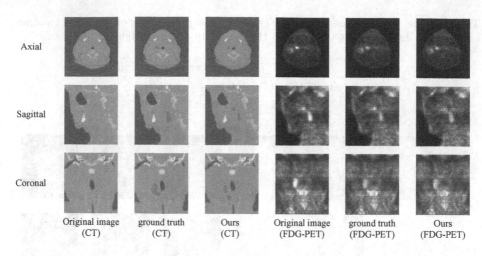

Fig. 6. Example of the original images, ground truths and predictions of the head and neck tumor segmentation.

To make full use of the features learned by different models and further improve the generalization ability of the model, we used the ensemble of the six models in Table 1 to train the test set and got the prediction results on the test set. The test set consisted of the CHUV and CHUP centers, where the CHUV dataset was brand-new data that has not been learned by the model. By comparing with ground truth, the DSC of our network model on the test set achieved 0.763 and the HD95 achieved 3.270. It showed that our model was effective and robust on datasets that the model had never learned before.

Although the model had achieved good results as a whole, it still had many shortcomings in segmenting small irregular tumors. Figure 6 showed the segmentation results of a case in the CHMR center. When facing the small irregular tumors, our proposed model was difficult to accurately describe the shape of tumors, and the segmentation results were very different from the ground truth.

4   Conclusion

In this paper, we proposed a network model which combined traditional CNN and transformer for automatic segmentation of head and neck tumors. In the U-shaped network, each encoder consisted of a convolutional block and a transformer block. The transformer block extracted remote dependencies from the feature map obtained by the convolution operation. To make better use of the remote dependency, global context information, and channel information of the feature map, a CCA block was proposed in the decoder for better tumor segmentation. Through the evaluation of the validation datasets, it was proved that our network model was effective. When we submitted our test results on the website, the DSC achieved 0.763, indicating that our model had a certain degree of robustness.

Although the model had achieved good results, it still had some limitations. First, the model had lots of parameters that had high requirements for experimental equipment. Secondly, the HD95 value of the proposed model was 3.270, while the best HD95 value in the challenge was 3.057, indicating that the proposed model was not effective in tumor boundary segmentation. Especially when segmenting irregular tumors, the segmentation result of this model was very different from the ground truth. In future research, we will pay attention to improving the segmentation performance of irregular tumors and tumor boundaries. Finally, the model had only been tested on the data set of head and neck cancer. We cannot guarantee that the ideal results can be obtained on other data sets.

References

1. Andrearczyk, V., et al.: Overview of the HECKTOR challenge at MICCAI 2021: automatic head and neck tumor segmentation and outcome prediction in PET/CT images. In: Andrearczyk, V., Oreiller, V., Hatt, M., Depeursinge, A. (eds.) HECK-TOR 2021. LNCS, vol. 13209, pp. 1–37. Springer, Cham (2022)
2. Oreiller, V., et al.: Head and Neck Tumor segmentation in PET/CT: the HECKTOR challenge. In: Medical Image Analysis (2021)
3. Dosovitskiy, A., et al.: An image is worth 16x16 words: transformers for image recognition at scale. arXiv:abs/2010.11929 (2021)
4. Gao, Y., Zhou, M., Metaxas, D.N.: UTNet: a hybrid transformer architecture for medical image segmentation. arXiv:abs/2107.00781 (2021)
5. Iantsen, A., Jaouen, V., Visvikis, D., Hatt, M.: Squeeze-and-excitation normalization for brain tumor segmentation. In: Crimi, A., Bakas, S. (eds.) BrainLes 2020. LNCS, vol. 12659, pp. 366–373. Springer, Cham (2021). https://doi.org/10.1007/978-3-030-72087-2_32
6. Cao, Y., Xu, J., Lin, S., Wei, F., Hu, H.: GCNet: non-local networks meet squeeze-excitation networks and beyond. In: 2019 IEEE/CVF International Conference on Computer Vision Workshop (ICCVW), pp. 1971–1980 (2019)
7. Wang, Q., Wu, B., Zhu, P., Li, P., Zuo, W., Hu, Q.: ECA-net: efficient channel attention for deep convolutional neural networks. In: 2020 IEEE/CVF Conference on Computer Vision and Pattern Recognition (CVPR), pp. 11531–11539 (2020)
8. Eelbode, T., et al.: Optimization for medical image segmentation: theory and practice when evaluating with dice score or Jaccard index. IEEE Trans. Med. Imaging **39**, 3679–3690 (2020)
9. Ye, C., Wang, W., Zhang, S., Wang, K.: Multi-depth fusion network for whole-heart CT image segmentation. IEEE Access **7**, 23421–23429 (2019)

A Coarse-to-Fine Framework for Head and Neck Tumor Segmentation in CT and PET Images

Chengyang An[iD], Huai Chen[iD], and Lisheng Wang[✉][iD]

Department of Automation, Institute of Image Processing and Pattern Recognition,
Shanghai Jiao Tong University, Shanghai 200240, People's Republic of China
lswang@sjtu.edu.cn

Abstract. Radiomics analysis can help patients suffered from head and neck (H&N) cancer customize tailoring treatments. It requires a large number of segmentation of the H&N tumor area in PET and CT images. However, the cost of manual segmentation is extremely high. In this paper, we propose a coarse-to-fine framework to segment the H&N tumor automatically in FluoroDeoxyGlucose (FDG)-Positron Emission Tomography (PET) and Computed Tomography (CT) images. Specifically, we trained three 3D-UNets with residual blocks to make coarse stage, fine stage and refined stage predictions respectively. Experiments show that such a training framework can improve the segmentation quality step by step. We evaluated our framework with Dice Similarity Coefficient (DSC) and Hausdorff Distance at 95% (HD95) of 0.7733 and 3.0882 respectively in the task 1 of the HEad and neCK TumOR segmentation and outcome prediction in PET/CT images (HECKTOR2021) Challenge and ranked second.

Keywords: Automatic segmentation · Head and neck cancer · Coarse-to-fine

1 Introduction

The conversion of digital medical images into mineable high-dimensional data, a process that is known as radiomics. The motivation is that biomedical images contain information that reflects the underlying pathophysiology, and these relationships can be revealed through quantitative image analysis [1]. For example, based on PET and CT images, it's helpful to better identify patients with poor prognosis in a non-invasive fashion, and use the acquired images for diagnosis and customized tailoring treatments to specific patients with H&N cancers. It will highly benefit from automated targeted quantitative assessments through radiomics models. However, radiomics models often requires the segmentation of the H&N tumor area in PET and CT images [2,3]. Manually delineating region of interest (ROI) by experts is costly and error prone due to fatigue and

V. Andrearczyk et al. (Eds.): HECKTOR 2021, LNCS 13209, pp. 50–57, 2022.
https://doi.org/10.1007/978-3-030-98253-9_3

other reasons [4]. Therefore, an effective automatic segmentation algorithm for H&N tumors is urgently needed.

In the past few years, deep learning techniques based on convolutional neural networks (CNN) and Transformers have achieved excellent results in many computer vision tasks for medical image analysis. In 2020, Vincent Andrearczyk et al. organized the first HECKTOR (HEad and neCK TumOR segmentation) challenge [5] and offered the opportunity to participants to develop automatic bi-modal approaches for the 3D segmentation of H&N tumors in FDG-PET and CT scans, focusing on oropharyngeal cancers. The purpose of this challenge is to determine the best method to utilize rich bi-modal information in H&N primary tumor segmentation. This valuable knowledge will be transferred to many other cancer types related to PET and CT imaging, making large-scale and reproducible radiomic research possible [6]. Andrei Iantsen et al. proposed an algorithm of Squeeze-and-Excitation Normalization and won the first place with a Dice Score of 0.759 in this challenge [7]. Chen Huai et al. proposed a framework that can iteratively refine the segmentation of H&N area and got the highest precision of 0.8479 in the challenge [8]. For the automatic segmentation problem in medical images, Fabian Isensee et al. proposed the nnUNet model and achieved the best results in many medical image segmentation challenges [9].

In this paper, We propose a framework for precise segmentation of H&N tumor areas in PET and CT images by three progressive steps and each step uses a 3D-UNet model. To achieve state of the art results, we combined the methods proposed by Andrei Iantsen et al. and Fabian Isensee et al. after this framework, and achieved Dice Similarity Coefficient (DSC) and Hausdorff Distance at 95% (HD95) of **0.7753** and **3.1432** respectively in the task 1 of the HECKTOR2021 Challenge.

2   Method

Our method is a three-progressive-step framework for H&N tumor segmentation, which is divided into three stages: coarse segmentation, fine segmentation and refined adjustment. The pre-processed CT and PET images are combined and input into the network. In the coarse segmentation stage, the recall rate is improved as much as possible to segment tumor areas. The corresponding tumor regions are cropped, and then the tumor regions selected in the previous step are finely segmented. Finally, in the refined adjustment stage, the segmentation results obtained in the second step are combined with the input data to make refined adjustments to the segmentation results.

2.1   Network Architecture

Inspired by the ResUNet structure [10], we designed a network for tumor segmentation in CT and PET two-modality images. The network structure is shown in Fig. 1. The number of channels in the middle feature map is 16, 32, 64, 128, 256, 128, 64, 32 and 16 respectively.

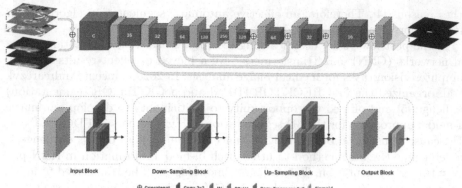

Fig. 1. Architecture of 3D-UNet. The input is a concatenated image of CT and PET images, and the final segmentation result is obtained after four times of down-sampling and up-sampling.

2.2 Coarse-to-Fine Framework

The framework is divided into 3 main parts. The illustration of our framework is shown in Fig. 2. At the coarse segmentation stage, we got the position and size of the tumor and selected the bounding box of each tumor for the fine segmentation stage. Then, at the fine segmentation stage, we input only the parts in bounding boxes of CT and PET images into the model. Finally, we concatenated the parts in bounding boxes of CT, PET and fine segmentation and input it into the model.

2.3 Training Details

Loss Function. Our loss function has two terms: F-Loss and Binary Cross-Entropy Loss (BCE Loss), i.e.

$$\ell = \lambda\ell_F + (1 - \lambda)\ell_{BCE} \tag{1}$$

In order to separate the tumor area completely in the coarse segmentation stage, we hope to increase the recall rate as much as possible while ensuring the basic segmentation shape at this stage. Inspired by Huai et al. [12], we used the F-Loss in our model. The F-Loss is defined as followed:

$$\ell_F(y, \hat{y}) = 1 - \frac{(1 + \beta^2) \sum\limits_{i=1}^{N} \hat{y}_i y_i}{\sum\limits_{i=1}^{N} \hat{y}_i + \beta^2 \sum\limits_{i=1}^{N} y_i + \epsilon} \tag{2}$$

where ϵ is set as 1e$-$8 to avoid the risk of being divided by 0. $\hat{y}$ is the predicted probability and y is the ground truth. N is the total number of voxels. y_i and $\hat{y}_i$

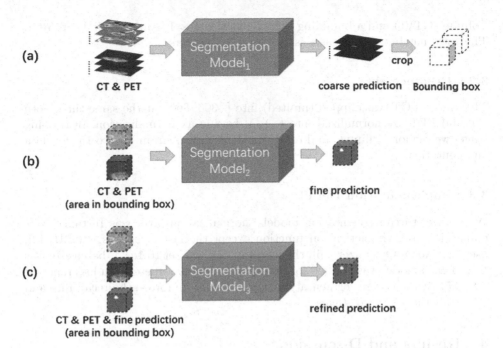

Fig. 2. An illustration of our framework. (a) **Coarse segmentation stage**, CT and PET images are concatenated and input into the segmentation model. We select the bounding boxes of tumor area according to the coarse segmentation. (b) **Fine segmentation stage**, the part of bounding box in the CT and PET images are input into the segmentation model. (c) **Refined adjustment stage**, we input the two-modal image and the fine segmentation obtained in the previous step into the segmentation model and get the final prediction result.

are the ith voxel in y and $\hat{y}$. The recall rate can be controlled on a reasonable segmentation effect by adjusting the β. The larger the β, the better the recall rate. When $\beta = 1$, F-Loss is equivalent to Dice Loss [11].

At the same time, we use BCE Loss to narrow the distance between the prediction result and the ground truth distribution. BCE Loss is defined as followed:

$$\ell_{BCE}(y, \hat{y}) = -(y \log(\hat{y}) + (1 - y) \log(1 - \hat{y})) \tag{3}$$

3   Experiments

3.1   Dataset

The challenge of HECKTOR2021 provided 224 cases from 5 centers for training. Each case includes one 3D FDG-PET volume (in SUV) registered with a 3D CT volume (from low-dose FDG-PET/CT) of the head and neck region, as well as a binary contour with the annotated ground truth of the primary Gross Tumor

Volume (GTVt) and a bounding box of $144 \times 144 \times 144$ mm^3 to indicate ROI. The total number of test cases is 101 from two centers.

3.2   Preprocessing

The value of CT was clipped (limited) into $[-300, 300]$, at the same time, both CT and PET are normalized into $[-1, 1]$ by min-ax normalization. In training stage, we randomly flipped and rotate on the XY plane from -15 to $15°$ for data augmentation.

3.3   Implementation Detail

We trained three segmentation models, normalization layer was Instance Normalization [13], the activation function except the last layer was PReLU [14], used the Adam Optimizer [15], the learning rate was set to 1e−4, batch size was 1, and each model was trained for 500 epochs. The parameter λ in loss function was 0.5. At the coarse segmentation stage, $\beta = 5$ to improve the recall rate and $\beta = 1$ for the next two stages.

4   Results and Discussion

We divide the training data provided by the challenge organizer into the training set and the validation set at a ratio of 4:1. Table 1 below is the results of the three stages of our framework on the validation set.

Table 1. Predictive indicators for each stage.

Stage	Recall	DSC
Coarse Segmentation Stage	0.9883	X
Fine Segmentation Stage	X	0.7913
Refined Adjustment Stage	X	0.7931

Experiments have proved that our proposed framework can optimize the segmentation results step by step. Figure 3 below shows the prediction results of each stage.

To achieve state of the art results, we have combined the model using Squeeze-and-Excitation Normalization [7] and nnUNet [9] model after this framework. The two methods are based on the original strategy. We evaluated our framework with DSC and HD95 of **0.7733** and **3.0882** respectively in the task 1 of the HECKTOR2021 Challenge. Some prediction results are shown in Fig. 4 below.

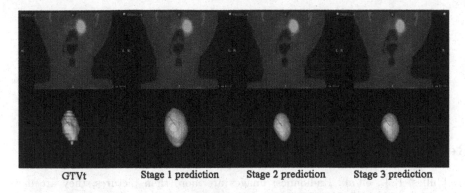

GTVt   Stage 1 prediction   Stage 2 prediction   Stage 3 prediction

Fig. 3. Predicted results of each stage in the framework.

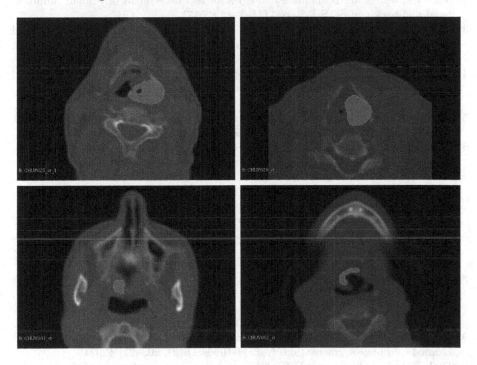

Fig. 4. Four prediction results of test cases.

The CPU we use is Intel(R) Core(TM) i9-10920X CPU @ 3.50 GHz, and the GPU is GeForce RTX 3090. We test in this environment. Table 2 below shows the inference time of each model in the three methods we integrated.

Table 2. Comparison of inference time.

Method	Time(s)
Ours	136
Iantsen et al. [7]	205 (5-folds)
Isensee et al. [9]	6000 (5-folds)

References

1. Gillies, R.J., et al.: Radiomics: images are more than pictures, they are data. Radiology **278**(2), 563–577 (2016)
2. Vallières, M., et al.: Radiomics strategies for risk assessment of tumour failure in head-and-neck cancer. Sci. Rep. **7**(1), 1–14 (2017)
3. Bogowicz, M., et al.: Comparison of PET and CT radiomics for prediction of local tumor control in head and neck squamous cell carcinoma. Acta Oncologica **56**(11), 1531–1536 (2017)
4. Andrearczyk, V., et al.: Automatic segmentation of head and neck tumors and nodal metastases in PET-CT scans. In: Medical Imaging with Deep Learning, pp. 33–43. PMLR (2020)
5. Oreiller, V., et al.: Head and neck tumor segmentation in PET/CT: the HECKTOR challenge. Med. Image Anal. **77**, 102336 (2022)
6. Andrearczyk, V., et al.: Overview of the HECKTOR challenge at MICCAI 2020: automatic head and neck tumor segmentation in PET/CT. In: Andrearczyk, V., Oreiller, V., Depeursinge, A. (eds.) HECKTOR 2020. LNCS, vol. 12603, pp. 1–21. Springer, Cham (2021). https://doi.org/10.1007/978-3-030-67194-5_1
7. Iantsen, A., Visvikis, D., Hatt, M.: Squeeze-and-excitation normalization for automated delineation of head and neck primary tumors in combined PET and CT images. In: Andrearczyk, V., Oreiller, V., Depeursinge, A. (eds.) HECKTOR 2020. LNCS, vol. 12603, pp. 37–43. Springer, Cham (2021). https://doi.org/10.1007/978-3-030-67194-5_4
8. Chen, H., Chen, H., Wang, L.: Iteratively refine the segmentation of head and neck tumor in FDG-PET and CT images. In: Andrearczyk, V., Oreiller, V., Depeursinge, A. (eds.) HECKTOR 2020. LNCS, vol. 12603, pp. 53–58. Springer, Cham (2021). https://doi.org/10.1007/978-3-030-67194-5_6
9. Isensee, F., et al.: nnU-Net: a self-configuring method for deep learning-based biomedical image segmentation. Nat. Methods **18**(2), 203–211 (2021)
10. Diakogiannis, Foivos I., et al.: ResUNet-a: A deep learning framework for semantic segmentation of remotely sensed data. ISPRS J. Photogrammetry Remote Sens. **162**, 94–114 (2020)
11. Sudre, C.H., Li, W., Vercauteren, T., Ourselin, S., Jorge Cardoso, M.: Generalised dice overlap as a deep learning loss function for highly unbalanced segmentations. In: Cardoso, M.J., et al. (eds.) DLMIA/ML-CDS -2017. LNCS, vol. 10553, pp. 240–248. Springer, Cham (2017). https://doi.org/10.1007/978-3-319-67558-9_28
12. Chen, H., Qian, D., Liu, W., Li, H., Wang, L.: An enhanced coarse-to-fine framework for the segmentation of clinical target volume. In: Shusharina, N., Heinrich, M.P., Huang, R. (eds.) MICCAI 2020. LNCS, vol. 12587, pp. 34–39. Springer, Cham (2021). https://doi.org/10.1007/978-3-030-71827-5_4

13. Ulyanov, D., et al.: Instance normalization: The missing ingredient for fast stylization. arXiv preprint arXiv:1607.08022 (2016)
14. He, Kaiming, et al.: Delving deep into rectifiers: surpassing human-level performance on imagenet classification. In: Proceedings of the IEEE International Conference on Computer Vision (2015)
15. Kingma, D.P., et al.: Adam: a method for stochastic optimization. arXiv preprint arXiv:1412.6980 (2014)

Automatic Segmentation of Head and Neck (H&N) Primary Tumors in PET and CT Images Using 3D-Inception-ResNet Model

Abdul Qayyum[1]([⊠]), Abdesslam Benzinou[1], Moona Mazher[2],
Mohamed Abdel-Nasser[2,3], and Domenec Puig[2]

[1] ENIB, UMR CNRS 6285 LabSTICC, 29238 Brest, France
qayyum@enib.fr
[2] Department of Computer Engineering and Mathematics, University Rovira i Virgili,
Tarragona, Spain
[3] Department of Electrical Engineering, Faculty of Engineering,
Aswan University, Aswan, Egypt

Abstract. In many computer vision areas, deep learning-based models achieved state-of-the-art performances and started to catch the attention in the context of medical imaging. The emergence of deep learning is significantly cutting-edge the state-of-the-art in medical image segmentation. For generalization and better optimization of current deep learning models for head and neck segmentation problems, head and neck tumor segmentation and outcome prediction in PET/CT images (HECKTOR21) challenge offered the opportunity to participants to develop automatic bi-modal approaches for the 3D segmentation of Head and Neck (H&N) tumors in PET/CT scans, focusing on oropharyngeal cancers. In this paper, a 3D Inception ResNet-based deep learning model (3D-Inception-ResNet) for head and neck tumor segmentation has been proposed. The 3D-Inception module has been introduced at the encoder side and the 3D ResNet module has been proposed at the decoder side. The 3D squeeze and excitation (SE) module is also inserted in each encoder block of the proposed model. The 3D depth-wise convolutional layer has been used in 3D inception and 3D ResNet module to get the optimized performance. The proposed model produced optimal Dice coefficients (DC) and HD95 scores and could be useful for the segmentation of head and neck tumors in PET/CT images. The code is publicly available (https://github.com/Res pectKnowledge/HeadandNeck21_3D_Segmentation).

Keywords: Head and neck segmentation · 3D-inception deep learning model · 3D-ResNet deep learning model · 3D squeeze and excitation module · 3D-depthwise module

1 Introduction

Head and Neck (H&N) cancers are among the most common cancers worldwide (5th leading cancer by incidence) [1]. Radiotherapy combined with cetuximab has been

V. Andrearczyk et al. (Eds.): HECKTOR 2021, LNCS 13209, pp. 58–67, 2022.
https://doi.org/10.1007/978-3-030-98253-9_4

established as a standard treatment [2]. However, locoregional failures remain a major challenge and occur in up to 40% of patients in the first two years after the treatment [3]. Recently, several radiomics studies based on Positron Emission Tomography (PET) and Computed Tomography (CT) imaging were proposed to better identify patients with a worse prognosis in a non-invasive fashion and by exploiting already available images such as these acquired for diagnosis and treatment planning [4–6].

Fluoro Deoxy Glucose (FDG)-Positron Emission Tomography (PET) and Computed Tomography (CT) imaging are the modalities of choice for the early staging and follow-up of Head and Neck (H&N) cancer [4]. Combined PET/CT imaging is broadly used in clinical practice for radiotherapy treatment planning, initial staging, and response assessment. The disease characteristics based on quantitative image biomarkers using medical images (i.e. radiomics) have great potential to optimize patient care, especially in (H&N) tumors. The quantitative evaluation of radiotracer endorsement in PET images and tissues density in CT images purposes at extracting clinically relevant features and building diagnostic, prognostic, and predictive models. The segmentation step of the radiomics workflow is the most time-consuming bottleneck and variability in traditional semi-automatic segmentation methods can significantly affect the extracted features. Under these circumstances, a fully automated segmentation is highly desirable to automate the whole process and facilitate its clinical routine usage.

The usage of PET information for automatic tumors segmentation has been explored for a long time in the literature. The different pre-deep learning-based methods such as thresholding, active contours, and mixture models have been used in-depth for the automatic segmentation of PET images [7]. Hatt et al. [8] at MICCAI 2016 proposed the first challenge on tumor segmentation in PET images. The basic standard and comparison between various methods for automatic tumor segmentation of PET images are highlighted and found in [9]. PET and CT images based on Multi-modal have been presented for various tasks such as lung cancer segmentation in [10–13] and bone lesion detection in [14]. Head and Neck Tumor segmentation challenge (HECKTOR) based on combined PET and CT images has been launched at MICCAI 2020 for assessing automatic algorithms for segmentation of Head and Neck (H&N) tumors. This challenge builds upon these works by comparing, on a publicly available dataset, recent segmentation architectures as well as the complementarity of the two modalities on a task of primary Gross Tumor Volume (GTVt) segmentation of H&N tumor in the oropharynx region. The challenge results were promising and provided some limitations on existing techniques that further motivate to improve the results using additional data curation, data cleaning, and new method generation. The problems and limitations in existing deep solutions are still need to solve and summaries as follows: (1) The existing techniques or methods are failed due to low FDG uptake on PET, a primary tumor that looks like a lymph node, abnormal uptake in the tongue and tumor present at the border of the oropharynx region. (2) The contours were mainly drawn based on the PET/CT fusion which is not sufficient to delineate the tumor. Other methods such as MRI with gadolinium or contrast CT are the gold standard to obtain the true contours for radiation oncology. Since the target clinical application is radiomics, however, the precision of the contours is not as important as for radiotherapy planning. Based on the aforementioned problems for tumor segmentation using combined CT and PET datasets, we have proposed a 3D-Inception-ResNet deep learning model for automatic segmentation of head and neck tumors. The main contributions in this paper are:

1. We have a 3D-Inception-ResNet deep learning model for head and neck tumor segmentation using CT/PET combined volume. The 3d-Inception module has been introduced at the encoder side and the 3D ResNet module has been proposed at the decoder side of the proposed model. The 3D squeeze and excitation (SE) module is also inserted in each encoder block of the proposed model.
2. The 3D depth-wise convolutional layer has been used in 3D inception and 3D ResNet module to get the optimized performance. The k-fold cross-validation dataset was used to train the proposed model.

2   Material and Methods

2.1   Head and Neck Tumor 2021(HECKTOR2021) Dataset Descriptions

The head and neck tumor dataset provided in MICCAI2021 consisted of the same patients that are published in HECKTOR20 MICCAI 2020 with an addition of 71 patients from a new center (CHUP), of which 23 were added to the training set, 48 to the testing set. The total number of training cases is 224 from 5 centers. The total number of test cases is 101 from two centers. A part of the test set will be from a center present in the training data (CHUP), another part from a different center (CHUV). For consistency, the GTVt (primary gross tumor volume) for these patients was annotated by experts following the same procedure as the curation performed for MICCAI 2020. We are only focused on task1 in this challenge. Task1 is about the segmentation of the head and neck region. All the dataset for 3D segmentation is provided in NIfTI volume along with the binary contour of the primary Gross Tumor Volume (GTVt). The bounding box for the training and testing dataset is provided in a CSV file. The evaluation (DSC scores) will be computed only within these bounding boxes at the original CT resolution. A detailed description of the dataset can be found [15, 16].

2.2   Proposed Method

The proposed 3D model based on the proposed 3D ResNet, 3D inception, and 3D Squeeze and excitation (SE) module is shown in Fig. 1. The proposed 3D model consisted of an encoder and decoder for 3D head and neck tumor segmentation. The input is combined with 3D PET/CT volume and the output will be a 3D segmentation map that consisted of tumor and background. The encoder module has four encoder blocks and each encoder block consisted of a $3 \times 3 \times 3$ Conv layer with 3D ResNet and SE block. The 3D maxpooling layer after each encoder block is used to downsample the spatial size of the input volume. Each decoder block consisted of a $3 \times 3 \times 3$ Conv layer with 3D inception module and a 3D up-sample layer used to increase the spatial size of the input volume. The feature maps are increased (24, 48, 96, 192, 384) in each encoder block and simultaneously, decreased (384, 192, 96, 48, 24) the feature maps in each decoder block.

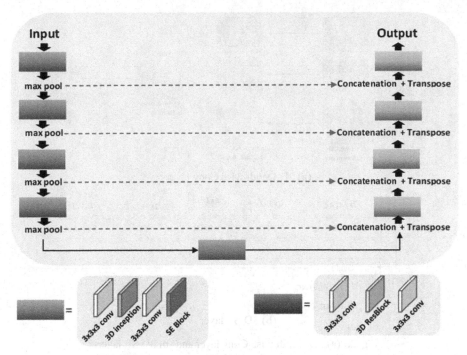

Fig.1. Proposed 3D segmentation of Head and Neck tumor from combined PET/CT 3D volume.

The detail of each 3D ResNet, SE, and 3D inception module are shown in Fig. 2 and Fig. 3. The network's left side works as an encoder to extract the features of different levels, and the right component of the network acts as a decoder to aggregate the features and the segmentation mask. Skip connection is used to combine coarse and fine feature maps for accurate segmentation. 3D depth-wise convolutional layer means that the computation is performed across the different channels (channel-wise). Inseparable convolution, the computation is factorized into two sequential steps: a channel-wise that processes channels independently and another 1x1xchannel Conv that merges the independently produced feature maps. The pointwise convolution combines linearly the output across all channels for every spatial location.

The squeeze operation performs based on the global average pooling (GAP) operation. The GAP module first squeezes the input into a scalar value per channel to capture the spatial content. The excitation function adaptively learns the inter-channel dependencies based on the fully connected layers. The sigmoid activation function and ReLU nonlinearity function are used as activation layers. The output of the 3D SE block is obtained by the channel-wise multiplication of input with the output of the sigmoid activation function. The 3D Resblock consisted of cascaded two blocks that include 3D Conv, 3D depth-wise Conv, and regularization layers such as ReLU activation and Instance3D normalization layer. The output of the two cascaded blocks is summed with the input that is passed another 3D Conv layer and residual output also passed another two regularization layers is shown in Fig. 3.

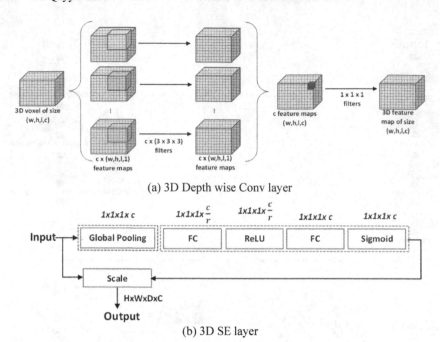

(a) 3D Depth wise Conv layer

(b) 3D SE layer

Fig. 2. (a) 3D Depth wise Conv layer and (b) 3D SE layer

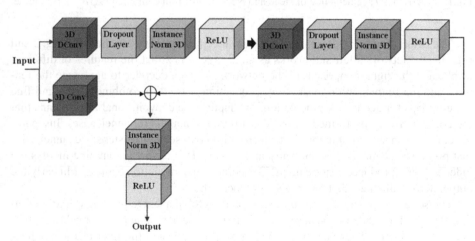

Fig. 3. 3D ResNet module. In this module, two cascaded blocks that include 3D Conv layer, Dropout, Instance 3D norm, and 3D Dconv (depth-wise convolutional layer) are added with input and passed to the regularization layers (IsntanceNorm 3D and ReLU).

In the GoogLeNet model, the concept of an "Inception module" has been presented to build a sparse, high-performance computing network architecture. The main idea presented in this module is to increase the depth and width of the network while keeping the computational budget constant. Furthermore, every convolutional layer is replaced

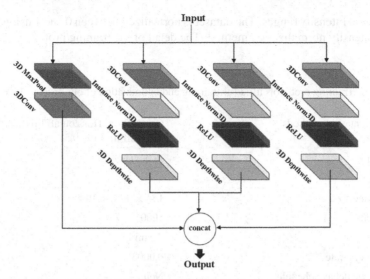

Fig. 4. 3D Inception module. In this module, four blocks are cascaded that include the 3D Conv layer, Dropout, Instance 3D norm, and 3D Dconv (depth-wise convolutional layer). The output of each block is cascaded channel-wise to get the multiscale feature maps information.

by an Inception module that consists of different layers with a various number of filters and is used in each module to avoid vanishing gradients during the training of deep models. During the training process, the accurate semantic segmentation is based on the local structural and global contextual information from medical images. The different architectures based on the multipath approach have been presented in the medical image segmentation that also used information at multi scales from given data [17].

Szegedy et al. [18] proposed 2D inception modules that capture features from input and aggregate these features at different scales for the classification task. We extended the 2D inception module into 3D with some extra 3D depth-wise layers for semantic segmentation for head and neck tumor segmentation. We proposed a 3D inception module in the decoder side of our proposed model. In the inception block, the features maps are aggregated from various branches using different kernels. The feature maps are aggregated by convolving with different kernels to minimize the number of parameters. Different sizes of the kernel make the network wider and have the capacity to learn more features. The proposed 3D inception module is shown in Fig. 4.

2.3 Training and Optimization Parameters

The proposed model is optimized using the Adam optimizer with a learning rate (0.0003). The dice and binary cross-entropy function in combined form were used as a loss function between ground-truth and predication mask. The 1000 number epochs with 2 batch sizes have been used for training and optimizing the proposed model. When a repeated validation dice score for continuous 20 epochs is reached, the early stopping criteria is used during the training of the proposed model. The PyTorch library has been used for the training, optimization, and development of the proposed model. The dataset cases have different intensity ranges. The dataset is normalized between 0 and 1 using the max and min intensity normalization method. The detail of the training protocol is shown in Table 1.

Table 1. Training protocols used for training and optimization of our proposed method.

Data augmentation methods	CenterCrop, HorizontalFlip (p = 0.5), VerticalFlip (p = 0.5)
Loss function	BCE+Dice
Batch size	2
Input volume size	$144 \times 144 \times 144$
Total epochs	1000
Optimizer	Adam
Initial learning rate	0.0003
Learning rate decay schedule	None
Stopping criteria, and optimal model selection criteria	The stopping criterion is reaching the maximum number of epoch (20)

3   Results and Discussion

3.1   Quantitative Results

Quantitative analysis of the proposed deep learning model using different proposed modules is given on the measurements of Dice coefficients (DC) and Hausdorff Distance (HD 95). The higher the DC, the better the segmentation results. Similarly, the lower the HD means the higher the model performance. The training dataset has been divided into 80% for training and 20% for validation and 5-fold cross-validation has been used for training and validation of our proposed model. In Table 2. the proposed 3D-inception-ResNet with 3D-SE model produced overall better DC and HD95 on validation fold1.

Table 3. showed that the proposed 3D-inception-ResNet-SE produced lower HD95 in fold1. The proposed model produced better DC and HD as compared to only the 3D ResNet module. The proposed model produced Dice score and HD95 score on the test dataset is shown in Table 4. The proposed model achieved lower dice on the test dataset as compared to the validation dataset. The performance is degraded due to test data samples may be different from validation data samples. The average performance on the test dataset for all participants was 75% Dice score means that our proposed method produced average performance on the test dataset.

Table 2. The Dice coefficient (DC) for 5-fold cross-validation using proposed 3D inception-ResNet and 3D-ResNet

Methods	Fold1	Fold2	Fold3	Fold4	Fold5
	DC				
Proposed 3D-inception-ResNet-SE	0.8112	0.7934	0.7899	0.8021	0.7945
Proposed 3D-ResNet-SE	0.7961	0.7781	0.7801	0.7632	0.7890

Table 3. The Hausdorff Distance (HD95) for 5-fold cross validation using proposed 3D inception-ResNet and 3D-ResNet.

Methods	Fold1	Fold2	Fold3	Fold4	Fold5
	HD95				
Proposed 3D-inception-ResNet-SE	10.25	11.56	10.99	12.66	11.66
Proposed 3D-ResNet-SE	11.91	13.77	12.17	11.98	12.36

Table 4. The Dice and HD95 on the test dataset.

Method	Dice	HD95
Proposed 3D-inception-ResNet-SE	0.7487	3.2700

3.2 Qualitative Results

The qualitative analysis using the proposed 3D model with and without inception block is shown in Fig. 5. The proposed model with inception block produced a better prediction map as compared to without inception block. The qualitative results showed the performance of our proposed 3D inception block for head and neck tumor segmentation. The proposed method without inception block produced overestimated pixels values at the boundary and some distorted pixels values from the center are shown in Fig. 5. We have shown the single slice from the whole volume using the CHGJ034 subject from the validation dataset.

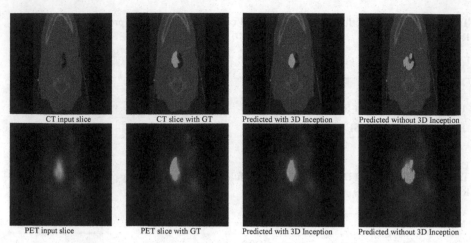

Fig. 5. The CT and PET input slice with ground truth (GT) and prediction using proposed with and without 3D inception block.

4   Conclusion and Future Work

In this paper, 3D deep learning model has been proposed for head and neck tumor segmentation. The proposed model produced optimal performance and produced average DC (0.81%) and HD95 (10.25) on the validation dataset. The results on the validation dataset are encouraging and may be useful for the real-time assessment of head and neck tumors in a combined PET/CT dataset. The proposed model could be used as the first step towards correct diagnoses and prediction of head and neck tumor segmentation. In the future, other 3D deep learning models (transformer-based) can be developed to further enhance head and neck tumor segmentation.

References

1. Parkin, D.M., Bray, F., Ferlay, J., Pisani, P.: Global cancer statistics 2002. CA Cancer J. Clin. **55**(2), 74–108 (2005)
2. Bonner, J.A., et al.: Radiotherapy plus cetuximab for locoregionally advanced head and neck cancer: 5-year survival data from a phase 3 randomised trial, and relation between cetuximab-induced rash and survival. Lancet Oncol. **11**(1), 21–28 (2010)
3. Chajon, E., et al.: Salivary gland-sparing other than parotid-sparing in definitive head-and-neck intensity-modulated radiotherapy does not seem to jeopardize local control. Radiat. Oncol. **8**(1), 1–9 (2013)
4. Vallieres, M., et al.: Radiomics strategies for risk assessment of tumour failure in head-and-neck cancer. Sci. Rep. **7**(1), 1–14 (2017)
5. Castelli, J., et al.: A PET-based nomogram for oropharyngeal cancers. Eur. J. Cancer **75**, 222–230 (2017)
6. Bogowicz, M., et al.: Comparison of PET and CT radiomics for prediction of local tumor control in head and neck squamous cell carcinoma. Acta Oncol. **56**(11), 1531–1536 (2017)
7. Foster, B., Bagci, U., Mansoor, A., Xu, Z., Mollura, D.J.: A review on segmentation of positron emission tomography images. Comput. Biol. Med. **50**, 76–96 (2014)

8. Hatt, M., et al.: The first MICCAI challenge on PET tumor segmentation. Med. Image Anal. **44**, 177–195 (2018)
9. Hatt, M., et al.: Classification and evaluation strategies of auto-segmentation approaches for PET: report of AAPM task group No. 211. Med. Phys. **44**(6), e1–e42 (2017)
10. Kumar, A., Fulham, M., Feng, D., Kim, J.: Co-learning feature fusion maps from PET-CT images of lung cancer. IEEE Trans. Med. Imaging **39**(1), 204–217 (2019)
11. Li, L., Zhao, X., Lu, W., Tan, S.: Deep learning for variational multimodality tumor segmentation in PET/CT. Neurocomputing **392**, 277–295 (2020)
12. Zhao, X., Li, L., Lu, W., Tan, S.: Tumor co-segmentation in PET/CT using multi-modality fully convolutional neural network. Phys. Med. Biol. **64**(1), 015011 (2018)
13. Zhong, Z., et al.: 3D fully convolutional networks for co-segmentation of tumors on PET-CT images. In: 2018 IEEE 15th International Symposium on Biomedical Imaging (ISBI 2018), pp. 228–231. IEEE, April 2018
14. Xu, L., et al.: Automated whole-body bone lesion detection for multiple myeloma on 68Ga-Pentixafor PET/CT imaging using deep learning methods. Contrast Media Mol. Imaging **2018**, article ID 2391925, 11 p. (2018). https://doi.org/10.1155/2018/2391925
15. Andrearczyk, V., et al.: Overview of the HECKTOR challenge at MICCAI 2021: automatic head and neck tumor segmentation and outcome prediction in PET/CT images. In: Andrearczyk, V., Oreiller, V., Hatt, M., Depeursinge, A. (eds.) HECKTOR 2021. LNCS, vol. 13209, pp. 1–37. Springer, Cham (2022)
16. Oreiller, V., et al.: Head and neck tumor segmentation in PET/CT: the HECKTOR challenge. Med. Image Anal. **77**, 102336 (2021) (under revision). - Refer to the challenge as "MICCAI 2021 HEad and neCK TumOR (HECKTOR) segmentation and outcome prediction challenge"
17. Kamnitsas, K., et al.: Efficient multi-scale 3D CNN with fully connected CRF for accurate brain lesion segmentation. Med. Image Anal. **36**, 61–78 (2017)
18. Szegedy, C., et al.: Going deeper with convolutions. In: Proceedings of the IEEE Conference on Computer Vision and Pattern Recognition, pp. 1–9 (2015)

The Head and Neck Tumor Segmentation in PET/CT Based on Multi-channel Attention Network

Guoshuai Wang[1,2], Zhengyong Huang[1,2], Hao Shen[1,2], and Zhanli Hu[1(✉)]

[1] Lauterbur Research Center for Biomedical Imaging, Shenzhen Institute of Advanced Technology, Chinese Academy of Sciences, Shenzhen 518055, China
zl.hu@siat.ac.cn
[2] University of Chinese Academy of Sciences, Beijing 100049, China

Abstract. Automatic segmentation of head and neck (H&N) tumors plays an important and challenging role in clinical practice and radiomics researchers. In this paper, we developed an automated tumor segmentation method based on combined positron emission tomography/computed tomography (PET/CT) images provided by the MICCAI 2021 Head and Neck Tumor (HECKTOR) Segmentation Challenge. Our model takes 3D U-Net as the backbone architecture, on which residual network is added. In addition, we proposed a multi-channel attention network (MCA-Net), which fuses the information of different receptive fields and gives different weights to each channel to better capture image detail information. In the end, our network scored well on the test set (DSC 0.7681, HD95 3.1549) (id: siat).

Keywords: Head and Neck tumor · Automatic segmentation · 3D U-Net · MCA-Net

1 Introduction

Head and neck squamous cell carcinoma (HNSCC) is the sixth most common cancer worldwide [1], with approximately 645,000 new cases of head and neck (H&N) cancer transpiring each year. The use of quantitative image biomarkers (i.e. radiomics) from medical images to predict pathological characteristics has shown eminent for optimizing patient care, notably for H&N tumors. Fludeoxyglucose (FDG)-Positron Emission Tomography (PET) and Computed Tomography (CT) are the preferred methods for the initial staging and follow-up of H&N cancer [2]. PET-CT plays an important role in the diagnosis and treatment of H&N tumors. Yet, an expensive and error-prone manual annotation process of the total tumor volume on medical images for radiomic analysis is needed. Manual annotation depends to a large extent on the oncologist's prior knowledge, with substantial intra- and inter-observer discrepancies in addition [3]. Consequently, it is of great significance to devise a diagnostic system that can automatically segment the H&N tumor target area, reduce the workload of oncologists and radiologists,

V. Andrearczyk et al. (Eds.): HECKTOR 2021, LNCS 13209, pp. 68–74, 2022.
https://doi.org/10.1007/978-3-030-98253-9_5

plus diminish the otherness and heterogeneity occasioned by contour mapping among different observers.

In the course of time, computer technology, artificial intelligence (AI), and deep learning (DL) have developed posthaste, the powerful modeling ability of convolutional neural network (CNN) has set off a new round of research upsurge. Deep learning-based image segmentation algorithms have also been introduced into the field of medical images. The automatic feature extraction capability of DL algorithms effectively overwhelms the deficiencies of traditional medical image segmentation algorithms that rely too much on the a priori knowledge of medical experts, nevertheless, the limitations of medical images (e.g., restrictions of imaging technology, sparse annotation data, etc.) determine that medical image segmentation faces with the need to increase the challenge.

The need for a standardized evaluation of automated segmentation algorithms and a comparative study of all current algorithms was highlighted in [4].The Head and Neck Tumor (HECKTOR) challenge [5] aims to advance automated segmentation algorithms by providing 3D PET/CT tumor datasets and real tumor segmentation labels annotated by medical experts, and this estimable knowledge will be utilized to numerous distinct cancer types relevant to PET/CT imaging, enabling large-scale and reproducible radiomics studies.

HECKTOR segmentation has attracted notable concentration in recent years. In [6], three DL algorithms consolidated with eight different loss functions were evaluated for PET image segmentation, and the performance of HECKTOR segmentation produced with several diverse combinations of bimodal images was compared in [7]. In view of the HECKTOR challenge, how to effectively integrate the bimodal synergistic information of PET and CT and how to achieve accurate (above medical expert labeling) automatic segmentation are still pressing problems. To address these problems, we propose a deep learning framework based on 3D U-Net [8] to perform prediction. The paper is structured as follows. See Sect. 2 for data description, network structure, selection of loss function, etc. The presentation and analysis of the experimental results are presented in Sect. 3. Finally, Sect. 4 concludes this paper.

2   Method

2.1   Data Preprocessing

The data sets were mainly from six different centers, including 224 patient samples from five different centers(CHGJ, CHUP, CHUM, CHMR, CHUS) in the training set and 101 patient samples from two different centers(CHUV, CHUP) in the test set. Each patient sample consists of CT and PET images and primary tumor volume (GTV) [2]. First, we sampled the data to the resolution of $1 \times 1 \times 1$ mm^3 and the size of $144 \times 144 \times 144$ by using the Trilinear interpolation. Then the Hounsfield Units of CT images are normalized to the range of $[-1, 1]$ and the PET images normalized using the mean and standard deviation:

$$x_i' = \frac{1}{\sigma_i}(x_i - \mu_i) \tag{1}$$

Where $\mu_i = E(x_i)$ and $\sigma_i = \sqrt{Var(x_i) + \varepsilon}$, ε is a small constant to prevent division by zero.

In addition, we perform random rotation (probability of rotation: 0.5, angular range for rotations: $0°$–$45°$) and mirror operation (probability of mirroring: 0.5) on the training set.

2.2 MCA Network

Attention mechanism was first proposed in the field of visual images, and Google Deep-Mind successfully used attention mechanism to realize image classification in RNN model in 2014 [9]. Subsequently, Bahdanau et al. successfully applied attention mechanism to the field of NLP [10]. Now, the attention mechanism based on CNN model has become a popular research topic. For example, Hu et al. proposed Squeeze-and-Excitation Networks (SE-Net) based on attention mechanism in 2017 [11] and X. Li et al. proposed Selective Kernel Networks (SK-Net) in 2019 [12]. Motivated by these works, we proposed Multi-channel Attention Network (MCA-Net) based on SK-Net and SE-Net. The architecture of MCA block is shown in Fig. 1.

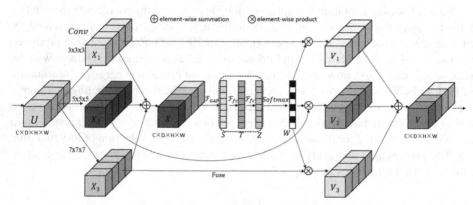

Fig. 1. The architecture of MCA block. The size of input X is D × H × W and C is the number of channels. In this paper, the feature map is split into three branches. The size of input is 144 × 144 × 144 and the initial C is 16. The size of output is the same as input.

As illustrated in Fig. 1, the given feature map $U = (u_1, u_2, \ldots, u_c)$ is split into multiple branches and each branch is implemented convolution operation respectively to obtain the information of different receptive fields. We implemented three conventional convolution with the size of kernel $3 \times 3 \times 3$, $5 \times 5 \times 5$ and $7 \times 7 \times 7$ with activation function of ReLU in this paper. We fuse the different information from three branches via an element-wise summation:

$$X = \sum_{i=1}^{n} X_i \tag{2}$$

Where n is the number of branches. Then we implement global average pooling and dimensionality reduction in each channel to obtain channel attention statistics $S \in \mathbb{R}^{C \times 1}$, and the i-th element of S is calculated by shrinking x_i:

$$s_i = \mathcal{F}_{GAP}(x_i) = \frac{1}{D \times H \times W} \sum_{j=1}^{D} \sum_{k=1}^{H} \sum_{l=1}^{W} x_i(j, k, l) \qquad (3)$$

$$V = \sum_{i=1}^{n} W \cdot X_i \qquad (4)$$

After S is obtained, $Z \in \mathbb{R}^{C \times 1}$ is obtained by two fully connected transformations without dimensionality reduction. $T \in \mathbb{R}^{C \times 1}$ is the result of the implicit layer and then Z is subjected to a softmax transformation to obtain the channel attention vector $W \in \mathbb{R}^{C \times 1}$, w_i represents the attention weight of each channel. Finally, we multiply the attention vector W with the result of the three convolutions and then add the results to get the final feature map V.

2.3 Network Architecture

In this work, our model structure is mainly based on 3D-UNet as backbone structure, on which residual networks [13] and our own MCA Networks are added. The overall architecture of the model is shown in Fig. 2:

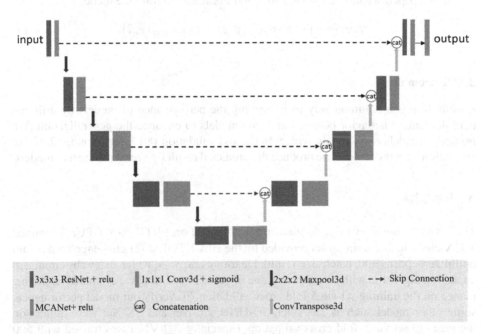

Fig. 2. The model architecture.

In our model, the input is PET/CT images with two channels and the initial number of convolution kernels is 16 and PET and CT contribute evenly to these channels. Double the number of convolution kernels after each pooling processing and in half after each deconvolution processing. Finally, we get the same size predicted image as input.

2.4 Loss Function

We pay attention to the similarity between predicted sample and label for image segmentation task. In general, we use the Dice coefficient to evaluate the segmentation result. To maximize the Dice coefficient, the Dice loss function [14] is usually used in training process. The Dice loss can be described as follows:

$$Loss_{Dice}(y, \hat{y}) = 1 - \frac{2 \sum_i^N y_i \hat{y}_i}{\sum_i^N y_i + \sum_i^N \hat{y}_i + \varepsilon} \tag{5}$$

Where y_i is the label and $\hat{y}_i$ is the predicted probability of i-th voxel. N is total number of voxels and ε is a small constant to avoid the zero division.

At the same time, we also added Focal loss function [15] to maximize the difference between background and segmentation target:

$$Loss_{Focal}(y, \hat{y}) = -(1 - \alpha)(1 - y)\hat{y}^\gamma \log(1 - \hat{y}) - \alpha y(1 - \hat{y})^\gamma \log \hat{y} \tag{6}$$

Where α is a balance weight factor, and the parameter α is set at 0.5. γ is the rate of weight decreases and the γ is set at 2.

In this paper, we utilize the Dice loss and Focal loss to train the model:

$$Loss(y, \hat{y}) = Loss_{Dice}(y, \hat{y}) + Loss_{Focal}(y, \hat{y}) \tag{7}$$

2.5 Ensembling

Ensembling is a common way of improving the performance of models by utilizing complementary behavior between multiple models to enhance the generalization. We trained 5 models respectively with 5-fold cross validation (80% for training, 20% for validation). On the test set, we produce the predicted result by averaging the five models.

3 Results

The network training was performed on NVIDIA GeForce RTX 3090 GPU. We trained MCA-Net with the training set provided by the HECKTOR 2021 challenge with Adam optimizer, epochs 300, batch size 1, initial learning rate 3e-4, which delays by simulated annealing algorithm with a minimum value of 1e-5. In addition, we evaluated its performance on the training set via 5-fold cross validation. To verify our model performance, some other model such as 3D-VNet, 3D-UNet, SE-Net and SK-Net were trained on the training set via 5-fold cross validation, especially, 3D-VNet was trained with 500 epochs every fold because if slow convergence. The segmentation of result could find in the Table 1. Compare to the 3D-VNet method, the proposed MCA-Net model improving segmentation performance by 4%–7% every fold and 7% in average result respectively.

The results show that proposed MCA-Net can efficiently segment H&N tumor with mean DSC of 0.811, more accurate compared to 3D-VNet (0.755), 3D U-Net (0.765), SE-Net (0.771) and SK-Net (0.789). So, we applied our MCA-Net model on the 101

Table 1. The result (DSC) of different methods in 5-fold cross validation

Method	Fold-0	Fold-1	Fold-2	Fold-3	Fold-4	Average
3D-VNet	0.749	0.744	0.743	0.778	0.763	0.755
3D-UNet	0.809	0.736	0.757	0.781	0.742	0.765
SE-Net	0.781	0.763	0.767	0.761	0.783	0.771
SK-Net	0.794	0.805	0.771	0.791	0.785	0.789
MCA-Net	0.809	0.776	0.821	0.838	0.814	0.811

test cases. At the same time, we implemented the above ensemble strategy by averaging the outputs of 5 models and gains the average DSC and HD95 scores with 0.7681, 3.1549 respectively. We found that our MCA-Net improved the head and neck tumor segmentation from PET and CT images. But it needs to be more GPU resources. At the same time, our MCA-Net could be applied to other similar segmentation tasks.

4 Discussion and Conclusion

In our MCA-Net, we implemented three conventional convolutions with the size of kernel $3 \times 3 \times 3, 5 \times 5 \times 5$ and $7 \times 7 \times 7$, which work better than two convolutions with the size of kernel $3 \times 3 \times 3, 5 \times 5 \times 5$. The reason is that more convolutional kernels of different sizes can obtain richer feature information, but the computational cost is also huge. As we continue to increase the number of branches in the convolution, we get a GPU memory shortage message, so in this paper we set the number of branches to 3 with the size of kernel $3 \times 3 \times 3, 5 \times 5 \times 5$ and $7 \times 7 \times 7$. To improve our segmentation performance, we also tried to apply some strategies, such as test time augmentation (TTA), which averages the predictions of the augmented test samples as the final prediction. But in our experiment, the TTA method cannot boost the performance of the model in the test set.

In the training process, we also found that the dice coefficient of some batch data was quite different from other data, which was close to 0. And we could call it an unusual phenomenon. We carefully check our data and try to change some other models (include 3D-Unet, 3D-Vnet). So, we suspect that the model in the poorest performance for the 'CHMR' center. Compared to other centers, the data of 'CHMR' center may influence our final model parameter and performance. Because of the number of data sets, we could not consider eliminating the data at that time. The processing of the data deserves further study in the future. We require more time to increase the capacity of our model network generalization.

In this paper, we proposed a multi-channel attention network to improve the segmentation of head and neck tumor, which has two advantages. First, compared with SE-Net, MCA-Net has multiple branches, which can extract richer feature information. Second, SE-Net and SK-Net reduce the dimension of the middle implicit layer when doing the fully connected transformation on the extracted attention vector, but MCA-Net does not, which avoids losing feature information. We achieved the average DSC and HD95 scores with 0.7681, 3.1549 respectively in the final rankings.

References

1. Hira, S., Vidhya, K., Wise-Draper, T.M.: Managing recurrent metastatic head and neck cancer. Hematol./Oncol. Clin. North Am. **35**, 1009–1020 (2021)
2. Andrearczyk, V., et al.: Overview of the HECKTOR challenge at MICCAI 2021: automatic head and neck tumor segmentation and outcome prediction in PET/CT images. In: Andrearczyk, V., Oreiller, V., Hatt, M., Depeursinge, A. (eds.) HECKTOR 2021. LNCS, vol. 13209, pp. 1–37. Springer, Cham (2022)
3. Gudi, S., et al.: Interobserver variability in the delineation of gross tumor volume and specified organs-at-risk during IMRT for head and neck cancers and the impact of FDG-PET/CT on such variability at the primary site. J. Med. Imaging Radiat. Sci. **48**(2), 184–192 (2017)
4. Hatt, M., et al.: Classification and evaluation strategies of auto-segmentation approaches for PET: report of AAPM task group no. 211. Med. Phys. **44**(6), e1–e42 (2017). https://doi.org/10.1002/mp.12124
5. Oreiller, V., et al.: Head and neck tumor segmentation in PET/CT: the HECKTOR challenge. Med. Image Anal. **77**, 102336 (2022). https://doi.org/10.1016/j.media.2021.102336
6. Shiri, I., et al.: Fully automated gross tumor volume delineation from PET in head and neck cancer using deep learning algorithms. Clin. Nucl. Med. **46**, 872–883 (2021)
7. Ren, J., et al.: Comparing different CT, PET and MRI multi-modality image combinations for deep learning-based head and neck tumor segmentation. Acta Oncologica (Stockholm, Sweden) **60**, 1399–1406 (2021)
8. Çiçek, Ö., Abdulkadir, A., Lienkamp, S., Brox, T., Ronneberger, O.: 3D U-Net: learning dense volumetric segmentation from sparse annotation. In: Ourselin, S., Joskowicz, L., Sabuncu, M.R., Unal, G., Wells, W. (eds.) MICCAI 2016. LNCS, vol. 9901, pp. 424–432. Springer, Cham (2016). https://doi.org/10.1007/978-3-319-46723-8_49
9. Mnih, V., Heess, N., Graves, A.: Recurrent models of visual attention. In: Advances in Neural Information Processing Systems, pp. 2204–2212 (2014)
10. Bahdanau, D., Cho, K., Bengio, Y.: Neural machine translation by jointly learning to align and translate. In: ICLR 2015, pp. 1–15 (2014)
11. Hu, J., Shen, L., Sun, G.: Squeeze-and-excitation networks. CoRR, vol.abs/1709.01507 (2017)
12. Li, X., Wang, W., Hu, X., Yang, J.: Selective kernel networks. In: 2019 IEEE/CVF Conference on Computer Vision and Pattern Recognition (CVPR), pp. 510–519 (2019). https://doi.org/10.1109/CVPR.2019.00060
13. He, K., et al.: Deep residual learning for image recognition. In: Proceedings of the IEEE Conference on Computer Vision and Pattern Recognition (2016)
14. Milletari, F., Navab, N., Ahmadi, S.-A.: V-net: fully convolutional neural networks for volumetric medical image segmentation. In: International Conference on 3D Vision, pp. 565–571. IEEE (2016)
15. Lin, T.-Y., Goyal, P., Girshick, R., He, K., Dollár, P.: Focal Loss for Dense Object Detection. arXiv preprint arXiv:1708.02002 (2017)

Multimodal Spatial Attention Network for Automatic Head and Neck Tumor Segmentation in FDG-PET and CT Images

Minjeong Cho[1], Yujin Choi[2], Donghwi Hwang[3,4], Si Young Yie[3,5,6], Hanvit Kim[7], and Jae Sung Lee[3,4,5,6,8,9](✉)

[1] School of Electrical and Electronics Engineering, Chung-Ang University, Seoul 06974, South Korea

[2] Department of Biomedical Engineering, Kyunghee University College of Electronics and Information, 1732 Deogyeong-daero, Giheung-gu, Yongin-si, Gyeonggi-do, South Korea

[3] Department of Nuclear Medicine, Seoul National University College of Medicine, Seoul 03080, South Korea
jaes@snu.ac.kr

[4] Department of Biomedical Sciences, Seoul National University College of Medicine, Seoul 03080, South Korea

[5] Interdisciplinary Program in Bioengineering, Seoul National University Graduate School, Seoul, Korea

[6] Integrated Major in Innovative Medical Science, Seoul National University Graduate School, Seoul, Korea

[7] Department of Electrical Engineering (EE), Ulsan National Institute of Science and Technology (UNIST), Ulsan, South Korea

[8] Artificial Intelligence Institute, Seoul National University, Seoul 08826, South Korea

[9] Brightonix Imaging Inc., Seoul 04782, Korea

Abstract. Quantitative positron emission tomography/computed tomography (PET/CT), owing to the functional metabolic information and anatomical information of the human body that it presents, is useful to achieve accurate tumor delineation. However, manual annotation of a Volume Of Interest (VOI) is a labor-intensive and time-consuming task. In this study, we automatically segmented the Head and Neck (H&N) primary tumor in combined PET/CT images. Herein, we propose a convolutional neural network named Multimodal Spatial Attention Network (MSA-Net), supplemented with a Spatial Attention Module (SAM), which uses a PET image as an input. We evaluated this model on the MICCAI 2021 HEad and neCK TumOR (HECKTOR) segmentation challenge dataset. Our method delivered a competitive cross-validation efficiency, with Dice Similarity Coefficient (DSC) 0.757, precision 0.788, and recall 0.785. When we tested out method on test dataset, we achieved an average DSC and Hausdorff Distance at 95% (HD95) of 0.766 and 3.155 respectively. Our team name is 'Heck_Uihak'.

Keywords: Multimodal image segmentation · PET-CT · Head and neck segmentation · HECKTOR 2021

© Springer Nature Switzerland AG 2022
V. Andrearczyk et al. (Eds.): HECKTOR 2021, LNCS 13209, pp. 75–82, 2022.
https://doi.org/10.1007/978-3-030-98253-9_6

1   Introduction

Positron Emission Tomography (PET) with fluorodeoxyglucose (FDG) is a useful imaging modality for detecting various cancers with high sensitivity and specificity. Although accurate tumor delineation is essential for quantifying reliable information of the tumor tissues, it remains a challenge in PET. Therefore, it may be necessary to utilize the complementary information provided by positron emission tomography/computed tomography (PET/CT) for accurate tumor segmentation [1].

PET/CT, an integrated imaging modality, can simultaneously acquire functional metabolic information and anatomical information of the human body. Accurate tumor delineation from PET and CT would allow for effective diagnosis, clinical decision-making, treatment planning, and personalized medicine. In this context, quantitative assessment of tumors provides valuable information and thus constitutes an important part of diagnostic procedures. Manual segmentation of Volume Of Interests (VOIs) in three dimensions is laborious, time-consuming, and error-prone, whereas automatic segmentation is desirable as it allows for a faster, objective, and accurate process.

Herein, we propose an automated segmentation method for Head and Neck (H&N) tumors with PET/CT using a deep-learning approach. To achieve the goal of the MICCAI Head and neck TumOR (HECKTOR) 2021 challenge [2, 3], we used a convolutional neural network named Multimodal Spatial Attention Network (MSA-Net), supplemented with a Spatial Attention Module (SAM) [4].

2   Method

2.1   Dataset

The data used in this study were provided in the MICCAI HECKTOR 2021 challenge. The dataset consisted of information of 325 patients from six hospitals; information of 224 of them formed the training dataset from five centers (CHGJ, CHMR, CHUM, CHUS, and CHUP), and that of the rest 101 formed the test dataset.

In the training dataset, the label was a CT resolution-based 3D binary description (0s and 1s) for the primary gross tumor volume (GTVt), and the 3D CT and FDG-PET input images were co-registered for each patient. Because the data were collected from different institutions, the voxel size of each image was different. Bounding boxes with dimensions of 144 mm × 144 mm × 144 mm were provided, indicating the locations of oropharyngeal tumors.

For five-fold cross-validations, HECKTOR 2021 training data were split into training and validation subsets. Although the data were collected from different centers, a total of 224 patients were divided into groups of 44, 45, 45, 45, and 45 patients for five-fold cross-validation. The validation subset was used to evaluate the performance of the model, whereas the training subset was used to train the model.

Information for the test dataset was collected from two different locations. A part of the test dataset was from a center that exists in the training data (CHUP), whereas the other part came from a different location (CHUV).

2.2   Data Preprocessing

As each patient image had a different voxel size, the original 3D CT and PET images were resampled to 1 mm × 1 mm × 1 mm pixel spacing with trilinear interpolation. The GTVt images were resampled to the same size spacing with nearest neighbor interpolation. Then, these images were cropped to a size of 144 mm × 144 mm × 144 mm using bounding box information.

The intensities of CT images were clipped in the range of [−1024, 1024] Hounsfield units. Then, these were mapped to the range of [−1, 1]. PET intensities were normalized using Z-score normalization.

2.3   Network Architecture

Our proposed MSA-Net consists of two main components: the SAM that consists of an image transformer network (ITN) [4, 5] and encoder–decoder CNN backbone, which is a UNet architecture [6] with the Squeeze-and-Excitation (SE) normalization layers [7, 8]. The SAM uses the input PET image to create a spatial attention map that could be used to find the tumor location. The tumor is extracted from the PET/CT data using the backbone network. The spatial attention map of the PET is downsampled with average pooling to multiply it at different scales and then multiplied by the PET/CT feature map generated at multiple stages of the encoder in the segmented backbone. The generated feature map is concatenated on the decoder. By doing this rather than multiplying the entire concatenation, the network can learn by comparing the feature map in which the location of the tumor is emphasized and the feature map in which it is not emphasized. Therefore, the PET/CT feature maps focus on the area where spatial interest of the PET is the strongest to produce the final segmentation. This model is illustrated in Fig. 1 (Fig. 2).

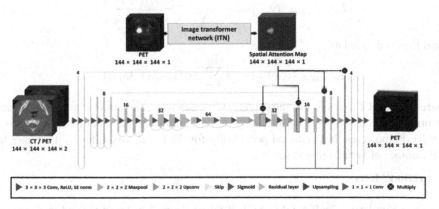

Fig. 1. Architecture of proposed network

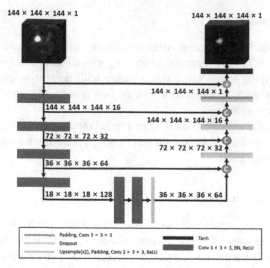

Fig. 2. Architecture of image transformer network

2.4 Training Scheme

Loss Function:
The network was trained using combination loss, which combines Soft Dice loss [9] and Focal loss [10] in (1).

$$L = L_{Soft\ Dice\ loss} + L_{Focal\ loss} \tag{1}$$

Soft dice loss is defined as

$$L_{Soft\ Dice\ loss}(y, \hat{y}) = 1 - \frac{2\sum_i^N y_i \hat{y}_i + \varepsilon}{\sum_i^N y_i + \sum_i^N \hat{y}_i + \varepsilon}. \tag{2}$$

Focal loss is defined as

$$L_{Focal loss}(y, \hat{y}) = -\frac{1}{N} \sum_i^N y_i (1 - \hat{y}_i)^\gamma \ln(\hat{y}_i), \tag{3}$$

where ε is set at 1 to avoid the zero division in cases when the tumor class is not present in training patches. In both definitions, $y_i \in \{0, 1\}$ indicates the label for the i-th voxel, $\hat{y}_i \in [0, 1]$ indicates the predicted probability for the $i - th$ voxel, and N indicates the total number of voxels. γ was set at 2.

Evaluation Metric:

(1) Dice Similarity Coefficient (DSC): It ranges from 0 to 1 and is used to determine how often the prediction and ground truth overlap.
    It is defined as

$$DSC(P, G) = \frac{2 \times |P \cap G|}{|P| + |G|}. \tag{4}$$

(2) Hausdorff Distance at 95% (HD95): The maximum Hausdorff distance between two sets is the distance between the closest points of the two sets. The maximum Hausdorff distance between sets P and G is a maximin function defined as

$$d_H(P, G) = \max\{d_{PG}, d_{GP}\} = \max\{\max_{p \in P} \min_{g \in G}(p, g), \max_{g \in G} \min_{p \in P}(p, g)\}. \quad (5)$$

95% HD is similar to maximum HD. However, it is based on the calculation of the 95th percentile of the distances between boundary points in P and G. This metric is used to eliminate the impact of a very small subset of the outliers.

Implementation:
We used Python and a trained network on a single GeForce RTX 3060 with 12 GB RAM using the PyTorch package, an open source deep-learning framework. The total number of epochs was 300. We used a batch size of 1 and the Adam optimizer [11]. After every 25 epochs, the cosine annealing schedule [12] was used to reduce the learning rate from 10^{-3} to 10^{-6}.

We flipped the input volume in left/right, superior/inferior, and anterior/posterior directions with a probability of 0.5 for data augmentation to prevent overfitting. Also, rotating input volume by a random angle between [0, 45] degrees with a probability of 0.5.

3 Results

We trained MSA-Net using the training dataset provided in the HECKTOR 2021 challenge and evaluated its performance on the training set with 5-fold cross-validation, as described earlier. We also performed training on the standard UNet for comparison with the proposed MSA-Net model. In other words, the UNet is trained with 5-fold cross-validation, and our proposed model also conducts 5-fold cross-validation. Table 1 presents the results of applying 5-fold cross-validation of UNet and MSA-Net. The DSC of our proposed model was 10.7% higher than that of the standard UNet.

Table 1. Mean segmentation results (DSC, precision, and recall) of the standard UNet and the proposed model in the five-fold cross-validation splits using the training image dataset.

	DSC		Precision		Recall	
	UNet	MSA-Net	UNet	MSA-Net	UNet	MSA-Net
fold-1	0.704	0.769	0.704	0.796	0.719	0.797
fold-2	0.676	0.755	0.748	0.802	0.704	0.751
fold-3	0.760	0.793	0.770	0.815	0.779	0.828
fold-4	0.625	0.704	0.674	0.730	0.657	0.758

(continued)

Table 1. (*continued*)

	DSC		Precision		Recall	
	UNet	MSA-Net	UNet	MSA-Net	UNet	MSA-Net
fold-5	0.654	0.764	0.724	0.797	0.675	0.793
Average	0.684	**0.757**	0.724	**0.788**	0.703	**0.785**

Our proposed model can catch tumors that are difficult to distinguish from the human eye, but the standard UNet cannot. Also, this model segments the shape of the tumor better, as shown by the predicted segmentations (Fig. 3 and Fig. 4).

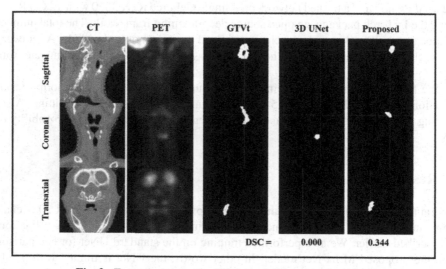

Fig. 3. Exemplary outputs of UNet and the proposed method

Our results for the test set were obtained using ensembles of five trained models. The test set predictions were made by averaging the predictions of the different models and applying a threshold operation with a value of 0.4. For the test set, our proposed network achieved DSC and HD95 of 0.766 and 3.155, respectively.

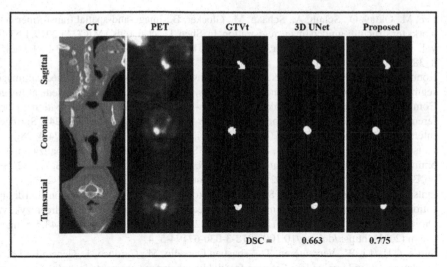

Fig. 4. Another exemplary outputs of UNet and the proposed method

4 Discussion and Conclusions

Herein, we propose a framework for automatically segmenting H&N tumors. The framework consists of two models. The first model, SAM, learns to highlight tumor areas spatially. Then, the final segmentation is performed through the second backbone network. The experimental results on the training dataset show that our proposed network has higher performance than the standard UNet. When our method was evaluated using test data, the DSC and HD95 of our network were 0.766 and 3.155.

Some data include tumors which are difficult to be distinguished due to their dimness and/or small size. The backbone model had difficulty to segment these tumors from the background. Meanwhile, since the spatial attention module could provide spatial information of area most likely to be tumors to the backbone model, performance of the proposed model outperformed the backbone model's.

References

1. Zhong, Z., et al.: 3D fully convolutional networks for co-segmentation of tumors on PET-CT images. In: 2018 IEEE 15th International Symposium on Biomedical Imaging (ISBI 2018), pp. 228–231. IEEE (2018)
2. Andrearczyk, V., et al.: Overview of the HECKTOR challenge at MICCAI 2021: automatic head and neck tumor segmentation and outcome prediction in PET/CT images. arXiv preprint arXiv:2201.04138 (2022)
3. Oreiller, V., et al.: Head and neck tumor segmentation in PET/CT: the HECKTOR challenge. Med. Image Anal. **77**, 102336 (2021)
4. Fu, X., et al.: Multimodal spatial attention module for targeting multimodal PET-CT lung tumor segmentation. IEEE J. Biomed. Health Inf. **25**, 3507–3516 (2021)

5. Lee, M., Oktay, O., Schuh, A., Schaap, M., Glocker, B.: Image-and-spatial transformer networks for structure-guided image registration. In: Shen, D., et al. (eds.) MICCAI 2019. LNCS, vol. 11765, pp. 337–345. Springer, Cham (2019). https://doi.org/10.1007/978-3-030-32245-8_38

6. Ronneberger, O., Fischer, P., Brox, T.: U-net: convolutional networks for biomedical image segmentation. In: Navab, N., Hornegger, J., Wells, W.M., Frangi, A.F. (eds.) Medical Image Computing and Computer-Assisted Intervention – MICCAI 2015: 18th International Conference, Munich, Germany, October 5-9, 2015, Proceedings, Part III, pp. 234–241. Springer International Publishing, Cham (2015). https://doi.org/10.1007/978-3-319-24574-4_28

7. Iantsen, A., Jaouen, V., Visvikis, D., Hatt, M.: Squeeze-and-excitation normalization for brain tumor segmentation. In: Crimi, A., Bakas, S. (eds.) BrainLes 2020. LNCS, vol. 12659, pp. 366–373. Springer, Cham (2021). https://doi.org/10.1007/978-3-030-72087-2_32

8. Iantsen, A., Visvikis, D., Hatt, M.: Squeeze-and-excitation normalization for automated delineation of head and neck primary tumors in combined PET and CT images. In: Andrearczyk, V., Oreiller, V., Depeursinge, A. (eds.) HECKTOR 2020. LNCS, vol. 12603, pp. 37–43. Springer, Cham (2021). https://doi.org/10.1007/978-3-030-67194-5_4

9. Sudre, C.H., Li, W., Vercauteren, T., Sebastien Ourselin, M., Cardoso, J.: Generalised dice overlap as a deep learning loss function for highly unbalanced segmentations. In: Jorge Cardoso, M., et al. (eds.) DLMIA/ML-CDS -2017. LNCS, vol. 10553, pp. 240–248. Springer, Cham (2017). https://doi.org/10.1007/978-3-319-67558-9_28

10. Lin, T.-Y., et al.: Focal loss for dense object detection. In Proceedings of the IEEE International Conference on Computer Vision. pp. 2980–2988 (2017)

11. Kingma, D.P., Ba, J.: Adam: a method for stochastic optimization (2014)

12. Loshchilov, I., Hutter, F.: Sgdr: stochastic gradient descent with warm restarts (2016)

PET Normalizations to Improve Deep Learning Auto-Segmentation of Head and Neck Tumors in 3D PET/CT

Jintao Ren[1], Bao-Ngoc Huynh[2], Aurora Rosvoll Groendahl[2], Oliver Tomic[2], Cecilia Marie Futsaether[2](✉), and Stine Sofia Korreman[1](✉)

[1] Department of Clinical Medicine, Aarhus University, Nordre Ringgade 1, 8000 Aarhus, Denmark
stine.korreman@oncology.au.dk

[2] Faculty of Science and Technology, Norwegian University of Life Sciences, Universitetstunet 3, 1433 Ås, Norway
cecilia.futsaether@nmbu.no

Abstract. Auto-segmentation of head and neck cancer (HNC) primary gross tumor volume (GTVt) is a necessary but challenging process for radiotherapy treatment planning and radiomics studies. The HEad and neCK TumOR Segmentation Challenge (HECKTOR) 2021 comprises two major tasks: auto-segmentation of GTVt in FDG-PET/CT images and the prediction of patient outcomes. In this paper, we focus on the segmentation part by proposing two PET normalization methods to mitigate impacts from intensity variances between PET scans for deep learning-based GTVt auto-segmentation. We also compared the performance of three popular hybrid loss functions. An ensemble of our proposed models achieved an average Dice Similarity Coefficient (DSC) of 0.779 and median 95% Hausdorff Distance (HD95) of 3.15 mm on the test set. Team: Aarhus_Oslo.

Keywords: Head and neck cancer · Deep learning · Gross tumor volume · Auto-segmentation

1 Introduction

Head and neck cancer (HNC) is the sixth most frequently occurring cancer globally [1]. For both radiotherapy treatment planning and radiomics research, determining the gross tumor volume (GTV) is critical. The task of outlining the primary tumor (GTVt) and the associated lymph nodes (GTVn) in HNC is known as GTV delineation. Manual delineation of GTV is a labor-intensive task that frequently suffers from inter-observer variations (IOV) due to the complicated

---

J. Ren and B.-N. Huynh—Authors contributed equally.

V. Andrearczyk et al. (Eds.): HECKTOR 2021, LNCS 13209, pp. 83–91, 2022.
https://doi.org/10.1007/978-3-030-98253-9_7

anatomical environment of the head and neck (H&N) region and irregular morphologies with ambiguous tumor boundaries. Auto-segmentation can potentially reduce labor requirements and increase delineation consistency. Recent advancements in deep learning have opened up the possibility of high quality automated GTV delineation.

MICCAI 2021 - HECKTOR2021 (HEad and neCK TumOR segmentation and outcome prediction in PET/CT images 2021) challenge provides a good platform for participants from different backgrounds around the world to compare a variety of approaches with the same data and evaluation criteria [2,3]. The challenge is composed of three tasks: 1. GTVt segmentation in Positron Emission Tomography/Computed Tomography (PET/CT) images. 2. Progression-Free Survival (PFS) prediction from PET/CT images and available clinical data. 3. PFS prediction with the addition of GTVt delineation ground truth. The focus of this study is task 1 - GTVt segmentation.

Fluorodeoxyglucose (FDG)-PET and CT are the two standard medical imaging modalities used for diagnosis and treatment planning of HNC. CT provides the anatomy, while PET shows metabolic tumor characteristics [4,5]. In the clinic, physicians use the fused FDG-PET/CT images to locate the target of interest and delineate the GTV. Previously, deep learning-based GTV auto-segmentation studies have shown that PET plays a more crucial role than CT in the GTV segmentation [6,7]. This is attributable to the fact that PET highlights high-metabolic activities from tumor cells. However, the limited spatial resolution and low signal-to-noise ratio of FDG-PET images [8] may also lead to erroneous predictions [7]. PET also has higher numerical intensity variations between different institutions than CT, which can be caused by differences in individual FDG dosages and scanner settings.

A variety of approaches have been proposed in HNC GTV segmentation to better utilize multi-modality information [9,10]. No-new-UNet (nnUNet) [11] has emerged as an out-of-the-box tool that provides close to state-of-the-art (SOTA) performance without requiring competence in conventional deep learning practice techniques (i.e., preprocessing, hyper-parameter tuning, structure adjustments, etc.). In this study, we use nnUNet as the deep learning pipeline for the segmentation experiments.

To address the above-listed problems of PET, we explored two normalization methods, namely PET-clip and PET-sin, on PET images to improve the accuracy of GTVt automatic segmentation. Furthermore, we also compared the impact of multiple loss functions on the segmentation task.

2   Materials and Methods

2.1   Data

HECKTOR2021 provided FDG-PET/CT images and patient clinical information of 224 oropharyngeal cancer patients from five clinical centers, with the delineated GTVt serving as the ground-truth. The total number of test cases was 101 from two centers, of which one was an external center. For each patient, a $144 \times 144 \times 144\,\mathrm{mm}^3$ bounding box was provided to crop the relevant H&N region from PET/CT. In addition, the PET intensities had been converted to Standardized Uptake Values (SUV).

Following instructions from HECTOR2021, we resampled PET/CT 3D images, along with the 3D delineated GTVt mask to a $1\,\mathrm{mm}^3$ isotropic grid using spline interpolation and nearest-neighbor interpolation, respectively. Subsequently, we cropped them into volumes of size $144 \times 144 \times 144\,\mathrm{mm}^3$ using the provided bounding box.

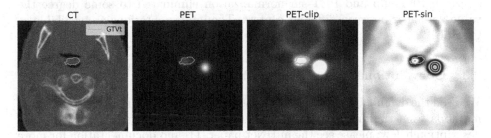

Fig. 1. An example image slice from CHUM054. From left to right: CT, original PET, PET-clip, and PET-sin. The ground truth GTVt is annotated in green. On the original PET, the SUV for GTVn is higher than GTVt. After PET was clipped, both GTVt and GTVn have similar SUV. GTVn on the PET-sin image shows an onion-like structure. (Color figure online)

2.2 PET-Clip and PET-Sin Normalization

For the segmentation task, one of the hurdles is the model's capability to generalize to the external center data. Many candidates performed well on their validation set but poorly on the external test set in the previous competition (HECKTOR2020 [12]). To minimize the generalization error, we assessed two normalization strategies on PET.

FDG-PET is an excellent imaging method for diagnosing cancer. However, FDG not only accumulates in tumors, but often also appears in inflammation, biopsy, wound repair, and benign regions [8]. Involved lymph nodes also often have very high SUV, but segmentation of these nodes was not included in this segmentation task. In some circumstances, if the lymph node areas have a higher SUV than tumor (Fig. 1), the model may confuse the GTVn with the GTVt. The PET-clip approach was designed to reduce the intensity discrepancies between GTVn and GTVt so that the model treats both areas equally. Previous research [13] suggests that SUV with a threshold of 5.5 seems to have little impact on the predictive value of metabolic tumor volume. We boldly use 5 as the threshold to better minimize the disparities across different data sources. Specifically, all PET voxels in the PET-clip group were clipped into the range of [0–5] SUV.

Another approach we attempted for PET normalization was the trigonometric sine function. Each PET voxel intensity x_n (in SUV) was transformed using $y_n = \sin(x_n)$, where y_n refers to the PET voxel intensity after sine transformation. PET-sin maps all the SUV values to a range between $[-1, 1]$. We noticed that GTVn on PET-sin has a multi-layer structure that resembles an ellipsoidal onion in the axial view. This is dependent on the specific SUV value and the GTV's shape. Thus, the sine-transformation mapped the SUV from the PET image into a narrower standardized value range while preserving the clear distinctions between the high and low SUV regions.

Both PET-clip and PET-sin normalization eliminated to some degree the value variances between different centers. The implementations are available at GitHub[1].

2.3 Convolutional Neural Network (CNN)

We used the default UNet network structure, pre-processing, and post-processing methods from the nnUNet 3D full-resolution pipeline. We will provide a brief description below; please see the nnUNet paper [11] and documentation for more details.

From the original size $2 \times 144 \times 144 \times 144$, the PET/CT images were cropped into $2 \times 128 \times 128 \times 128$ patches with overlaps. To remove any extraneous information, CT intensities were clipped to the 5 to 95 percentile of the Hounsfield units (HU). Then, each image (including PET-clip and PET-sin) was standardized using z-scores (subtraction of the mean and division by the standard deviation). To avoid overfitting, image augmentation was used with random rotation in all directions between $[-30, 30]$ degrees and random scaling between $[0.7, 1.4]$. All 224 patients were divided into five groups at random for five-fold cross-validation and trained five times, with one fold of patients being left out for validation and the others for training each time.

2.4 Experiments

We used the combined CT and PET with Dice + Cross-Entropy (CE) loss as the baseline model; CT and PET-clip with Dice + CE loss as 'PET-clip'; CT and PET-sin with Dice + CE loss as 'PET-sin'.

Previous studies [10,14] on GTV segmentation have shown the necessity of employing loss functions which address the class imbalance problem. However, these studies all used different training setups (network structures, data pipelines, etc.). In order to understand the differences in performance across different loss functions, we also evaluated Dice + Focal ($\gamma = 2$) loss and Dice + TopK ($K = 10$) loss functions [15] using identical training setups as baseline.

For evaluation, we used the Dice Similarity Coefficient (DSC) and 95% Hausdorff Distance (HD95) as primary metrics, as per the challenge setup. The mean DSC and median HD95 of five-fold cross-validation were used to evaluate the

---

[1] https://github.com/BBQtime/HNSCC-ct-pet-tumor-segmentation.

training performance. For test set segmentation, we calculated the mean of predicted softmax probability maps from five-fold cross-validation models before applying a step function with a 0.5 threshold.

Each model was trained for 1000 epochs using the Stochastic gradient descent (SGD) optimizer on an NVIDIA RTX Quadro 8000 (48 GB) GPU with a batch size of 2. The initial learning rate was 0.01 and decayed with the Poly learning-rate scheduler [16]. For each fold of the five-fold cross-validation, the training time was about 50 h.

3   Results

Across the five-fold cross-validation results, there were no substantial differences between the overall performance of models using the different PET normalization schemes, as shown in Table 1.

Table 1. Average scores from five-fold Cross-validation between different experiment groups. Columns 1, 2, 3 applied Cross-entropy + Dice loss; Columns 1, 4, 5 applied only z-score normalization, whereas columns 2 and 3 used clip and sin normalization prior to z-score normalization.

	1. Baseline	2. PET-clip	3. PET-sin	4. Dice + Focal	5. Dice + TopK
DSC	0.77	0.77	0.76	0.76	0.75
HD95 (mm)	2.72	2.72	2.93	2.92	3.14

PET-clip achieved the same overall performance as baseline, while PET-sin slightly decreased the scores Table 1. In addition, all combinations of loss functions using the Dice loss (baseline Dice + CE, Dice + Focal, and Dice + TopK) achieved comparable performances; nevertheless, models utilizing the baseline configuration performed slightly better.

However, in some cases, after applying either the PET-clip or the PET-sin, the segmentation result improved. For instance, Fig. 2 illustrates a false positive (top) and a false negative (bottom) segmentation predicted by the baseline model. Both of these effects were reduced after using the two proposed PET normalization methods, thereby increasing the model performance.

Table 2. Model evaluation on the external test set (DSC rounded up by 4 decimal places to show differences).

	1. Baseline	2. PET-clip	3. PET-sin	4. Ensemble 2 + 3	5. Ensemble 1 + 2 + 3
DSC	0.7671	0.7747	0.7726	0.7787	0.7790
HD95 (mm)	3.15	3.27	3.27	3.15	3.15

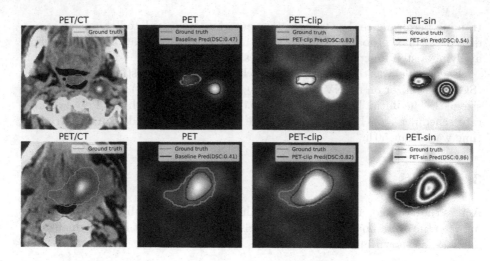

Fig. 2. Top row: On the PET original image, GTVt has a lower SUV as compared to GTVn, and the baseline model predicted a false positive region on GTVn giving a DSC = 0.47. On the PET-clip image, GTVt SUV was at the same level as GTVn, and the PET-clip model ignored GTVn giving a high DSC = 0.83. Bottom row: The false-negative region was reduced after application of either PET-clip or PET-sin.

For the evaluation on test data, we made five trials following the challenge rule. They were baseline, PET-clip, PET-sin, the ensemble of two models (baseline + PET-clip), and the ensemble of three models (baseline + PET-clip + PET-sin). Among these, the ensemble of three models achieved the best DSC and HD95, although the improvement was marginal (Table 2).

4   Discussion and Conclusion

In this study, two PET image normalization techniques have been proposed, aiming to improve the GTVt segmentation performance. Through the PET-clip, all of the high SUV regions on the PET image were trimmed to the same level as GTVt. As a result, the model cannot easily identify GTVt from non-GTVt regions depending on PET intensity alone; therefore, spatial anatomical information from CT may have a higher weight for the task. On the other hand, the PET-sin turned the heatmap-based PET image into a more structure-based image. Changes in the SUV gradient in the GTV gave rise to an onion-like layer structure. While we chose 1 as the scale factor for SUV in the sine function, a larger factor value will produce more "onion rings", which may help on GTV localization. However, this will introduce additional noise into the image on small SUV uptake regions, thus reducing the model's prediction accuracy. It would be interesting to investigate further by making this parameter learnable. A more in-depth investigation into visualizing activation maps would assist in understanding the effects of both approaches.

According to the cross-validation scores, neither of the two PET normalization approaches substantially impacted the segmentation results. According to the external test set results, these approaches led to a small improvement in DSC compared to the baseline method. The adapted deep learning pipeline with default z-score normalization appears to a certain degree to be sufficient to compensate for the PET intensity variations between different centers. However, we cannot conclude that the two introduced normalization methods reduced the generalization error for external center data since we split folds randomly rather than grouping by distinct centers during cross-validation training. A leave-one-center-out cross-validation study should be conducted to explore the problem further.

From an information theory perspective, PET manipulations do not supply additional information to the data but rather discard some information (PET-clip). This experiment suggests, in agreement with earlier research [6,7], that PET is significant in determining the location of a tumor, but not so in precisely defining the GTVt contour. A cascade model, which first provides a preliminary contour using PET and then refines it with CT, could be helpful for this GTVt segmentation task.

For cross-validation, all three hybrid loss functions performed similarly. The baseline Dice + CE loss appears to be sufficient for this GTVt segmentation task. We made no substantial adjustments to the Focal loss and TopK loss parameters, which might have improved their performance. Nevertheless, doing so will be a tedious task.

Finally, the best test set DSC and HD95 were achieved using the ensemble of baseline, PET-clip, and PET-sin. It must be admitted that an ensemble approach is widely used throughout our experiment, and each independent model itself was obtained by an ensemble through five-fold cross-validation. An ensemble approach has the advantage of greatly reducing variance and improving the model's robustness. The disadvantage is that it necessitates the training of multiple models, which takes a long time.

In conclusion, there was no negative impact of PET-clip or PET-sin normalization on the segmentation outcomes. However, whether these changes improve the model's generalization ability should be explored further. The sum of the Dice and CE loss functions is sufficient for the task with the proposed pipeline. An ensemble of baseline, PET-clip, and PET-sin models achieved the highest DSC and HD95 on the external test data.

Acknowledgments. This work is supported by Aarhus University Research Foundation (No. AUFF-F-2016-FLS-8-4); Danish Cancer Society (No. R231-A13856, No. R191-A11526) and DCCC Radiotherapy - The Danish National Research Center for Radiotherapy.

References

1. Economopoulou, P., Psyrri, A.: Head and neck cancers: Essentials for clinicians, chap. 1 (2017)
2. Oreiller, V., et al.: Head and Neck Tumor Segmentation in PET/CT: The HECKTOR Challenge. Medical Image Analysis (2021)
3. Andrearczyk, V., et al.: Overview of the HECKTOR challenge at MICCAI 2021: automatic head and neck tumor segmentation and outcome prediction in PET/CT images. In: Andrearczyk, V., Oreiller, V., Hatt, M., Depeursinge, A. (eds.) HECKTOR 2021. LNCS, vol. 13209, pp. 1–37. Springer, Cham (2022)
4. Boellaard, R., et al.: FDG PET/CT: EANM procedure guidelines for tumour imaging: version 2.0. Eur. J. Nuclear Med. Mol. Imaging **42**(2), 328–354 (2014). https://doi.org/10.1007/s00259-014-2961-x
5. Jensen, K., et al.: The Danish Head and Neck Cancer Group (DAHANCA) 2020 radiotherapy guidelines. Radiotherapy Oncol. **151**, 149–151 (2020). https://doi.org/10.1016/j.radonc.2020.07.037
6. Groendahl, A.R., et al.: A comparison of methods for fully automatic segmentation of tumors and involved nodes in PET/CT of head and neck cancers. Phys. Med. Biol. **66**(6), 065012 (2021). https://doi.org/10.1088/1361-6560/abe553
7. Ren, J., Eriksen, J.G., Nijkamp, J., Korreman, S.S.: Comparing different CT, PET and MRI multi-modality image combinations for deep learning-based head and neck tumor segmentation. Acta Oncologica, pp. 1–8, July 2021. https://doi.org/10.1080/0284186x.2021.1949034
8. Rosenbaum, S.J., Lind, T., Antoch, G., Bockisch, A.: False-positive FDG PET uptake-the role of PET/CT. Eur. Radiol. **16**(5), 1054–1065 (2005). https://doi.org/10.1007/s00330-005-0088-y
9. Guo, Z., Guo, N., Gong, K., Li, Q., et al.: Gross tumor volume segmentation for head and neck cancer radiotherapy using deep dense multi-modality network. Phys. Med. Biol. **64**(20), 205015 (2019)
10. Iantsen, A., Visvikis, D., Hatt, M.: Squeeze-and-excitation normalization for automated delineation of head and neck primary tumors in combined PET and CT images. In: Andrearczyk, V., Oreiller, V., Depeursinge, A. (eds.) HECKTOR 2020. LNCS, vol. 12603, pp. 37–43. Springer, Cham (2021). https://doi.org/10.1007/978-3-030-67194-5_4
11. Isensee, F., Jaeger, P.F., Kohl, S.A., Petersen, J., Maier-Hein, K.H.: nnu-net: a self-configuring method for deep learning-based biomedical image segmentation. Nat. Methods **18**(2), 203–211 (2021)
12. Andrearczyk, V., et al.: Overview of the HECKTOR challenge at MICCAI 2020: automatic head and neck tumor segmentation in PET/CT. In: Andrearczyk, V., Oreiller, V., Depeursinge, A. (eds.) HECKTOR 2020. LNCS, vol. 12603, p. Overview of the HECKTOR challenge at MICCAI 2020: automatic head and neck tumor segmentation in PET/CT-21. Springer, Cham (2021). https://doi.org/10.1007/978-3-030-67194-5_1
13. Castelli, J., Depeursinge, A., De Bari, B., Devillers, A., De Crevoisier, R., Bourhis, J., Prior, J.O.: Metabolic tumor volume and total lesion glycolysis in oropharyngeal cancer treated with definitive radiotherapy: which threshold is the best predictor of local control? Clin. Nucl. Med. **42**(6), e281–e285 (2017)
14. Ma, J., Yang, X.: Combining CNN and hybrid active contours for head and neck tumor segmentation in CT and PET images. In: Andrearczyk, V., Oreiller, V., Depeursinge, A. (eds.) HECKTOR 2020. LNCS, vol. 12603, pp. 59–64. Springer, Cham (2021). https://doi.org/10.1007/978-3-030-67194-5_7

15. Ma, J., Chen, J., Ng, M., Huang, R., Li, Y., Li, C., Yang, X., Martel, A.L.: Loss odyssey in medical image segmentation. Medical Image Analysis, p. 102035 (2021)
16. Chen, L.C., Papandreou, G., Kokkinos, I., Murphy, K., Yuille, A.L.: DeepLab: semantic image segmentation with deep convolutional nets, atrous convolution, and fully connected CRFs. IEEE Trans. Pattern Anal. Mach. Intell. **40**(4), 834–848 (2017)

The Head and Neck Tumor Segmentation Based on 3D U-Net

Juanying Xie[✉] [iD] and Ying Peng

School of Computer Science, Shaanxi Normal University, Xi'an 710119,
People's Republic of China
xiejuany@snnu.edu.cn

Abstract. Head and neck cancer is one of the common malignancies. Radiation therapy is primary treatment of this type of cancer. Mapping the target area of the head and neck tumor is the key step to make the appropriate radiotherapy schedule. However, it is a very time consuming and boring work. Therefore, automatic segmenting the head and neck tumor is of the very significant work. This paper adopts the U-Net network used in medical image segmentation commonly to carry out the automatic segmentation to head and neck tumors based on the dual-modality PET-CT images. The 5-fold cross validation experiments are carried out. The average experimental results are 0.764, 7.467, 0.839, and 0.797 in terms of Dice score, HD95, recall, and precision, respectively. The mean of Dice and the median of HD95 on the test set are 0.778 and 3.088, respectively.

Keywords: Head and neck cancers · U-Net network · Image segmentation · PET-CT images

1 Introduction

Head and neck cancers are the most common cancers in the world. It is the cancer whose numbers of the primary sites and of the pathological types are both ranked in the first place compared to other cancers [1]. Radiation therapy is one of the most effective treatments for the head and neck cancers. However, the radiation therapy relies heavily on the labeled tumor locations on medical images. Manual labeling tumor locations is a time-consuming and laboring work. Furthermore, the accuracy of the manual labeling depends on the subjective judgment relying on the professional experience of the physician, which is very unstable [2]. The different individual tumors are often very different in shapes and sizes. The traditional segmentation methods are usually inefficient and inaccurate, so that they often cannot meet the clinical requirements. Therefore, it is very significant to study the automatic segmentation of the head and neck tumors.

The most widely used instrument in imaging head and neck is the PET/CT scanner in clinical practice. It integrates PET imaging and CT imaging into the same device. PET and CT images characterize lesions from different but complementary ways. The former provides physiological and metabolic information to locate the tumor, while the latter provides anatomical information indicating the location and boundaries of normal

© Springer Nature Switzerland AG 2022
V. Andrearczyk et al. (Eds.): HECKTOR 2021, LNCS 13209, pp. 92–98, 2022.
https://doi.org/10.1007/978-3-030-98253-9_8

tissue. These two types of modes together can provide more accurate references for labeling tumors.

In recent years, deep learning has been widely used in image classification, segmentation and object detection due to its high robustness and efficiency. Therefore, this paper adopts the U-Net [3] commonly used in medical image segmentation to carry out the automatic segmentation of the head and neck tumors based on the dual-modality PET-CT images.

2 Data and Methods

2.1 Dataset

The dataset used in the experiments in this paper are from the Head and Neck Tumor Segmentation Challenge named HECTOR2021 [4, 5]. The training set contains 224 cases in total from 5 centers, such as CHGJ, CHMR, CHUM, CHUS, and CHUP. The testing set contains 101 cases in total from 2 centers including CHUP and CHUV. Each case includes both CT and PET images, GTVt (primary Gross Tumor Volume, that is, the label of each case) in NIfTI format, as well as a bounding box to locate the oropharynx region. In addition, each case also contains the patient's gender and age and other information. The test data do not include GTVt information.

2.2 Data Preprocessing

In order to improve the quality of the data so that the network can learn the features of the data easily, the following preprocessing operations are performed to the training set of HECKTOR 2021 in this paper. First, the voxel spacing of all images is resampled to 1.0 mm × 1.0 mm × 1.0 mm. That is, the third-order spline interpolation method is used for CT and PET images, and the nearest neighbor interpolation method is used for the labeling images. Second, the gray values of CT images are constrained in the range of [−200, 200], so as to exclude the gray values not related to the tumor while strengthening the tumor-related information. The gray value of a pixel point is set to 200 when it is greater than 200, and similarly the gray value of a pixel point is set to −200 when its gray value is less than −200; otherwise the gray value of the pixel point remains unchanged. Finally, the Z-score normalization is used for CT and PET images. That is the gray value minus the mean will be divided by the standard deviation.

2.3 Network Structure

The network we adopted is the 3D U-Net in [3]. The architecture is shown in Fig. 1. The basic module of the encoder and decoder is Conv-SE Norm-ReLU [6, 7]. The operation of downsampling in the encoder is achieved by max pooling. The upsampling operation in the decoder is achieved by using the transpose convolution of 2 × 2 × 2. There are initial 32 feature maps in this network architecture, and they are doubled along each downsampling operation in the encoder, but the maximum feature maps is no more than 320. They will be halved by each transposed convolution in the decoder [8]. At the end of the decoder the spatial size achieved is the same as the initial input size. Then there is a 1 × 1 × 1 convolution into 2 channels and a softmax function followed.

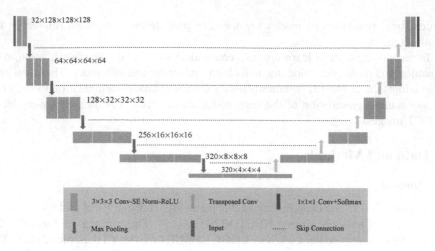

Fig. 1. The architecture of the 3D U-Net.

2.4 Implementation Details

The experiments are implemented in the environment of Linux operating system, and the NVIDIA GeForce RTX 3090 GPU with 24 GB memory is used to accelerate training. The program is implemented using nnUNet [8], and the nnUNet repository address is: https://github.com/MIC-DKFZ/nnUNet. The detailed configuration of the experimental environment is shown in Table 1.

Table 1. Environment and requirements of experiments.

Operating system	Linux
CPU	Intel(R) Xeon(R) Silver 4214R CPU @ 2.40 GHz
GPU	NVIDIA GeForce RTX 3090
CUDA version	11.1
Programming language	Python 3.7
Deep learning framework	Pytorch 1.8.0

The loss function used in experiments is the sum of Dice loss and focal loss. The training process is optimized using SGD optimizer. The initial learning rate and nesterov momentum are 0.01 and 0.99, respectively. We use 'polyLR' [9] schedule to adjust the learning rate dynamically. The learning rate is updated in (1).

$$\alpha = \alpha_0 * (1 - \frac{e}{N})^{0.9} \tag{1}$$

Where α_0 is initial learning rate, e is an epoch counter, and N is the number of total epochs.

The 5-fold cross-validation experiments are carried out. The given training data by the HECKTOR2021 Challenge organization are randomly partitioned into 5 folds, where each fold is excluded from others, and 4 folds together from the 5 ones are used as the training data to train our model, till each fold is used as the test fold. We trained the models from scratch without using any pre-trained weights. There are 5 models achieved in total. The mean results of 5 models are recorded. It took about 48 h to train one model.

To avoid overfitting due to inadequate training for the model, the following data augmentation methods are used to augment the training data, such as random rotation from $-30°$ to $30°$, random scaling with parameter of 0.7–1.4, mirroring, adding Gaussian noise of 0–0.1 and Gamma correction of 0.7–1.5. Table 2 shows the details of all parameters used in experiments.

Table 2. Training protocols of experiments.

Initialization of the network	He normal initialization
Patch size	$128 \times 128 \times 128$
Batch size	2
Alpha and gamma	0.25, 2.0
Optimizer	SGD
Initial learning rate	0.01
Learning rate decay schedule	polyLR
Nesterov momentum	0.99
Epoch	1000
Data augmentation methods with related parameters	Random rotation ($-30°$–$30°$), random scaling (0.7–1.4), mirroring, adding Gaussian noise (0–0.1) and Gamma correction (0.7–1.5)

3 Results

The experimental results are evaluated in terms of four metrics, including Dice, HD95, recall, and precision. The results of the 5-fold cross-validation experiments are shown in Table 3. The bold fonts in Table 3 indicate the best results of the related metrics. The pretty good result and the bad result of our models are shown in Fig. 2, where the green line means the ground truth, and purple line indicates the segmentation results of our achieved models. We adopt our achieved five models to predict the given test data by the HECKTOR2021 Challenge organization. The results on the test data are presented in Table 4.

Table 3. The results of 5-fold cross-validation experiments in terms of Dice, HD95, recall and precision.

	Dice	HD95	Recall	Precision
Fold_1	0.731	9.324	0.859	0.788
Fold_2	0.745	8.787	0.789	0.766
Fold_3	0.764	**5.933**	0.832	**0.818**
Fold_4	**0.803**	6.733	**0.869**	0.816
Fold_5	0.775	6.558	0.846	0.795
Average	0.764	7.467	0.839	0.797

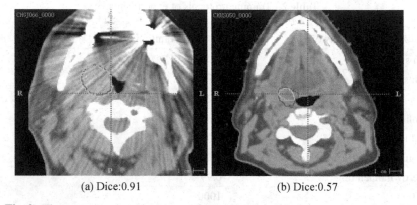

(a) Dice:0.91   (b) Dice:0.57

Fig. 2. The pretty good and bad cases of 5-fold cross-validation experimental results.

Table 4. The results of the five models on test data.

Data	Dice	HD95
Test data	0.778	3.088

4   Discussions

The results in Table 3 show that best results of the 5-fold cross-validation experiments on the given training data are 0.803, 5.933, 0.869 and 0.818 in terms of Dice, HD95, recall and precision, respectively. They are from different models due to the randomly partition of the 5 folds. The average results of 5-fold cross validation experiments on given training data are 0.764, 7.467, 0.839 and 0.797, respectively, in terms of Dice, HD95, recall and precision.

The results in Fig. 2 show that our models can predict the head and neck tumor well for some cases. Although the segmentation results are not perfect in some cases, our prediction results include the ground truth, that is, our predicting one is a little bigger than the true tumors as shown in Fig. 2b. Therefore we can say our model is practical for clinic practice because it will not lead the missing diagnosis.

The results in Table 4 show that results of our 5 models are 0.778 and 3.088 in terms of the mean of Dice and the median of HD95 for the given test data.

5 Conclusions

The U-shaped network was adopted in this paper for segmenting the head and neck tumors using the dual-modality PET-CT images. Although it works well, it still suffers from the insufficient generalization in segmenting the data from unseen centers. We will continue to study how can solve this challenging work and we also hope to work together with those scholars in the related research fields from all over the world to overcome the challenges in this field.

Acknowledgements. This work is supported in part by the National Natural Science Foundation of China under grant No. 62076159, 61673251 and 12031010, and is also by the Fundamental Research Funds for the Central Universities under grant No. GK202105003, and the Innovation Funds of Graduate Programs at Shaanxi Normal University under grant No. 2015CXS028 and 2016CSY009.

We also acknowledge the HECKTOR2021 challenge organization committee for their providing the competition platform and inviting us submitting this paper for our having won the first place in this competition.

References

1. O'rorke, M.A., Ellison, M.V., Murray, L.J., Moran, M., James, J., Anderson, L.A.: Human papillomavirus related head and neck cancer survival: a systematic review and meta-analysis. Oral Oncol. **48**(12), 1191–1201 (2012). https://doi.org/10.1016/j.oraloncology.2012.06.019
2. Gudi, S., et al.: Interobserver variability in the delineation of gross tumor volume and specified organs-at-risk during IMRT for head and neck cancers and the impact of FDG-PET/CT on such variability at the primary site. J. Med. Imaging Radiat. Sci. **48**(2), 184–192 (2017). https://doi.org/10.1016/j.jmir.2016.11.003
3. Çiçek, Ö., Abdulkadir, A., Lienkamp, S.S., Brox, T., Ronneberger, O.: 3D U-net: learning dense volumetric segmentation from sparse annotation. In: Ourselin, S., Joskowicz, L., Sabuncu, M.R., Unal, G., Wells, W. (eds.) MICCAI 2016. LNCS, vol. 9901, pp. 424–432. Springer, Cham (2016). https://doi.org/10.1007/978-3-319-46723-8_49
4. Andrearczyk, V., et al.: Overview of the HECKTOR challenge at MICCAI 2021: automatic head and neck tumor segmentation and outcome prediction in PET/CT images. In: Andrearczyk, V., Oreiller, V., Hatt, M., Depeursinge, A. (eds.) HECKTOR 2021. LNCS, vol. 13209, pp. 1–37. Springer, Cham (2022)
5. Oreiller, V., et al.: Head and neck tumor segmentation in PET/CT: the HECKTOR challenge. Med. Image Anal. **77**, 102336 (2021, under revision)
6. Iantsen, A., Jaouen, V., Visvikis, D., Hatt, M.: Squeeze-and-excitation normalization for brain tumor segmentation. In: International MICCAI Brainlesion Workshop (2020)
7. Iantsen, A., Visvikis, D., Hatt, M.: Squeeze-and-excitation normalization for automated delineation of head and neck primary tumors in combined PET and CT images. In: Andrearczyk, V., Oreiller V., Depeursinge A. (eds.) HECKTOR 2020. LNCS, vol. 12603, pp. 37–43. Springer, Cham (2021). https://doi.org/10.1007/978-3-030-67194-5_4.

8. Isensee, F., Jäger, P.F., Kohl, S.A., Petersen, J., Maier-Hein, K.H.: nnU-Net: a self-configuring method for deep learning-based biomedical image segmentation. Nat. Methods **18**(2), 203–211 (2021). https://doi.org/10.1038/s41592-020-01008-z
9. Chen, L.C., Papandreou, G., Kokkinos, I., Murphy, K., Yuille, A. L.: DeepLab: semantic image segmentation with deep convolutional nets, atrous convolution, and fully connected CRFs. IEEE Trans. Pattern Anal. Mach. Intell., **40**(4), 834–848 (2017). https://doi.org/10.1109/TPAMI.2017.2699184

3D U-Net Applied to Simple Attention Module for Head and Neck Tumor Segmentation in PET and CT Images

Tao Liu[ID], Yixin Su[ID], Jiabao Zhang[ID], Tianqi Wei[ID], and Zhiyong Xiao[✉]

School of Artificial Intelligence and Computer Science, Jiangnan University,
Wuxi 214122, People's Republic of China
zhiyong.xiao@jiangnan.edu.cn

Abstract. Accurate and prompt automatic segmentation of medical images is crucial to clinical surgery and radiotherapy. In this paper, we proposed a new automatic tumor segmentation method for the MICCAI 2021 Head and Neck Tumor segmentation challenge (HECKTOR), using positron emission tomography/computed tomography (PET/CT) images. The main structure of our model is a 3D U-net network with SimAM attention module, which takes advantage of the energy function to give unique weights to disparate pixels. This module enabled us to effectively extract the feature of medical images without adding additional parameters. In the test set consisting of 101 patients, our team, '*C235*', obtained result with the Dice Similarity coefficient (DSC) of 0.756 and Hausdorf Distance at 95% (HD95) 3.269. The full implementation based on PyTorch and the trained models are available at https://github.com/TravisL24/HECKTOR

Keywords: Head and neck cancer · 3D U-net · SimAM · Segmentation

1 Introduction

Radiotherapy is an effective treatment for head and neck cancers which are among the most common cancers worldwide. The use of traditional computed tomography/magnetic resonance imaging (CT/MRI) for staging diagnosis has certain errors and limitations. Positron emission tomography/computed tomography (PET/CT) images can reflect the proliferative activity of tumor cells in each region of the tumor by measuring Standardized Uptake Value (SUV) in each region of the tumor at one time. It can effectively provide the distribution of bioactive tumor tissues, improve the accuracy of tumor diagnosis, and become an indispensable tool for modern radiotherapy.

Recently, convolution neural network (CNN) had developed rapidly in the field of 3D medical image segmentation. There are two mainstream solutions for medical 3D tumor segmentation: The first is to use a 2D convolutional neural network. Dong et al. [1] proposed a brain tumor segmentation network based on UNet, which slices 3D MRI images into multiple 2D images, enhances the images, and uses 2D convolution. In [2], Zhu et al. used a two steps approach. At the beginning, the 3D image is selected by the

© Springer Nature Switzerland AG 2022
V. Andrearczyk et al. (Eds.): HECKTOR 2021, LNCS 13209, pp. 99–108, 2022.
https://doi.org/10.1007/978-3-030-98253-9_9

ResNet-based classification network to select the axial slices that may contain tumors. Then, 2D Unet segments these slices and generates a binary output mask. Although the above methods all slice the 3D image, the slice does not contain the spatial information of the image [3]. When the slices are recombined into a 3D segmentation mask, it is prone to have low segmentation accuracy due to spatial misalignment.

The second is to use a 3D convolutional neural network. Iantsen et al. proposed a U-shaped network that combines residual layers and 'Squeeze and Excitation' (SE) normalization [4]. In [5], Yuan proposed to integrate information across different scales by using a dynamic Scale Attention Network (SA-Net), based on a U-Net architecture. Although the above methods have achieved a certain improvement in accuracy, they all contain a large number of redundant features, which fails to make the model light-weight enough. He et al. proposed multi-view fusion convolution and multi-view hierarchical split block (MVHS block) [6]. These modules can efficiently obtain multi-scale and multi-view information, reduce redundant characteristic information, and improve network performance.

In this paper, we proposed a model based on 3D U-Net supplemented with a Simple Attention Module (SimAM) to address the goal of the MICCAI 2021 Head and Neck Tumor (HECKTOR) segmentation challenge [7].

In this paper, we proposed a model based on 3D U-Net supplemented with a Simple Attention Module (SimAM) to address the goal of the Head and Neck Tumor segmentation challenge (HECKTOR) challenge.

2 Materials and Methods

The chapter is organized as follows. The detail of the relevant background and dataset is described in Sect. 2.1. The core module of the model is explained in detail in Sect. 2.2. In Sect. 2.3, the model structure used in this article is explained. In Sect. 2.4–2.6, the details of data preprocessing, training scheme, and the final prediction scheme are introduced.

2.1 Dataset

In view of the high cost of medical image annotation, there are relatively few publicly available medical segmentation datasets, which are mainly concentrated in the field of single-modality and CT/MRI modalities. Most of the medical image datasets of PET/CT modalities are private, that is the reason why Andrearczyk et al. [7] organized the HECKTOR challenge and offered the opportunity to participants to develop automatic bi-modal approaches for the 3D segmentation of H&N tumors in PET/CT scans, focusing on oropharyngeal cancers.

The training data comprise 224 cases from five centers which can be split in any manner for cross-validation. The total number of test cases is 101 from two centers. A part of the test set will be from a center present in the training data, while another part from a different center. All primary gross tumor volume (GTVt) for patients were reannotated by experts in order to evaluate the methods by Dice Similarity Coefficient (DSC) and Hausdorf Distance at 95% (HD95) [8] (Fig. 1).

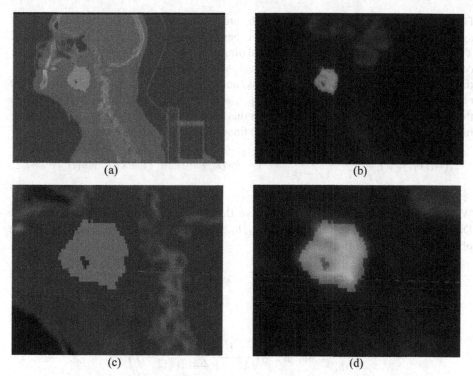

Fig. 1. Examples of 2D sagittal slices of HECKTOR images from patient 'CHGJ007'. (a) and (b) are the original size, (c) and (d) are the images resampled according to the official bounding box. The red area is the tumor annotation area. (Color figure online)

2.2 SimAM

The core component of the model used for this challenge is the full 3D weight attention module proposed in [9]. Different from the common channel-wise attention and spatial-wise attention mechanisms, the core idea of SimAM is to generate the corresponding weight for each pixel in each channel and spatial position, which can complete the learning of the associated information between channel and spatial at the same time. In visual neuroscience, the neurons with the largest amount of information are usually those show different discharge patterns from surrounding neurons. Webb et al. [10] proposed that active neurons have an inhibitory effect on surrounding neurons, which indicates we should give active neurons a higher priority in visual processing. In visual tasks, we take each pixel in the feature functions as a neuron, so as to make linear separability for each neuron, and find the target neuron by measuring the linear separability of the target neuron and other neurons. Based on this theory, the energy function can be defined for each neuron as follows:

$$e_t(w_t, b_t, y, x_i) = \left(y_t - \hat{t}\right)^2 + \frac{1}{M-1} \sum_{i=1}^{M-1} (y_o - \widehat{x_i})^2 \tag{1}$$

Here, $\hat{t} = w_t t + b_t$ and $\hat{x_i} = w_t x_i + b_t$ are linear transforms of t and x_i, where t and x_i are the target neuron and other neurons in a single channel of the input feature $X \in R^{C \times D \times H \times W}$, i is the index of spatial dimension. $M = D \times H \times W$ is the number of neurons on that channel. w_t and b_t are weight and bias the transform. The Eq. 1 attains the minimal value when the $\hat{t}$ equals to y_t, and all other $\hat{x_i}$ are y_o, where y_t and y_o are two different values. After minimizing Eq. 1, we can find the linear separability between the target neuron and all other neurons in the same channel. For simplicity, we employ binary labels for y_t and y_o(1 and -1). The final energy function is given by:

$$e_t\left(w_t, b_t, y, x_i\right) = \frac{1}{M-1}\sum\nolimits_{i=1}^{M-1}(-1-(w_t x_i + b_t))^2 + (1-(w_t t + b_t))^2 + \lambda w_t^2$$

(2)

We use the least squares method to solve the minimization of the energy function. Derivatives of the energy function $e_t(w_t, b_t, y, x_i)$ with respect to w_t and b_t can be obtained by Eq. 3 and Eq. 4.

$$\frac{\partial e_t}{\partial w_t} = 2\left(\left(t^2 + \frac{1}{M-1}\sum\nolimits_{i=1}^{M-1}x_i^2 + \lambda\right)w_t + \left(t + \frac{1}{M-1}\sum\nolimits_{i=1}^{M-1}x_i\right)b_t \right.$$
$$\left. + \frac{1}{M-1}\sum\nolimits_{i=1}^{M-1}x_i - t\right)$$

(3)

$$\frac{\partial e_t}{\partial b_t} = 4b_t + 2w_t\left(\frac{1}{M-1}\sum\nolimits_{i=1}^{M-1}x_i + t\right)$$

(4)

When Eq. 3 and Eq. 4 are equal to 0, a fast closed-form solution can be obtained, w_t and b_t can be achieved by:

$$w_t = -\frac{2(t-\mu_t)}{(t-\mu_t)^2 + 2\sigma_t^2 + 2\lambda}$$

(5)

$$b_t = -\frac{1}{2}(t+\mu_t)w_t$$

(6)

$\mu_t = \frac{1}{M-1}\sum_{i=1}^{M-1}x_i$ and $\sigma_t^2 = \frac{1}{M-1}\sum_{i=1}^{M-1}(x_i - \mu_t)^2$ are the mean and variance of all neurons except t in the channel. For the mean and variance are calculated by all neurons in a channel, we can assume that all neurons are independent and identically distributed, which can greatly reduce the calculation cost. Therefore, we can substitute Eq. 5 and Eq. 6 back to Eq. 2, and the final energy function can be simplified as follows:

$$e_t^* = \frac{4(\hat{\sigma}^2 + \lambda)}{(t-\hat{\mu})^2 + 2\hat{\sigma}^2 + 2\lambda}$$

(7)

where $\hat{\mu} = \frac{1}{M}\sum_{i=1}^{M}x_i$ and $\hat{\sigma}^2 = \frac{1}{M}\sum_{i=1}^{M}(x_i - \hat{\mu})^2$. Equation 7 shows that the smaller the energy, the greater the difference between the neuron and the surrounding, and more weight should be given to the visual segmentation task. After using sigmoid to scale the weight, the following expression can be obtained:

$$\tilde{X} = sigmoid\left(\frac{1}{E}\right) \odot X$$

(8)

where E groups all e_t^* across channel and spatial dimensions. The implementation module of Eq. 8 is shown in Fig. 2(a).

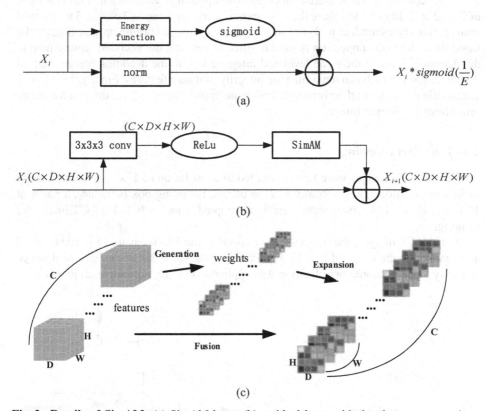

(a)

(b)

(c)

Fig. 2. Details of SimAM: (a) SimAM layer, (b) residual layer with the shortcut connection, (c) Transformation of feature map in SimAM. The features generate weights through the energy function. And the generated weights are expanded and fused with the original features to form new features. Different colors indicate that unique weight scalars are employed for each channel, for spatial location or for each point on the features. (Color figure online)

The energy function in Eq. 7 realizes the mean operation in Squeeze-and-Excitation (SE) block [11] and the standard deviation operation in Style-based Recalibration Module (SRM) [12], which is equivalent to deriving a parameter-free combination of SE and SRM attention module. This is also the main factor in achieving satisfactory accuracy when the model is quite lightweight.

2.3 Network Architecture

Our model adopts a 3D U-net model [13] using SimAM attention module, as shown in Fig. 3. Convolutional blocks are stacks of $3 \times 3 \times 3$ convolutions and ReLU activations followed by SimAM layers. The residual block adds a shortcut connection based on the convolution block, as shown in Fig. 2(b).

Firstly, the deployment of $7 \times 7 \times 7$ convolutions can greatly increase the receptive field of the model without additional computational cost. Then, in the encoder, three $3 \times 3 \times 3$ convolution blocks are connected to max pooling operation with the kernel size of $2 \times 2 \times 2$. In order to ensure the consistency of feature images, $3 \times 3 \times 3$ transposed convolution corresponding to downsampling is also deployed during upsampling. At the same time, the skip connection is used to directly transmit the encoded feature map to the decoding layer of the same level and integrate it with the decoding feature map of the lower level, which can maximize the integrity of image features. Finally, $1 \times 1 \times 1$ convolution and sigmoid activation function are applied to transform the feature image into binary prediction image.

2.4 Data Preprocessing

All PET and CT images were first resampled to a resolution of $1 \times 1 \times 1 \, mm^3$ using triple interpolation. Then according to the official bounding box boundary, a patch of $144 \times 144 \times 144$ voxels were directly resampled from a whole PET/CT image for training.

For the CT images, the images were clipped in the HU intensity of $[-1024, 1024]$ and mapped to the range of $[-1, 1]$. For PET images, since there is no standard image intensity value, z-score normalization was applied to directly normalize all pixels.

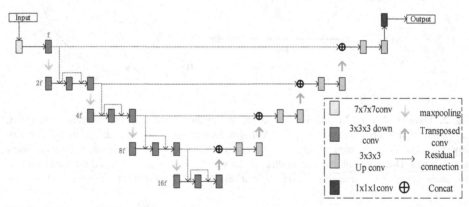

Fig. 3. 3D U-net with SimAM attention module. 'f' represents the number of convolution kernels. In this paper, f is equal to 4. One of the reasons for the small number of model parameters.

2.5 Training Scheme

According to the comparison of loss function combination summarized in [14], our model applies the weightless combination of Soft Dice Loss [15] and the Focal Loss [16]. They are defined as:

$$L_{Dice}(P, G) = 1 - \frac{2\sum_i^N p_i g_i + 1}{\sum_i^N p_i^2 + \sum_i^N g_i^2 + 1} \qquad (7)$$

$$L_{Focal} = -\frac{1}{N} \sum_i^N g_i(1-p_i)^\gamma ln(p_i) \tag{8}$$

where $p_i \in P$ - the predicted probability for the $i-th$ voxel, $g_i \in G$ - the label for the $i-th$ voxel. In addition, we performed $+1$ operation on the numerator and denominator of Soft Dice Loss to avoid division by 0, and the γ value in the Focal Loss is set at 2.

In the input stage, we choose to directly concatenate PET and CT images into a two-channel tensor as input. On the one hand, the rich feature information from different modalities can be fully exploited in all layers, from the first layer to the last one [17]. On the other hand, the existing layer-level fusion strategies mostly use dense connections to exchange complex features between different modalities, which will increase a large number of additional parameters. In addition, due to the low resolution of PET images, the tumor boundary is not clear. In the training process, we need to combine the boundary feature information of the CT image, which manifests the decision-level fusion strategy is also inappropriate.

In the training stage, the data were randomly divided into training and validation dataset according to the ratio of 8:2. The model was trained for 1000 epochs using Adam optimizer on one GPU NVIDIA GeForce GTX 1080 Ti (11 GB) with a batch size of 4. The cosine annealing schedule was applied to reduce the learning rate from 10^{-3} to 10^{-6} within every 25 epochs.

2.6 Ensembling

We randomly divided the whole training set according to the ratio of 8:2 to obtain 5 different splits, then, obtained 5 models after training, and predicted the test set respectively. The final prediction on the test set were generated by averaging the predicted values of each model and applying a threshold operation with a value equal to 0.5.

3 Result and Discussion

Our validation results in the Hector challenge are summarized in Table 1. From the fluctuation of the Dice Similarity coefficient (DSC) and the Hausdorf Distance at 95% (HD95) coefficient, the models trained by different splits have little difference in the segmentation of the tumor body position, but have a large difference in the tumor boundary segmentation. The ensemble of models effectively improved the accuracy of tumor body segmentation and ensured the relative refinement of tumor boundaries. With the ensemble of five models, our proposed 3D U-net model combined with SimAM attention mechanism achieved the result with the DSC of 0.756 and HD95 coefficient of 3.269.

Since the testing set is not made public, we will re-divide the public training set for comparison experiments. The training set of HECKTOR 2021 is based on HECKTOR 2020 with 23 additional cases. Therefore, we adopt the newly added case data as the testing set for the comparison experiment, and the rest is consistent with the training set in 2020. The model "SENorm" is the result of utilizing the model and the opensource code published by Iantsen [4]. Our model is also re-trained and evaluated on the newly divided datasets. In addition, we use Recall, Precision, DSC and HD95 as evaluation metrics, because HECKTOR 2021 and HECKTOR 2020 use different evaluation standards.

Table 1. The performance results on test set. The row 'Single' refers to the performance of a single model on the test set. The row 'Ensemble' indicates the test set result of the ensemble of five models. DSC: Dice Similarity coefficient, HD95: Hausdorf Distance at 95%

Model mode	DSC	HD95
Single	0.735 ± 0.015	3.154 ± 0.752
Ensemble	0.756	3.269

In Table 2, we conducted ablation experiments to verify the function of each module. The SimAM module reduces the HD95 coefficient by 9.32%., which shows that SimAM can adapt to the direct input of PET/CT medical images and learn the spatial relationship information between different channels.

Table 2. Ablation experiment. 'Res' represents the residual block.

Model	Parameter	DSC	Recall	Precision	HD95
Unet	0.55M	0.7737	0.7697	0.8362	5.8922
Unet + Res	0.56M	0.7812	0.8081	0.7975	5.8475
Unet + Res + SimAM (ours)	0.56M	0.7884	0.7666	0.8602	5.3024

In Table 3, it can find that the ensemble of multiple models can effectively improve the accuracy of the model. In addition, our model still maintains a higher DSC coefficient when reducing more parameters, which shows that the model can accurately detect the main position of the tumor. From the perspective of Recall and Precision, although our model has a high accuracy in identifying the tumor, it fails to detect the whole extent of the tumor, which is further confirmed in the HD95 coefficient. The higher HD95 coefficient indicates that the boundaries of tumor segmentation cannot live up to our expectations. This shows that the tumor location information in the PET image is fully learned, but the tumor boundary information in the CT image is not. In subsequent

Table 3. Comparison of model segmentation results. 'Single' represents the average of the segmentation results of a single model. 'Ensemble' represents the segmentation result of the ensemble of multiple models. The number in the 'parameter' indicates the number of ensemble models.

	Model	Parameter	DSC	Recall	Precision	HD95
Single	SENorm	21.75M	0.7779	0.8049	0.8079	5.1254
	Ours	0.56M	0.7884	0.7666	0.8602	5.3024
Ensemble	SENorm	21.75M * 8	0.7911	0.8105	0.8198	4.7047
	Ours	0.56M * 4	0.7999	0.7972	0.8460	5.2036

research, different modal fusion methods can be considered to improve the accuracy of tumor boundary segmentation.

4 Conclusion

In this paper, we propose a lightweight automatic segmentation model for segmentation of head and neck tumors in PET and CT modalities. The core of the model is the improved 3D SimAM attention mechanism. Different from the existing channel or spatial attention mechanism, this module can derive 3D attention weights through the feature map of the defined energy function, whose advantage is to complete the attention integration of channels and spaces without adding additional parameters. In this HECKTOR2021 challenge, we obtain the result with the DSC of 0.756 and the HD95 coefficient of 3.269. Although there is a large gap in accuracy compared with the first place of this challenge this year, we believe that a lightweight and relatively accurate model is what is needed in practical applications.

In the future, we will continue to explore the limits of the model in multi-modal segmentation. In the follow-up work, it is urgent for us to use different modal fusion methods to more fully learn the spatial information between PET/CT images to improve the accuracy of segmentation. Moreover, the generalization of the model should also be studied on other large-scale multi-modal tumor segmentation datasets.

References

1. Dong, H., Yang, G., Liu, F., Mo, Y., Guo, Y.: Automatic brain tumor detection and segmentation using u-net based fully convolutional networks. In: Valdés Hernández, M., González-Castro, V., (eds.) Medical Image Understanding and Analysis. MIUA 2017. Communications in Computer and Information Science, vol. 723, pp. 506–517. Springer, Cham. https://doi.org/10.1007/978-3-319-60964-5_44
2. Zhu, S., Dai, Z., Wen, N.: Two-stage approach for segmenting gross tumor volume in head and neck cancer with CT and PET imaging. In: Andrearczyk, V., Oreiller, V., Depeursinge, A. (eds.) HECKTOR 2020. LNCS, vol. 12603, pp. 22–27. Springer, Cham (2021). https://doi.org/10.1007/978-3-030-67194-5_2
3. Xiao, Z., Du, N., Liu, J., Zhang, W.: SR-NET: a sequence offset fusion net and refine net for undersampled multislice MR image reconstruction. Comput. Meth. Prog. Biomed. **202**, 105997 (2021)
4. Iantsen, A., Visvikis, D., Hatt, M.: Squeeze-and-excitation normalization for automated delineation of head and neck primary tumors in combined PET and CT images. In: Andrearczyk, V., Oreiller, V., Depeursinge, A. (eds.) HECKTOR 2020. LNCS, vol. 12603, pp. 37–43. Springer, Cham (2021). https://doi.org/10.1007/978-3-030-67194-5_4
5. Yuan, Y.: Automatic head and neck tumor segmentation in PET/CT with scale attention network. In: Andrearczyk, V., Oreiller, V., Depeursinge, A. (eds.) HECKTOR 2020. LNCS, vol. 12603, pp. 44–52. Springer, Cham (2021). https://doi.org/10.1007/978-3-030-67194-5_5
6. Xiao, Z., He, K., Liu, J., Zhang, W.: Multi-view hierarchical split network for brain tumor segmentation. Biomed. Sig. Process. Control **69**, 102897 (2021)
7. Andrearczyk, V., et al.: Overview of the HECKTOR challenge at MICCAI 2021: automatic head and neck tumor segmentation and outcome prediction in PET/CT images. In: Andrearczyk, V., Oreiller, V., Hatt, M., Depeursinge, A. (eds.) HECKTOR 2021. LNCS, vol. 13209, pp. 1–37. Springer, Cham (2022)

8. Oreiller, V., et al.: Head and Neck Tumor Segmentation in PET/CT: The HECKTOR Challenge, Medical Image Analysis (2021)

9. Yang, L., Zhang, R., Li, L., Xie, X..: SimAM: a simple, parameter-free attention module for convolutional neural networks. In: Proceedings of the 38th International Conference on Machine Learning, in Proceedings of Machine Learning Research, vol. 139, pp. 11866–11867 (2021)

10. Webb, B.S., Dhruv, N.T., Solomon, S.G., Tailby, C., Lennie, P.: Early and late mechanisms of surround suppression in striate cortex of Macaque. J. Neurosci. 25(50), 11666–11675 (2005)

11. Hu, J., Shen, L., Sun, G.: Squeeze-and-excitation networks. In: IEEE Conference on Computer Vision and Pattern Recognition, pp. 7132–7141 (2018b)

12. Lee, H., Kim, H.-E., Nam, H.: SRM: a style-based re-calibration module for convolutional neural networks. In: IEEE International Conference on Computer Vision, pp. 1854–1862 (2019)

13. Çiçek, Ö., Abdulkadir, A., Lienkamp, S.S., Brox, T., Ronneberger, O.: 3D U-Net: learning dense volumetric segmentation from sparse annotation. In: Ourselin, S., Joskowicz, L., Sabuncu, M.R., Unal, G., Wells, W. (eds.) MICCAI 2016. LNCS, vol. 9901, pp. 424–432. Springer, Cham (2016). https://doi.org/10.1007/978-3-319-46723-8_49

14. Andrearczyk, V.,et al.: Overview of the HECKTOR challenge at MICCAI 2020: automatic head and neck tumor segmentation in PET/CT. In: Andrearczyk, V., Oreiller, V., Depeursinge, A. (eds.) HECKTOR 2020. LNCS, vol. 12603, pp. 1–21. Springer, Cham (2021). https://doi.org/10.1007/978-3-030-67194-5_1

15. Milletari, F., Navab, N., Ahmadi, S.-A.: V-net: fully convolutional neural networks for volumetric medical image segmentation. In: International Conference on 3D Vision, pp. 565–571. IEEE (2016)

16. Lin, T.-Y., Goyal, P., Girshick, R., He, K., Doll´ar, P.: Focal Loss for Dense Object Detec-tion. arXiv preprint arXiv:1708.02002 (2017)

17. Zhou, T., Ruan, S., Canu, S.: A review: deep learning for medical image segmentation using multi-modality fusion. Array 3–4, 100004 (2019)

Skip-SCSE Multi-scale Attention and Co-learning Method for Oropharyngeal Tumor Segmentation on Multi-modal PET-CT Images

Alessia De Biase(✉)(ID), Wei Tang(ID), Nikos Sourlos(ID), Baoqiang Ma(ID),
Jiapan Guo(ID), Nanna Maria Sijtsema(ID), and Peter van Ooijen(ID)

University Medical Center Groningen (UMCG), 9700 Groningen, RB,
The Netherlands
{a.de.biase,w.tang,n.sourlos,b.ma,j.guo,n.m.sijtsema,
p.m.a.van.ooijen}@umcg.nl

Abstract. One of the primary treatment options for head and neck
cancer is (chemo)radiation. Accurate delineation of the contour of the
tumors is of great importance in the successful treatment of the tumor
and in the prediction of patient outcomes. With this paper we take part
in the HECKTOR 2021 challenge and we propose our methods for automatic tumor segmentation on PET and CT images of oropharyngeal cancer patients. To achieve this goal, we investigated different deep learning
methods with the purpose of highlighting relevant image and modality
related features, to refine the contour of the primary tumor. More specifically, we tested a Co-learning method [1] and a 3D Skip Spatial and
Channel Squeeze and Excitation Multi-Scale Attention method (Skip-
scSE M), on the challenge dataset. The best results achieved on the test
set were 0.762 mean Dice Similarity Score and 3.143 median of the Hausdorf Distance at 95%.

Keywords: Head and neck tumor segmentation · Oropharynx ·
Neural networks · Deep learning · AI · Head and neck cancer ·
PET-CT · Multi-modal · HECKTOR2021 · Automatic segmentation ·
Radiotherapy treatment planning · Co-learning ·
Squeeze-and-excitation · Normalization

1 Introduction

Head and neck cancer is a very common cancer. It accounts for 3% of all cancers
worldwide and for over 1.5% of all cancer deaths in the US [2]. Radiotherapy is
an effective treatment in these cases [3], therefore, to deliver a high dose in the
tumor area, without damaging surrounding normal tissues, proper delineation
of the tumor contour needs to be performed. This delineation of the gross tumor

Aicrowd Group Name: "umcg".

© Springer Nature Switzerland AG 2022
V. Andrearczyk et al. (Eds.): HECKTOR 2021, LNCS 13209, pp. 109–120, 2022.
https://doi.org/10.1007/978-3-030-98253-9_10

volume (GTVt) is usually performed manually by experts. The method of manual delineation of contours is susceptible to inter- and intra-observer variability, and so error-prone [4]. Adding more modalities can help to decrease those variations. The difficulty of manually segmenting head and neck tumors lies in the fact that adjacent normal tissues are very similar to tumors in that region and cannot be easily distinguished even by experienced clinicians, especially in CT images [5]. Moreover, this method is extremely time consuming, since radiation oncologists have to do the contouring in 3 perpendicular cross sections, starting in the axial slices and then going in the sagittal and coronal views. This method is intractable in 3D and semi-automated methods are used in practice. These are the reasons why AI methods for automatic segmentation have gained interest in the recent years.

There are many methods for automatic lesion segmentation using just one modality [6], but they showed lower performance compared to multi-modality models. In the context of tumor segmentation, the increased performance of the combination of both Positron Emission Tomography (PET) and Computer Tomography (CT) images is due to the fact that each modality can provide complementary information about the anatomy (CT) and the metabolic activity (PET) of the tumor. On the other hand, the fusion of PET and CT scans, in current methods, does not reach a good balance between feature map interactions across multi-modal channels and single modality-related feature maps enhancement.

To address this problem we first trained a Multi-Modal Co-Learning model, on the provided dataset, using as input PET and CT images separately. The chosen model was first proposed to solve the problem of Liver Lesion Segmentation on PET-CT images by Z. Xue et al. in [1], demonstrating to reach significantly higher accuracy than the baseline models. To further strengthen our results, we designed a 3D Skip-scSE multi-Scale attention model to capture the more important information between encoder and decoder feature maps, as well as to get global context information with different scale features.

HECKTOR (HEad and neCK TumOR) 2021 challenge is the successor of the HECKTOR 2020, both aimed to improve the automatic segmentation methods for head and neck cancer based on PET and CT images [7,8]. In addition, this year the challenge has been extended to include outcome prediction in patients using both image and clinical data.

In this paper, we present our methods to participate in the first task of the challenge, which is the automatic tumor segmentation on oropharyngeal cancer patients, based on PET-CT images.

2   Data and Methods

2.1   Dataset Description

Our dataset contains in total 325 cases of aligned PET-CT images from a total of 6 centers. The total number of training cases are 224 from 5 centers (CHGJ, CHMR, CHUM, CHUS and CHUP) whereas the number of test cases that are used in the final ranking are 101 from 2 centers (CHUP - which is also present in the training set - and CHUV). All cases were annotated by experts and only the annotations of the training cases were provided.

Compared to last year's challenge, 76 new cases from another clinical center (CHU Milétrie, Poitiers, France) were added to the provided dataset. Each center has its own set of parameters for image acquisition, and the vendors differ between some centers. These factors, along with the presence of lymph nodes with high SUV values in PET images, are the main contributors to why this segmentation task is challenging.

For the segmentation task, CSV files containing bounding boxes of the tumor region for each patient were also provided. The quality of the performance of the algorithms is assessed only within the volume of the bounding boxes at the original CT resolution, based on the Dice Similarity Coefficient (DSC) and of the Hausdorf distance at 95% (HD95) between the ground-truth annotations and the predictions of the implemented algorithms. The final ranking is based on the mean DSC and the median HD95 calculated on all test set images.

Since the evaluation results are calculated on bounding boxes of $144 \times 144 \times 144$ mm^3 that contain the tumor region, the images were first resampled to $1 \times 1 \times 1$ mm^3 isotropic resolution and then cropped to that volume size. As a pre-processing technique we used a z score normalization based on the formula

$$z_{score} = (image - mean)/std \tag{1}$$

To validate the performance of different models we chose to create an internal test set of 45 patients from the provided 224 training cases. For the remaining 179 cases, we split them in training and validation sets, either randomly or by keeping-one-center-out each time. Both our training and test sets include cases from all the 5 centers. At the end, an ensemble of a particular model trained on each of these splits was used to get the final predictions of images in the organizers' test set. No data augmentation was performed in any of the experiments.

2.2 Methods: An Overview

Two different methods were used to address the tumor segmentation task: the Co-learning Multi-Modal PET-CT and the Skip-scSE Multi-Scale Attention. Both methods train on 3D images of size $144 \times 144 \times 144$. The first one takes as input PET and CT as two separate images, while the second one concatenates them in the channel dimension (input size is $2 \times 144 \times 144 \times 144$). There are 4 downsampling and 4 upsampling steps in both methods. For the Co-learning method the architecture of each encoder and decoder block is the same of the V-Net. While for the Skip-scSE Multi-Scale Attention network the output is a binary mask of size $144 \times 144 \times 144$, for the Co-learning Multi-Modal network there are two segmentation outputs of same size which are then averaged in one final binary mask prediction.

2.3   Co-learning Method with Multi-modal PET-CT

The key features of the Co-Learning method are the *Shared Down-sampling Block* (SDB), the *Feature Co-learning Block* (FCB) and the *Up-sampling block* (US). PET and CT images are down-sampled by two separate encoder blocks. An encoder step refers to an encoder block for both PET and CT at a same resolution. Whenever we refer to a "branch", we indicate a sequence of encoder blocks or decoder blocks belonging to the same encoder or decoder respectively. After each encoder step, the same shared down-sampling block (SDB) allows the two modalities to share single modality-related information gained in the exact same areas of the image. These blocks help to map the PET-CT feature maps to the same spatial Region of Interest (ROI), which can preserve the spatial and structural consistencies between them. There are two decoder branches: one that uses the output from the SDB from each stage of the encoder as input of each decoder; the other one that uses the outputs of the FCB from each stage of the encoder, as input to each stage of the second decoder, as shown in Fig. 1. The Feature Co-learning blocks have as inputs a pair of feature maps from the CT encoder and the PET encoder and aim to combine the information learnt from each modality. The hierarchical skip connections of the Feature Co-learning module allow feature fusion at different resolutions. Information at different scales is then used for the reconstruction of the prediction in the second branch. Finally, the US block ensures consistency in resolution in the second decoder branch.

The loss function used in [1] is a weighted sum of two Focal Tversky loss functions (L_{ft_i}, with $i = 1, 2$ indicating the decoder branch) and a similarity loss function L_{sim}. L_{ft_i} has as input the output of each i decoder branch. L_{sim} constrains the output prediction of the two decoder branches using the average L_1 distance between the probability output of the first and of the second decoder path, as shown in the equation:

$$L_{co-learn} = \sum_{i=1}^{2} \lambda_i * L_{ft_i} + \mu * L_{sim} \qquad (2)$$

The Focal Tversky loss [9] has the advantage of avoiding the excessive decrease in performance (common problem with large class imbalances), and of choosing whether precision or sensitivity is more important. The hyperparameters λ_1 and λ_2 weight which of the two branches has more influence on the output. Values of λ, which gave comparable results for both precision and recall, were $\lambda_1 = 0.4$ and $\lambda_2 = 0.5$.

After applying the z-score normalization on PET and CT images, scaling in the range $[-1, 1]$ was needed. We trained the Co-learning model for 800 epochs, with a batch size 1 and $\mu = 0.5$. The optimal learning rate to train our model on that dataset was $lr = 1e^{-5}$ using Adam as optimizer and the MultiStepLR scheduler. The two outputs from the two branches are then averaged, resulting in one final prediction.

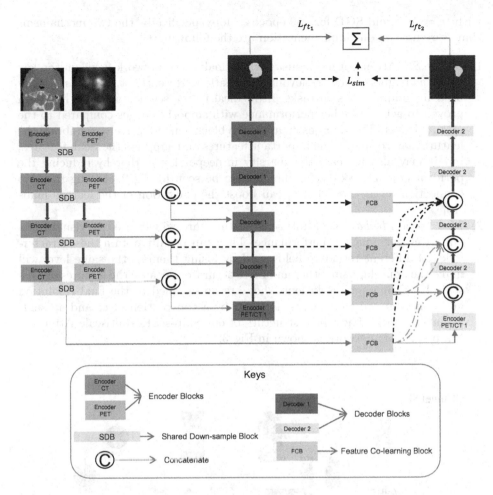

Fig. 1. Co-learning method framework. For more information about the structure details see [1].

2.4   3D Skip-ScSE Multi-scale Attention Model

The current standard U-Net skip connection network mainly has the following shortcomings: First, the direct connection of the encoder and decoder features cannot map the gap information between 2 different features. Second, it ignores the remote feature dependence and global context information at different scales. In response to these problems, we adapted the attention mechanism in the skip connection to capture the differences between encoder and decoder in a 3D U-Net, highlight the key information, and then we used different scales of feature information from the decoder, combined with attention mechanism to supervise the whole learning process. This model was trained by dice and bce loss function, with $lr = 1e^{-4}$ and SGD for 250 epochs. More specifically, the two mechanisms that were used in our implementation are the following:

1) *Skip-scSE*. Attention mechanism, which makes the network focus on the key parts and ignores the irrelevant information. Recently it has been widely used in computer vision tasks. In medical image segmentation, it has been proven to achieve higher performance with smoother edges compared to the normal U-Net. The scSE, as an attention block, can adaptively recalibrate the feature map to signify the important features and suppress the weak features, but it may also increase the sparsity in deeper layers thereby reducing the performance. The skip-scSE module can be seen in Fig. 2. Adding the skip connection in the scSE block, can boost the excitation of the feature maps [10].

2) *Multi-scale Mechanism.* Multi-scale features are widely used in visual tasks and often get amazing performance. They can usually obtain the characteristics of different receptive fields, and by fusing them at the same level will make the boundary smoother and more accurate. Because the features of each scale have different resolutions, they are up-sampled to the final resolution and concatenated to a tensor, followed by a convolution layer and an scSE attention block. The whole structure of our Skip-scSE-Multiscale Attention (Skip-scSE-M) network is shown in Fig. 3.

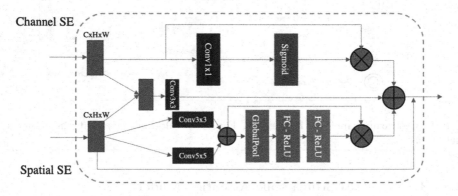

Fig. 2. Skip connection with scSE (skip-scSE) module

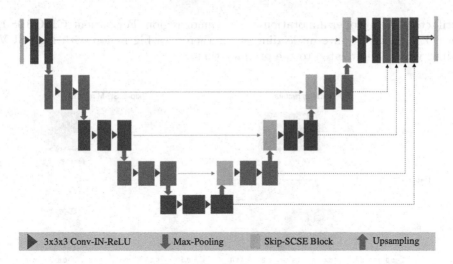

Fig. 3. The architecture of the proposed 3D Skip-scSE-M model

3   Results

All the models were trained on a TITAN V100 GPU on the Peregrine Cluster of the University of Groningen using the PyTorch framework in Python.

In Table 1, only the best results obtained from the models described in the Data and Method section above, are reported. The results shown in the table are obtained from the models trained and validated on different splits of the training set (made of 179 patients) and tested on the same internal test set (of 45 patients). For the Co-learning model we trained and validated with 5-fold cross validation and we chose the best 5 models according to the lowest validation loss value. In the Skip-scSE-M training, we used leave-one-center-out cross validation and we selected the 5 best models based on the highest DSC on each validation set. Finally we tested on the internal test set and we averaged the results for both methods.

The Skip-scSE-M model resulted in higher performance compared to the Co learning method in all metrics except precision. In Fig. 4, the distributions of Dice Similarity Score, Precision and Recall calculated on the internal test set are shown, using boxplots. The results from the Co-learning model have a more uniform distribution than the ones from the Skip-scSE-M model, showing the higher stability of the second method on the test set, compared to the first one. The median value of the metrics does not differ much between the two models. The Skip-scSE-M model distributions show two outliers for each one of the metrics. There is a common outlier for all metrics: patient with ID **CHGJ089**. The second one shown in the boxplots is patient with **CHUM013** for Dice Score and Recall and patient with ID **CHMR013** for Precision. Visualizing the CT and the PET for these patients we noticed that in two out of three cases, the lower performance can be justified by a dataset-related issue: presence of metal artifacts and improper annotation of the tumor region. For patient **CHGJ089**, the PET image is quite misleading, since there is a big region with high SUV values which is unrelated to the primary tumor.

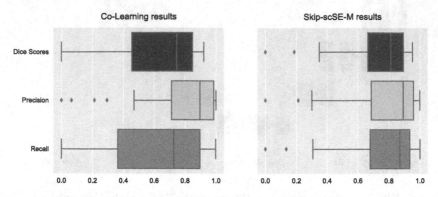

Fig. 4. Comparison of the distribution of evaluation metrics on the internal test set between the Co-learning model and the Skip-scSE-M model. The average value of these metrics is shown in Table 1

Table 1. Comparison of performance of the Co-learning and the Skip-scSE-M models on the internal test set of 45 patients. For the DSC, Precision and Recall the average values are reported, for the HD the median value in mm is reported.

Model	Data split	DSC	HD (mm)	Precision	Recall
Co-learning	Random 5-fold CV	0.634	7.071	0.792	0.601
Skip-scSE	LOCO CV	0.744	5.000	0.793	0.759

In Table 2 we show a comparison of the evaluation metrics among different centers. The center with lowest performance according to the Skip-scSE-M model is **CHMR** while the best is **CHUP**. The DSC values are consistent with the ones obtained in last year papers [7]. The Co-learning method has significantly lower performance in DSC on the **CHUS** and **CHUM** centers compared to the other method.

Table 2. Comparison of performance of the Co-learning and the Skip-scSE-M models on internal test set across different centers. For the DSC the average values are reported, for the HD the median value in mm is reported for each center.

Model	Co-learning		Skip-scSE-M	
Center	DSC	HD (mm)	DSC	HD (mm)
CHGJ (n = 11)	0.703	5.00	0.791	3.46
CHMR (n = 4)	**0.660**	18.12	0.652	7.32
CHUM (n = 11)	0.578	6.32	0.723	4.12
CHUP (n = 5)	0.736	13.34	**0.831**	9.43
CHUS (n = 14)	0.578	10.80	0.720	7.17

To have a visual comparison of the metric differences between the two used methods, we ordered the patients in the internal test set by DSC and plotted their scores, as shown in Fig. 5. The curves of the Co-leaning and the Skip-scSE-M method show a bigger gap when the Dice Score is lower than 0.8, while a smaller one when the value increases. In the second plot we would expect a decreasing pattern of HD values, however the Co-learning method curve has an irregular pattern compared to the Skip-scSE-M curve.

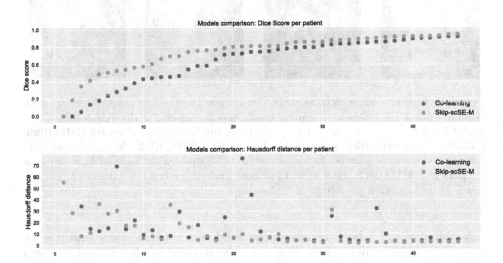

Fig. 5. Visual metric comparison between the Co-learning model and the Skip-scSE-M model

In Table 3 the submission results obtained in the challenge test data are shown. The highest Dice Score is 0.762 and it was performed testing the Skip-scSE-M model trained and validated using leave one center out cross validation on the entire training set (224 patients). The results are obtained based on an

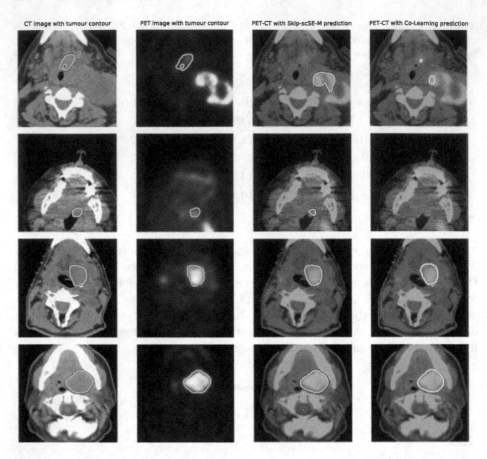

Fig. 6. Comparison of predictions of the two models. From left to right: CT image with ground truth, PET image with ground truth, and predictions of the Skip-scSE-M, and of the Co-learning models on a PET/CT. From top to bottom: Patients **CHGJ089**, **CHUM013**, **CHUS027**, **CHGJ016**. For patient **CHGJ089** Co-learning method detects a very small part of the tumor, while the Skip-scSE-M method fails, whereas for patient **CHUM013** the first method fails while the second one does not. For the rest 2 patients similar performance is observed.

ensembling of the 5 best models on the validation sets. In the second row we report the mean value of Dice Score calculated between the original contours and the average between the probability output of the Co-Learning method from Table 3 and the Skip-scSE-M model submission results.

Table 3. Submission results on the challenge test data. The Ensemble method (**Skip-scSE-M + Co-learning**) was obtained averaging the probability results of the Skip-scSE-M and of the Co-learning method.

Model	Mean DSC	Median HD
Skip-scSE-M	**0.762**	**3.143**
Skip-scSE-M + Co-learning	0.706	3.291

4   Discussion and Conclusion

In this paper we explored different deep learning techniques with the goal of overcoming some of the issues emerged in the last year challenge and, meanwhile, gave our valuable contribution to the difficult task of automatic tumor segmentation in the head and neck region.

To perform this task, we used two different algorithms, namely the Co-learning and the Skip-scSE-M methods, and we compared their results on the challenge dataset.

PET and CT images are used separately in the Co-learning method inspired by [1]. The reason why we chose this approach is to use deep learning structures to extract both relevant information from each modality, separately and combined. The Co-learning approach has many limitations, especially related to the learning phase. The loss function does not always decrease during training, even after many epochs. The training performance heavily depends on random seed selection, showing its weakness in the tumor segmentation task. To address this problem, different optimizers, schedulers and learning rates were tested. In the original paper, the deep learning framework has, as input, a specific region of interest corresponding to the liver region, while in our task we did not have such small focus. This is one of the reasons why this method did not reach high performance in the whole dataset.

We also tried different attention methods. Finally, we chose the Skip-scSE-M model because it can provide a more stable and robust performance. But for some examples, it is still not very predictable.

The limitations of our algorithms can also be related to the imaging data quality and characteristics. Our methods may not perform well on some cases in which we have low FDG uptake on PET, or when the primary tumor can be confused with a lymph node, or when the tumor is present at the border of the oropharynx region [7]. Unfortunately, images with these problems are common in real clinical practice, this is why, during training, we decided not to remove them from our dataset. Larger studies on how to deal with the most common modality related issues are essential before actually implementing this algorithm in real clinical practice.

Suggestions for future works are assessing the quality of the dataset before training, using pre-processing techniques for solving metal artifacts problems and add variability in the annotations.

Acknowledgement. We would like to thank the Center for Information Technology of the University of Groningen for their support and for providing access to the Peregrine high performance computing cluster.

References

1. Xue, Z., et al.: Multi-modal co-learning for liver lesion segmentation on PET-CT images. IEEE Trans. Med. Imaging. https://doi.org/10.1109/TMI.2021.3089702
2. Chow, L.Q.M.: Head and Neck Cancer. N Engl. J. Med. **382**(1), 60–72 (2020). PMID: 31893516. https://doi.org/10.1056/NEJMra1715715
3. Yeh, S.A.: Radiotherapy for head and neck cancer. Semin. Plast. Surg. **24**(2), 127–136 (2010). https://doi.org/10.1055/s-0030-1255330
4. Gudi, S., et al.: Interobserver variability in the delineation of gross tumour volume and specified organs-at-risk during IMRT for head and neck cancers and the impact of FDG-PET/CT on such variability at the primary site. J. Med. Imaging Radiat. Sci. **48**(2), 184–192 (2017)
5. Andrearczyk, V., et al.: Automatic segmentation of head and neck tumors and nodal metastases in PET-CT scans. In: Medical Imaging with Deep Learning (MIDL) (2020)
6. Moe, Y.M., et al.: Deep learning for automatic tumour segmentation in PET/CT images of patients with head and neck cancers. Medical Imaging with Deep Learning (2019)
7. Andrearczyk, V., et al.: Overview of the HECKTOR challenge at MICCAI 2021: automatic head and neck tumor segmentation and outcome prediction in PET/CT images. In: Andrearczyk, V., Oreiller, V., Hatt, M., Depeursinge, A. (eds.) HECK-TOR 2021. LNCS, vol. 13209, pp. 1–37. Springer, Cham (2022)
8. Oreiller, V., et al.: Head and Neck Tumor Segmentation in PET/CT: The HECK-TOR Challenge, Medical Image Analysis (2021). (under revision)
9. Abraham, N., Khan, N.M.: A novel Focal Tversky loss function with improved attention U-Net for lesion segmentation, arXiv preprint arXiv:1810.07842 (2018)
10. Islam, M., Wijethilake, N., Ren, H.: Glioblastoma multiforme prognosis: MRI missing modality generation, segmentation and radiogenomic survival prediction. Comput. Med. Imaging Graph. **91**, 101906 (2021)

Head and Neck Cancer Primary Tumor Auto Segmentation Using Model Ensembling of Deep Learning in PET/CT Images

Mohamed A. Naser(✉) ⓘ, Kareem A. Wahid ⓘ, Lisanne V. van Dijk ⓘ, Renjie He ⓘ, Moamen Abobakr Abdelaal ⓘ, Cem Dede ⓘ, Abdallah S. R. Mohamed ⓘ, and Clifton D. Fuller ⓘ

Department of Radiation Oncology, The University of Texas MD Anderson Cancer, Houston, TX 77030, USA
manaser@mdanderson.org

Abstract. Auto-segmentation of primary tumors in oropharyngeal cancer using PET/CT images is an unmet need that has the potential to improve radiation oncology workflows. In this study, we develop a series of deep learning models based on a 3D Residual Unet (ResUnet) architecture that can segment oropharyngeal tumors with high performance as demonstrated through internal and external validation of large-scale datasets (training size = 224 patients, testing size = 101 patients) as part of the 2021 HECKTOR Challenge. Specifically, we leverage ResUNet models with either 256 or 512 bottleneck layer channels that demonstrate internal validation (10-fold cross-validation) mean Dice similarity coefficient (DSC) up to 0.771 and median 95% Hausdorff distance (95% HD) as low as 2.919 mm. We employ label fusion ensemble approaches, including Simultaneous Truth and Performance Level Estimation (STAPLE) and a voxel-level threshold approach based on majority voting (AVERAGE), to generate consensus segmentations on the test data by combining the segmentations produced through different trained cross-validation models. We demonstrate that our best performing ensembling approach (256 channels AVERAGE) achieves a mean DSC of 0.770 and median 95% HD of 3.143 mm through independent external validation on the test set. Our DSC and 95% HD test results are within 0.01 and 0.06 mm of the top ranked model in the competition, respectively. Concordance of internal and external validation results suggests our models are robust and can generalize well to unseen PET/CT data. We advocate that ResUNet models coupled to label fusion ensembling approaches are promising candidates for PET/CT oropharyngeal primary tumors auto-segmentation. Future investigations should target the ideal combination of channel combinations and label fusion strategies to maximize segmentation performance.

Keywords: PET · CT · Tumor segmentation · Head and neck cancer · Oropharyngeal cancer · Deep learning · Auto-contouring

© Springer Nature Switzerland AG 2022
V. Andrearczyk et al. (Eds.): HECKTOR 2021, LNCS 13209, pp. 121–133, 2022.
https://doi.org/10.1007/978-3-030-98253-9_11

1 Introduction

Oropharyngeal cancer (OPC) is a type of head and neck squamous cell carcinoma that affects a large number of individuals across the world [1]. Radiation therapy is an effective component of OPC treatment but is highly dependent on accurate segmentation of gross tumor volumes [2], i.e., visible gross disease that is informed by clinical examination and radiographic findings. Importantly, precise tumor delineation is crucial to ensure adequate radiation therapy dose to target volumes while minimizing dose to surrounding healthy tissues. The combination of computed tomography (CT) with positron emission tomography (PET) allows for sufficient anatomic detail in determining tumor location coupled to underlying physiologic information [3]. However, tumor segmentation in OPC has long been seen as an inefficient and potentially inconsistent process as multiple studies have demonstrated high human inter- and intra-observer segmentation variability [4, 5]. Therefore, developing automated tools, such as those based on deep learning [6–9], to reduce the variability in OPC PET/CT tumor segmentation while retaining reasonable performance is imperative for improving the radiation therapy workflow.

The annual Medical Image Computing and Computer Assisted Intervention Society (MICCAI) Head and Neck Tumor Segmentation Challenge (HECKTOR) has provided an avenue to systematically evaluate different OPC primary tumor auto-segmentation methodologies through the release of high-quality, multi-institutional training and testing PET/CT data. We previously participated in the 2020 HECKTOR challenge and achieved reasonable results using deep learning approaches [10]. Subsequently, we improve upon our previous approach through various architectural modifications, ensembling of independent models' predictions, and additional provided training/testing data, that ultimately leads to improved segmentation performance. This work presents the results of our OPC primary tumor auto-segmentation model based on a ResUnet deep learning model applied to the 2021 HECKTOR Challenge PET/CT training and testing data.

2 Methods

We developed deep learning models (Sect. 2.3) for auto-segmentation of primary tumors of OPC patients using co-registered ^{18}F-FDG PET and CT imaging data (Sect. 2.1). The ground truth manual segmentations of the tumors and the normalized imaging data (Sect. 2.2) were used to train the models (Sect. 2.4). The performance of the trained models for auto-segmentation were validated using a 10-fold cross-validation approach (Sect. 2.5).

2.1 Imaging Data

The data set used in this study, which was released through AIcrowd [11] for the HECKTOR Challenge at MICCAI 2021 [12–14], consists of co-registered ^{18}F-FDG PET and CT scans for 325 OPC patients (224 patients used for training and 101 patients used for testing, previously partitioned by the HECKTOR Challenge organizers). All imaging data in the training set (224 patients) was paired with ground truth manual segmentations of the OPC primary tumors derived from clinical experts (HECKTOR Challenge organizers). All training and testing data were provided in Neuroimaging Informatics Technology Initiative (NIfTI) format.

2.2 Image Processing

All images (i.e., PET, CT, and tumor segmentation masks) were cropped to fixed bounding box volumes, provided with the imaging data (Sect. 2.1) by the HECKTOR Challenge organizers [11], of size $144 \times 144 \times 144$ mm^3 in the x, y and z dimensions. To mitigate the variable resolution of the PET and CT images, the cropped images were resampled to a fixed image resolution of 1 mm in the x, y, and z dimensions. We used spline interpolation of order 3 for resampling the PET/CT images and nearest-neighbor interpolation for resampling the segmentation masks. We based our cropping and resampling work on the code provided by the HECKTOR Challenge organizers (https://github.com/vor eille/hecktor). The CT intensities were truncated to the range of $[-200, 200]$ Hounsfield Units (HU) to increase soft tissue contrast and then were normalized to a $[-1, 1]$ scale. The intensities of PET images were normalized with z-score normalization ([intensity-mean]/standard deviation). We used the Medical Open Network for AI (MONAI) [15] software transformation packages to rescale and normalize the intensities of the PET/CT images. Image processing steps used in this manuscript are displayed in Fig. 1.

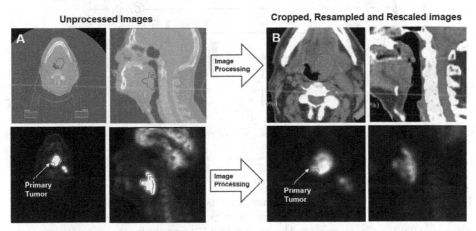

Fig. 1. An illustration of the workflow used for image processing. (A) Overlays of the provided ground truth tumor segmentation masks (red outline) and the original CT (top) and PET (bottom) images. (B) Overlays of the provided ground truth tumor segmentation masks (red outline) and the processed CT (top) and PET (bottom) images. (Color figure online)

2.3 Model Architecture

A deep learning convolutional neural network model based on the ResUnet architecture included in the MONAI software package was used for the analysis. As shown in Fig. 2, the network consisted of 4 convolution blocks in the encoding and decoding branches and a bottleneck convolution block between the two branches. All convolution layers used a kernel size of 3 except one convolution layer in the bottleneck, which used a kernel size of 1. The number of output channels for each convolution layer is given above each layer, as shown in Fig. 2. Each convolution block in the encoding branch was composed of a two-strided convolution layer and a residual connection that contained a two-strided convolution layer and a one-strided convolution layer. In the bottleneck, the residual connection contained two one-strided convolution layers. In the decoding branch, each block contained a two strided convolution transpose layer, a one strided convolution layer and a residual connection. Batch normalization and parametric ReLU (PReLU) activation functions were used throughout the architecture. The PET/CT images acted as two channel inputs to the model, while a two-channel output provided the tumor segmentation mask (i.e., 0 = background, 1 = tumor). The architecture shown in Fig. 2 corresponds to a ResUnet with a maximum of 512 channels in the bottleneck layer (512 Model) where the number of channels in the convolution layers was (32, 64, 128, 256, and 512). We also implemented a model using a maximum of 256 channels in the bottleneck layer (256 Model), which has the same structure as the 512 Model, but the number of channels in the convolution layers was (16, 32, 64, 128, and 256).

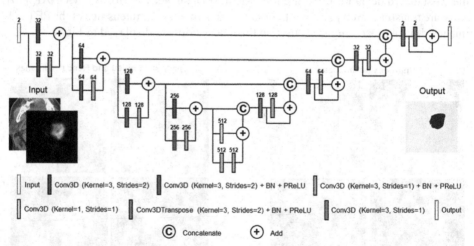

Fig. 2. Schematic of the ResUnet architecture used for the segmentation model. The number of channels (32, 64, 128, 256, and 512) is given above each block. The batch normalization and the parametric ReLU layers are annotated by BN and PReLU, respectively. The channels given in the figure are for the 512 model, while for the 256 model the channels are (16, 32, 64, 128, and 256).

2.4 Model Implementation

We used a 10-fold cross-validation approach where the 224 patients from the training data were randomly divided into ten non-overlapping sets. Each set (22 patients) was used for model validation while the remaining 202 patients in the remaining sets were used for training, i.e., each set was used once for testing and nine times for training.

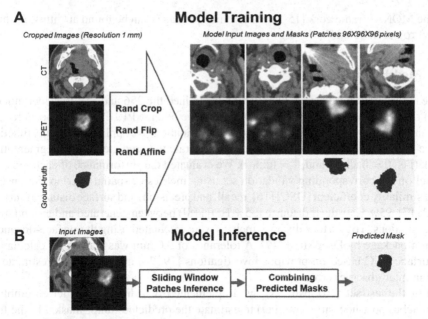

Fig. 3. An illustration of the workflow of the training and inference phases of the segmentation model. (A) Data transformation and augmentation is used to produce input data to the model. An example of four patches of CT and PET images and the corresponding ground truth tumor segmentation with at 50% representation of the tumor in these patches used for training the segmentation model (four patches of images per patient - $96 \times 96 \times 96$ voxels each). (B) Segmentation model prediction using sliding window inferences ($96 \times 96 \times 96$ voxels each) and combining the predicted masks from all patches to provide the final mask.

The processed PET, CT, and tumor masks (2.2) were randomly cropped to four random fixed-sized regions (patches) of size (96, 96, 96) per patch per patient. The random spatial cropping considered the patch center of mass as foreground (i.e., a tumor - positive) or background (i.e., non-tumor - negative) with a 50% probability for both the positive and negative cases as shown in Fig. 3A. We used a batch size of 2 patients' images and, therefore, a total of 8 patches of images. The shape of the input tensor provided to the network (2.3) for a batch size of 2, patches per image of 4, a two-channel input (PET/CT), and patch size of (96, 96, 96) is (8, 2, 96, 96, 96). The tumor mask was used as the ground truth target to train the segmentation model. The shape of the target tensor provided was (8, 1, 96, 96, 96). To minimize overfitting, in addition to the random spatial cropping to patch the images and masks, we implemented additional data augmentation to both image and mask patches which includes random horizontal flips of 50%, and random affine transformations with an axial rotation range of 12° and scale range of 10%. We used Adam as the optimizer and Dice loss as the loss function. The model was trained for 700 iterations with a learning rate of 2×10^{-4} for the first 550 iterations and 1×10^{-4} for the remaining 150 iterations. The image processing (2.2), data augmentation, network architecture, and loss function were used from the software packages provided by the MONAI framework [15]; code for these packages can be found at "https://github.com/Project-MONAI/".

2.5 Model Validation

For each validation fold (i.e., 22 patients), we trained the 256 and 512 ResUnet models (2.3) on the remaining 202 patients. Therefore, we obtained 10 different models for the 256 and 512 networks each from 10-fold cross-validation. We applied an argmax function to the two-channel output of each model to generate the predicted tumor segmentation mask (i.e., 0 = background, 1 = tumor). We evaluated the performance of each separate model on the corresponding validation set using metrics of spatial overlap (Sørensen–Dice similarity coefficient [DSC] [16], recall, and precision) and surface distance (surface DSC [17], 95% Hausdorff distance [95% HD] [18]) between generated and ground truth segmentations. The surface distance metrics were calculated using the surface-distances Python package by DeepMind [17]. A tolerance of 3.0 mm was chosen for calculation of surface DSC based on previous investigations [19, 20] as a reasonable estimate of human inter-observer error.

For the test set (101 patients), we implemented two different model ensembling approaches post-hoc (after training) to estimate the predicted tumor masks. In the first approach, we use used the Simultaneous Truth and Performance Level Estimation (STAPLE) algorithm [21] as a method to fuse labels generated by applying the 10 models produced during the 10-fold cross-validation on the test data set, i.e., generate the consensus predicted masks from the different generated predicted masks (STAPLE approach). The STAPLE algorithm was derived from publicly available Python code (https://github.com/fepegar/staple). In the second approach, we implemented a simple threshold of agreement based on all cross-validation fold models at the voxel level (AVERAGE approach). The total number of cross-validation models used in thresholding could be modulated as a parameter for this approach. For our purposes, we selected a threshold of 5 cross-validation folds as a proxy for majority voting, i.e. at least 5 cross-validation models must consider a voxel to be a tumor (label = 1) for that final voxel label to be considered as a tumor (label = 1). Majority voting in this context was chosen since it is common in other model ensembling approaches [22].

3 Results

The performances of the segmentation models are illustrated in Fig. 4, which shows Boxplots of the DSC, recall, precision, surface DSC, and 95% HD distributions obtained using the 10-fold cross-validation approach described in (2.5). The mean ± standard deviation values of the DSC, recall, precision, surface DSC, and 95% HD achieved by the 256 and 512 Models are 0.771 ± 0.039 and 0.768 ± 0.041, 0.807 ± 0.042 and 0.793 ± 0.038, 0.788 ± 0.038 and 0.797 ± 0.038, 0.892 ± 0.042 and 0.890 ± 0.044, and 6.976 ± 2.405 and 6.807 ± 2.357 respectively. The mean and median values of these metrics

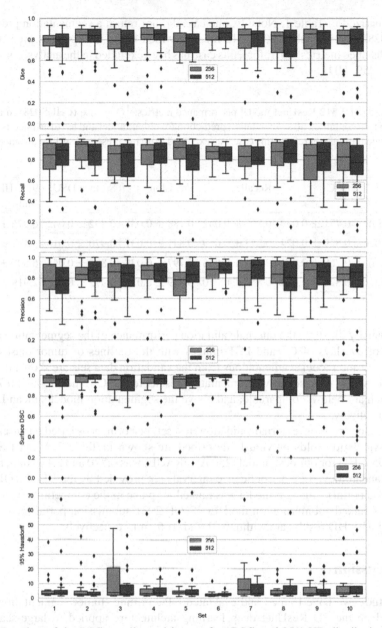

Fig. 4. Boxplots of the DSC, recall, precision, surface DSC, and 95% HD distributions for the 10-fold cross-validation data sets (Set 1 to Set 10–22 patients each*) used for the 256 and 512 ResUnet models. The lines inside the boxes refer to the median values. The stars refer to significant differences in the results by the two models (p-value < 0.05) using two-sided Wilcoxon signed-rank test. [1]One patient in Set 1 (CHUS028) did not return a segmentation prediction for either model and was thus excluded from the analysis of surface distance metrics (surface DSC, 95% HD).

are summarized in Table 1. Notably, one case did not return a segmentation prediction (CHUS028) for either the 256 or 512 models, which led to the spurious prediction of surface distance metrics. Therefore, this case has been excluded in the analysis of surface DSC and 95% HD.

Table 1. 256 and 512 ResUnet model performance metrics. [1]One case (CHUS028) in a cross-validation fold contained no segmentation prediction for either model and led to erroneous surface distance calculations; therefore, this case was excluded from the presented surface distance metric results.

Model	DSC	Recall	Precision	Surface DSC[1]	95% HD[1] (mm)
256 (mean)	0.771 ± 0.039	0.807 ± 0.042	0.788 ± 0.038	0.892 ± 0.042	6.976 ± 2.405
256 (median)	0.829 ± 0.024	0.873 ± 0.039	0.841 ± 0.037	0.970 ± 0.016	3.192 ± 0.816
512 (mean)	0.768 ± 0.041	0.793 ± 0.038	0.797 ± 0.038	0.890 ± 0.044	6.807 ± 2.357
512 (median)	0.828 ± 0.024	0.854 ± 0.040	0.849 ± 0.038	0.972 ± 0.013	2.919 ± 0.391

To visually illustrate the internal validation performance of the segmentation model, samples of overlays of CT and PET images with the outlines of tumor masks using ground truth and model segmentations from the validation data sets are shown in Fig. 5. The figure shows representative segmentation results for DSC values of 0.54, 0.77, and 0.96 which are below, comparable, and above the segmentation model's mean DSC of 0.77, respectively.

Finally, our models' external validation (test set) performances based on ensembling of cross-validation folds previously described are shown in Table 2. Mean DSC and median 95% HD for our best model (256 AVERAGE) was 0.770 and 3.143 mm, respectively (standard deviation or confidence intervals not provided by the HECKTOR 2021 submission portal). Our best model was ranked 8th place in the competition. Compared to the top ranked submission on the HECKTOR 2021 submission portal (pengy), our DSC and 95% HD results are within 0.01 and 0.06 mm, respectively.

4 Discussion

In this study, we have trained and evaluated OPC primary tumor segmentation models based on the 3D ResUnet deep learning architecture applied to large-scale and multi-institutional PET/CT data from the 2021 HECKTOR Challenge. Moreover, we investigate a variety of architectural modifications (512 vs. 256 channels in bottleneck layer) and ensembling techniques (STAPLE vs. AVERAGE) for test set predictions. Our approaches yield high and consistent segmentation performance in internal validation (cross-validation) and external validation (independent test set), thereby providing further empirical evidence for the feasibility of deep learning-based primary tumor segmentation for fully-automated OPC radiotherapy workflows.

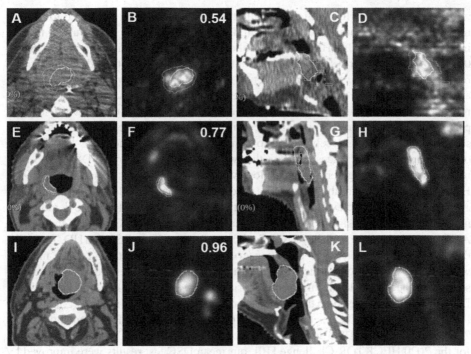

Fig. 5. Illustrative examples overlaying the ground truth tumor segmentations (red) and predicted tumor segmentations (yellow) on the CT images (first and third columns) and PET images (second and forth columns) with different 3D volumetric DSC values (below, equivalent, and above the mean estimated DSC value of 0.77) given at the right top corners of the PET images in the second column. (Color figure online)

Table 2. Test set results for ensemble models. Metrics are reported from the HECKTOR 2021 submission portal.

Model	Mean DSC	Median 95% HD (mm)
256 STAPLE	0.763	3.433
512 STAPLE	0.759	3.291
256 AVERAGE	0.770	3.143
512 AVERAGE	0.761	3.155

Through internal validation procedures on the training set (10-fold cross-validation), we attain mean DSC, recall, and precision values of 0.771, 0.807, and 0.788 for the 256 model and 0.768, 0.793, and 0.797, for the 512 model, respectively. While the 512 model offers a greater number of channels that could provide greater contextual information, maximum DSC performance is achieved with the 256 model. This may indicate the 256 model led to less over-fitting on the training data evaluation procedure. Interestingly, there was a tradeoff between recall and precision for the two models tested, with the 256 model offering higher recall at the cost of precision compared to the 512 model. Regardless, both these internal validation results improve upon our 3D models implemented in the 2020 HECKTOR Challenge, which only achieved a mean DSC of 0.69 [10]. These improved results potentially highlight the utility and importance of residual connections in a Unet architecture for this task. Moreover, image processing approaches that improve target class balance (i.e., tumor vs non-tumor) in the provided images significantly improve model sensitivity. Finally, we have further investigated the performance of our models using surface distance metrics, as these metrics have been suggested to be more closely linked to clinically meaningful endpoints [23, 24]. We observe minimal differences between the two models for the surface DSC, with both models showing strong performance. However, the 512 model has a slightly lower 95% HD, which may be favorable when more precise tumor boundary definitions are desired.

When models were evaluated on the test data (external validation), we demonstrate high performance consistent with the internal validation results. Generally, the AVERAGE method outperformed the STAPLE method in terms of both DSC and HD. Typically, the AVERAGE method led to more conservative estimates than the STAPLE method, which could indicate ground truth segmentations in the test set tended to be more conservative when considering tumor boundaries. Interestingly, while a tradeoff between DSC and HD exists based on the channel number for the STAPLE method (256 = better DSC, 512 = better HD), this tradeoff is not present with the AVERAGE method, as the 256 model has better DSC and HD compared to the 512 model. Compared to our entry for the 2020 HECKTOR Challenge [10], our mean DSC test results were improved by a sizable degree from the original DSC of 0.637 (0.133 increase for our best model). Moreover, we also improve upon the performance of the winning submission in the 2020 HECKTOR Challenge [25], which achieved a DSC of 0.759 (0.011 increase for our best model). Our positive results may in part be due to the inclusion of ensembling coupled to our improved network modifications (as indicated in the internal validation). The utility of ensembling for PET/CT OPC tumor segmentation has been previously noted since the winning entry in the 2020 challenge used an ensembling approach based on leave-one-center-out cross-validation models to yield the best performing DSC results [25]. Therefore, our results further incentivize the ensembling of model predictions for OPC tumor segmentation data. While our results were not ranked particularly highly within the 2021 HECKTOR leaderboard (8th place), it is worth noting our best model, and most of the models within the top 10–15 entries, scored highly similarly for both DSC and 95% HD. This may indicate a theoretical upper limit on this segmentation task, regardless of model implementations.

In recent years, there has been increasing evidence suggesting the utility of applying deep learning for fully-automated OPC tumor auto-segmentation in various imaging modalities [20, 26–28]. PET/CT has recently shown excellent performance when used as inputs to deep learning models, partly due to the large and highly curated datasets provided by the HECKTOR Challenge [12]. While direct comparison of performance metrics between segmentation studies is often ill-advised, the HECKTOR Challenge offers a systematic method for directly compare segmentation methods with each other. Moreover, since it has been suggested that the mean interobserver DSC for head and neck tumors in human experts is approximately 0.69 [29], our results indicate the potential for further testing to develop auto-segmentation workflows. However, it should be noted that before any definitive statements could be said about the clinical value of an auto-segmentation tool, the dosimetric impact and clinical acceptability of auto-segmeneted structures should be thoroughly evaluated through further studies [24].

One limitation of our study is the reliance of our loss function purely on DSC as an optimization metric. We have chosen the DSC loss since it has provided excellent results in previous investigations and due to its widespread acceptance. However, other loss functions such, as cross entropy [10] and focal loss [25], can be combined with the DSC loss for model optimization which may require further investigation. Moreover, additional measures of spatial similarity, such as surface DSC and 95% HD, are relevant in auto-segmentation for radiotherapy applications [24], and therefore may be attractive candidates for use in model loss optimization [30]. The importance of additional measures of spatial similarity seems to have been noted by the HECKTOR Challenge organizers, as the 95% HD has now become a metric used in the leaderboard to rank contestant performance. An additional limitation of our study is we have only tested a few label fusion approaches as ensembling techniques for our models. For example, we have selected STAPLE as a label fusion method because of its general ubiquity and widely available implementations. However, STAPLE has been criticized in the past [31]; therefore, additional label fusion approaches may be necessary to test in this framework [32]. Moreover, for the AVERAGE ensembling method, the specific threshold in the number of cross-validation models used to determine final label fusion can be seen as an additional parameter to tune. While we have chosen a 5-model threshold as a proxy for majority voting, alternative thresholding strategies can lead to more conservative or liberal estimates of tumor segmentation.

5   Conclusion

This study presented the development and validation of deep learning models using a 3D ResUnet architecture to segment OPC primary tumors in an end-to-end automated workflow based on PET/CT images. Using a combination of pre-processing steps, architectural design decisions, and model ensembling approaches, we achieve promising internal and external validation segmentation performance, with external validation mean DSC and median 95% HD of 0.770 and 3.143 mm, respectively, for our best model. Our method notably improves upon our previous iteration of our model submitted in the 2020 HECKTOR Challenge. Future studies should seek to further optimize these methods for improved OPC tumor segmentation performance in forthcoming iterations of the HECKTOR Challenge.

Acknowledgements. M.A.N. is supported by a National Institutes of Health (NIH) Grant (R01 DE028290-01). K.A.W. is supported by a training fellowship from The University of Texas Health Science Center at Houston Center for Clinical and Translational Sciences TL1 Program (TL1TR003169), the American Legion Auxiliary Fellowship in Cancer Research, and a NIDCR F31 fellowship (1 F31 DE031502-01). C.D.F. received funding from the National Institute for Dental and Craniofacial Research Award (1R01DE025248-01/R56DE025248) and Academic-Industrial Partnership Award (R01 DE028290), the National Science Foundation (NSF), Division of Mathematical Sciences, Joint NIH/NSF Initiative on Quantitative Approaches to Biomedical

Big Data (QuBBD) Grant (NSF 1557679), the NIH Big Data to Knowledge (BD2K) Program of the National Cancer Institute (NCI) Early Stage Development of Technologies in Biomedical Computing, Informatics, and Big Data Science Award (1R01CA214825), the NCI Early Phase Clinical Trials in Imaging and Image-Guided Interventions Program (1R01CA218148), the NIH/NCI Cancer Center Support Grant (CCSG) Pilot Research Program Award from the UT MD Anderson CCSG Radiation Oncology and Cancer Imaging Program (P30CA016672), the NIH/NCI Head and Neck Specialized Programs of Research Excellence (SPORE) Developmental Research Program Award (P50 CA097007) and the National Institute of Biomedical Imaging and Bioengineering (NIBIB) Research Education Program (R25EB025787). He has received direct industry grant support, speaking honoraria and travel funding from Elekta AB.

References

1. Hay, A., Nixon, I.J.: Recent advances in understanding colorectal cancer. F1000Research **7**, 1528 (2018)
2. Njeh, C.F.: Tumor delineation: The weakest link in the search for accuracy in radiotherapy. J. Med. Phys./Assoc. Med. Phys. India **33**, 136 (2008)
3. Foster, B., Bagci, U., Mansoor, A., Xu, Z., Mollura, D.J.: A review on segmentation of positron emission tomography images. Comput. Biol. Med. **50**, 76–96 (2014)
4. Segedin, B., Petric, P.: Uncertainties in target volume delineation in radiotherapy–are they relevant and what can we do about them? Radiol. Oncol. **50**, 254–262 (2016)
5. Rasch, C., Steenbakkers, R., van Herk, M.: Target definition in prostate, head, and neck. In: Seminars in Radiation Oncology, pp. 136–145. Elsevier (2005)
6. Tajbakhsh, N., Jeyaseelan, L., Li, Q., Chiang, J.N., Wu, Z., Ding, X.: Embracing imperfect datasets: a review of deep learning solutions for medical image segmentation. Med. Image Anal. **63**, 101693 (2020)
7. Zhou, T., Ruan, S., Canu, S.: A review: deep learning for medical image segmentation using multi-modality fusion. Array **3**, 100004 (2019)
8. Bakator, M., Radosav, D.: Deep learning and medical diagnosis: a review of literature. Multimodal Technol. Interact. **2**, 47 (2018)
9. Naser, M.A., Deen, M.J.: Brain tumor segmentation and grading of lower-grade glioma using deep learning in MRI images. Comput. Biol. Med. **121**, 103758 (2020). https://doi.org/10.1016/j.compbiomed.2020.103758
10. Naser, M.A., Dijk, L.V., He, R., Wahid, K.A., Fuller, C.D.: Tumor segmentation in patients with head and neck cancers using deep learning based-on multi-modality PET/CT images. In: Andrearczyk, V., Oreiller, V., Depeursinge, A. (eds.) HECKTOR 2020. LNCS, vol. 12603, pp. 85–98. Springer, Cham (2021). https://doi.org/10.1007/978-3-030-67194-5_10
11. AIcrowd MICCAI 2020: HECKTOR Challenges
12. Andrearczyk, V., et al.: Overview of the HECKTOR challenge at MICCAI 2020: automatic head and neck tumor segmentation in PET/CT. In: Andrearczyk, V., Oreiller, V., Depeursinge, A. (eds.) HECKTOR 2020. LNCS, vol. 12603, pp. 1–21. Springer, Cham (2021). https://doi.org/10.1007/978-3-030-67194-5_1
13. Andrearczyk, V., et al.: Overview of the HECKTOR challenge at MICCAI 2021: automatic head and neck tumor segmentation and outcome prediction in PET/CT images. In: Andrearczyk, V., Oreiller, V., Hatt, M., Depeursinge, A. (eds.) HECKTOR 2021. LNCS, vol. 13209, pp. 1–37. Springer, Cham (2022)
14. Oreiller, V., et al.: Head and neck tumor segmentation in PET/CT: the HECKTOR challenge. Med. Image Anal. **77**, 102336 (2021)

15. The MONAI Consortium: Project MONAI (2020). https://doi.org/10.5281/zenodo.4323059
16. Dice, L.R.: Measures of the amount of ecologic association between species. Ecology **26**, 297–302 (1945)
17. Nikolov, S., et al.: Clinically applicable segmentation of head and neck anatomy for radiotherapy: deep learning algorithm development and validation study. J. Med. Internet Res. **23**, e26151 (2021)
18. Taha, A.A., Hanbury, A.: Metrics for evaluating 3D medical image segmentation: analysis, selection, and tool. BMC Med. Imaging **15**, 1–28 (2015)
19. Blinde, S., et al.: Large interobserver variation in the international MR-LINAC oropharyngeal carcinoma delineation study. Int. J. Radiat. Oncol. Biol. Phys. **99**, E639–E640 (2017)
20. Wahid, K.A., et al.: Evaluation of deep learning-based multiparametric MRI oropharyngeal primary tumor auto-segmentation and investigation of input channel effects: results from a prospective imaging registry. Clin. Transl. Radiat. Oncol. **32**, 6–14 (2022). https://doi.org/10.1016/j.ctro.2021.10.003
21. Warfield, S.K., Zou, K.H., Wells, W.M.: Simultaneous truth and performance level estimation (STAPLE): an algorithm for the validation of image segmentation. IEEE Trans. Med. Imaging. **23**, 903–921 (2004)
22. Zhou, Z.H.: Ensemble learning. In: Machine Learning. Springer, Singapore (2021). https://doi.org/10.1007/978-981-15-1967-3_8
23. Kiser, K.J., Barman, A., Stieb, S., Fuller, C.D., Giancardo, L.: Novel autosegmentation spatial similarity metrics capture the time required to correct segmentations better than traditional metrics in a thoracic cavity segmentation workflow. J. Digit. Imaging **34**(3), 541–553 (2021). https://doi.org/10.1007/s10278-021-00460-3
24. Sherer, M.V., et al.: Metrics to evaluate the performance of auto-segmentation for radiation treatment planning: a critical review. Radiother. Oncol. **160**, 185–191 (2021)
25. Iantsen, A., Visvikis, D., Hatt, M.: Squeeze-and-excitation normalization for automated delineation of head and neck primary tumors in combined PET and CT images. In: Andrearczyk, V., Oreiller, V., Depeursinge, A. (eds.) HECKTOR 2020. LNCS, vol. 12603, pp. 37–43. Springer, Cham (2021). https://doi.org/10.1007/978-3-030-67194-5_4
26. Outeiral, R.R., Bos, P., Al-Mamgani, A., Jasperse, B., Simões, R., van der Heide, U.A.: Oropharyngeal primary tumor segmentation for radiotherapy planning on magnetic resonance imaging using deep learning. Phys. Imaging Radiat. Oncol. **19**, 39–44 (2021)
27. Andrearczyk, V., et al.: Automatic segmentation of head and neck tumors and nodal metastases in PET-CT scans. In: Medical Imaging with Deep Learning, pp. 33–43. PMLR (2020)
28. Moe, Y.M., et al.: Deep learning for automatic tumour segmentation in PET/CT images of patients with head and neck cancers. arXiv Prepr. arXiv:1908.00841 (2019)
29. Gudi, S., et al.: Interobserver variability in the delineation of gross tumour volume and specified organs-at-risk during IMRT for head and neck cancers and the impact of FDG-PET/CT on such variability at the primary site. J. Med. Imaging Radiat. Sci. **48**, 184–192 (2017). https://doi.org/10.1016/j.jmir.2016.11.003
30. Karimi, D., Salcudean, S.E.: Reducing the Hausdorff distance in medical image segmentation with convolutional neural networks. IEEE Trans. Med. Imaging. **39**, 499–513 (2019)
31. Van Leemput, K., Sabuncu, M.R.: A cautionary analysis of staple using direct inference of segmentation truth. In: Golland, P., Hata, N., Barillot, C., Hornegger, J., Howe, R. (eds.) MICCAI 2014. LNCS, vol. 8673, pp. 398–406. Springer, Cham (2014). https://doi.org/10.1007/978-3-319-10404-1_50
32. Robitaille, N., Duchesne, S.: Label fusion strategy selection. Int. J. Biomed. Imaging **2012**, 1–13 (2012)

Priori and Posteriori Attention for Generalizing Head and Neck Tumors Segmentation

Jiangshan Lu[1], Wenhui Lei[2], Ran Gu[1], and Guotai Wang[1(✉)]

[1] School of Mechanical and Electrical Engineering, University of Electronic Science and Technology of China, Chengdu, China
guotai.wang@uestc.edu.cn
[2] Shanghai Jiao Tong University, Shanghai, China

Abstract. Head and neck cancer is one of the most common cancers in the world. The automatic segmentation of head and neck tumors with the help of computer is of great significance for treatment. In the context of the MICCAI 2021 HEad and neCK tumOR (HECKTOR) segmentation challenge, we propose a combination of a priori and a posteriori attention to segment tumor regions from PET/CT images. Specifically, 1) According to the imaging characteristics of PET, we use the normalized PET as an attention map to emphasize the tumor area on CT as a priori attention. 2) We add channel attention to the model as a posteriori attention. 3) For the test set contains unseen domains, we use Mixup to mix the PET and CT in the train set to simulate unseen domains and enhance the generalization of the network. Our results on the test set are produced with the use of an ensemble of multiple models, and our method ranked third place in the MICCAI 2021 HECKTOR challenge with DSC is 0.7735 and HD95 is 3.0882.

Keywords: Head and neck tumor · Attention · Automatic segmentation

1 Introduction

In traditional clinical applications, for the segmentation of tumor regions, it is generally the clinician's manual delineation of the tumor gross target volume (GTV), which is tedious, time-consuming and easy to bring in inter-observer interaction. Differences within observers are highly subjective, and the results drawn by different clinicians may not be the same. Therefore, the automatic segmentation of tumors is very important for preoperative decision-making. It can effectively help clinicians determine the boundaries of the tumor and provide other important information about the tumor.

Head and Neck cancers are among the most common cancers worldwide (5th leading cancer by incidence) [1]. Radiotherapy combined with cetuximab has been established as standard treatment. However, locoregional failures remain a

© Springer Nature Switzerland AG 2022
V. Andrearczyk et al. (Eds.): HECKTOR 2021, LNCS 13209, pp. 134–140, 2022.
https://doi.org/10.1007/978-3-030-98253-9_12

major challenge and occur in up to 40% of patients in the first two years after the treatment. Recently, several radiomics studies based on Positron Emission Tomography (PET) and Computed Tomography (CT) imaging were proposed to better identify patients with a worse prognosis in a non-invasive fashion and by exploiting already available images such as those acquired for diagnosis and treatment planning.

Delineating the boundaries of tumors from PET/CT scans with the help of computers is of great significance in clinical practice for radiotherapy planning and response evaluation. This method of automatically segmenting tumor boundaries can effectively save clinicians' time, and greatly improves the accuracy of segmentation, providing a reliable solution for tumor treatment.

The MICCAI 2021 Head and Neck Tumor segmentation challenge (HECK-TOR) aims at evaluating automatic algorithms for the automatic segmentation of Head and Neck primary tumors in FDG-PET and CT image [2]. The data contains the same patients from MICCAI 2020 with the addition of a cohort of 71 patients from a new center (CHUP), of which 23 were added to the training set, 48 to the testing set [3]. The total number of training cases is 224 from 5 centers. The total number of test cases is 101 from two centers. A part of the test set will be from a center present in the training data (CHUP), another part from a different center (CHUV). For consistency, the GTVt (primary gross tumor volume) for these patients were annotated by experts, as shown in Fig. 1. The methods are evaluated using the Dice Similarity Coefficient (DSC) and Hausdorf Distance at 95% (HD95).

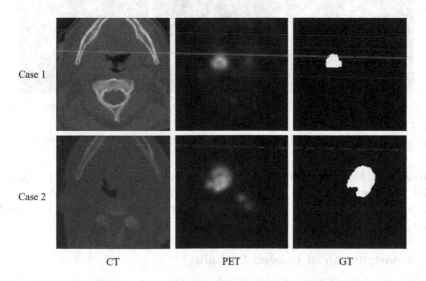

Fig. 1. Visual examples of CT and PET image and the corresponding ground truth.

This paper describes our approach based on the combination of prior attention and posterior attention, and constructs an unseen domain to improve the generalization of the model. Finally, the test is completed in the test set.

2   Method

Our method is generated through the ensemble of multiple training models. Firstly, the first part includes 8 pre-training models published by the winner of HECKTOR 2020 [4]; the second part is to use the data of five centers to leave a cross validation training of five models; the third part is the model trained by the enhanced data after PET and CT modal fusion.

2.1   Attention Model

Prior Attention Model. The tumor area on the PET image own higher pixel intensity than that of the normal area because of its high metabolic rate. While on the CT image, the HU value intensity difference between the tumor and the normal area is limited. However, it could effectively reflect the anatomical position and restrain false positive predictions. To combine the characteristics of PET and CT images, we adopt a fusion strategy to multiply the normalized PET and CT images as prior attention information, as shown in Fig. 2.

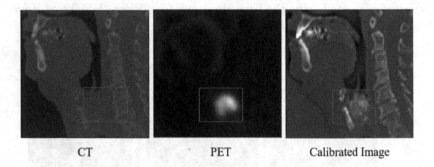

CT   PET   Calibrated Image

Fig. 2. Visual examples of CT and PET image and the priori image.

Posterior Attention Model. As shown in Fig. 3, our backbone network is the U-Net architecture [5], and uses the Squeeze-and-Excitation attention mechanism [6] as a posterior attention strategy to enhance the characteristics of the tumor area and suppress the characteristics of the background.

2.2   Construction of Unseen Domain

For the test set containing unseen domains, in order to improve the generalization of the model, the PET image is multiplied by a random number α between 0 and 1, the CT image is multiplied by $1-\alpha$, while the two multiplied values are added together to construct unseen data, thereby simulating a domain that has never been seen before, as shown in Fig. 4.

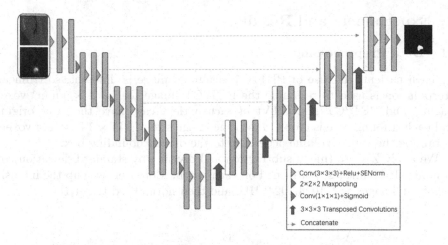

Fig. 3. The network of training a segmentation model.

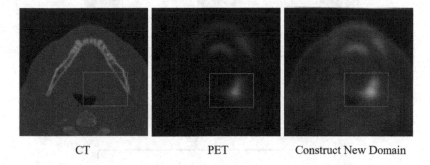

CT   PET   Construct New Domain

Fig. 4. Visual examples of CT and PET image and construct new domain image.

2.3 Model Ensemble

For the ensemble of models, the first part includes 8 pre-training models published by the winner of HECKTOR 2020; the second part is the model of the five centers we trained, using the method of leaving one center for cross validation, that is, using the data of four centers for training and the data of the remaining fifth center for validation; the third part includes 1) the calibrated data and constructed new domain data and the original PET and CT modals are input to the network as four channels to train the model; 2) the calibrated data and constructed new domain data are used as new samples to train the model. This can provide a priori attention for the model and improve the generalization of the network. Finally, our prediction of the test is generated by averaging the prediction results of all models and truncating with a threshold of 0.5, as shown in Fig. 5.

3   Experiments and Results

3.1   Data Preprocessing

For each patient, the size of PET/CT scanning image is different, so trilinear interpolation is used to resample the PET/CT image to $1 \times 1 \times 1\,\text{mm}^3$ voxel spacing, and PET/CT and GTVt of each patient case have the same origin. And each training case is cropped into the image size of $144 \times 144 \times 144$ voxels containing the tumor volume according to the official bounding box.

We apply Z-score (mean subtraction and division by standard deviation) to normalize PET images, performed on each patch. Moreover, we clip the intensities of CT images to [−1024,1024] HU, and then normalized to [−1,1].

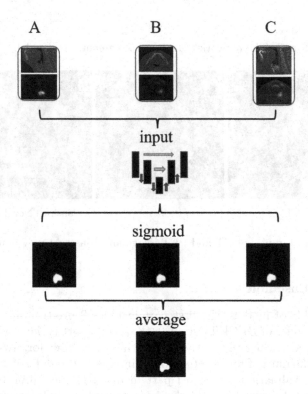

Fig. 5. Overall framework of our proposed method. The a model includes 8 pre-training models published by the winner of HECKTOR 2020; The B model includes five centers with one cross validation training model; The C model is to input the a priori data after PET and CT fusion and the data for constructing a new domain into the network as additional channels and additional samples together with the original data for training.

3.2   Network Training

The model was implemented in the framework of pytorch, and was trained on Ubuntu with two GPUs NVIDIA GeForce RTX 1080 Ti with 11 GB RAM. Adam is select to optimize the model parameters. We use cosine annealing schedule with the learning rate of l0–3, and a decay of 10–6 every 25 epochs. We train the segmentation model with 800 epochs with a batch size of 2, the loss function is the sum of Dice Loss and the Focal Loss.

3.3   Experimental Results

The results of our multiple model ensemble methods in the test sets of 53 CHUV centers and 48 CHUP centers, as shown in Fig. 6. Our method gains the average DSC and HD95 with 0.7735, 3.0882 respectively, as shown in Table 1.

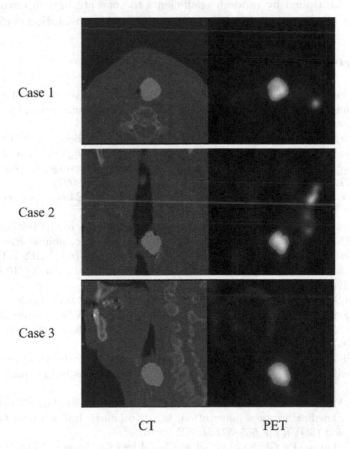

Fig. 6. Visual examples of segmentation result from test set.

Table 1. Quantitative results on the testing set

Participants	DSC	HD95	Rank
Pengy	0.7785	3.0882	1
SJTU_EIEE	0.7733	3.0882	2
HiLab(**Ours**)	0.7735	3.0882	3

4 Conclusion

In this study, we multiply the normalized PET and CT images to construct a priori attention image that fuses PET and CT image features, and add SE attention mechanism to U-Net as a posteriori attention strategy. At the same time, in order to construct the unseen domain, the normalized PET and CT images are multiplied by random coefficients to generate unseen data. Finally, we ensemble all the above models to improve its final segmentation performance.

References

1. Bray, F., Ferlay, J., Soerjomataram, I., et al.: Global cancer statistics 2018: GLOBO-CAN estimates of incidence and mortality worldwide for 36 cancers in 185 countries. CA: Cancer J. Clin. **68**(6), 394–424 (2018)
2. Andrearczyk, V., et al.: Overview of the HECKTOR challenge at MICCAI 2021: automatic head and neck tumor segmentation and outcome prediction in PET/CT images. In: Andrearczyk, V., Oreiller, V., Hatt, M., Depeursinge, A. (eds.) HECK-TOR 2021. LNCS, vol. 13209, pp. 1–37. Springer, Cham (2022)
3. Oreiller, V., et al.: Head and neck tumor segmentation in PET/CT: The HECKTOR challenge. Med. Image Anal. **77**, 102336 (2022)
4. Iantsen, A., Visvikis, D., Hatt, M.: Squeeze-and-excitation normalization for auto-mated delineation of head and neck primary tumors in combined PET and CT images. In: Andrearczyk, V., Oreiller, V., Depeursinge, A. (eds.) HECKTOR 2020. LNCS, vol. 12603, pp. 37–43. Springer, Cham (2021). https://doi.org/10.1007/978-3-030-67194-5_4
5. Ronneberger, O., Fischer, P., Brox, T.: U-net: convolutional networks for biomedical image segmentation. In: Navab, N., Hornegger, J., Wells, W.M., Frangi, A.F. (eds.) MICCAI 2015. LNCS, vol. 9351, pp. 234–241. Springer, Cham (2015). https://doi.org/10.1007/978-3-319-24574-4_28
6. Hu, J., Shen, L., Sun, G.: Squeeze-and-excitation networks. In: Proceedings of the IEEE Conference on Computer Vision and Pattern Recognition, pp. 7132–7141 (2018)
7. Milletari, F., Navab, N., Ahmadi, S.A.: V-net: fully convolutional neural networks for volumetric medical image segmentation. In: 2016 Fourth International Conference on 3D Vision (3DV), pp. 565–571. IEEE (2016)
8. Lin, T.Y., Goyal, P., Girshick, R., et al.: Focal loss for dense object detection. In: Proceedings of the IEEE International Conference on Computer Vision, pp. 2980–2988 (2017)

Head and Neck Tumor Segmentation with Deeply-Supervised 3D UNet and Progression-Free Survival Prediction with Linear Model

Kanchan Ghimire[1]([✉]), Quan Chen[1,2], and Xue Feng[1,3]

[1] Carina Medical, Lexington, KY 40513, USA
kghimire@carinaai.com
[2] Department of Radiation Medicine, University of Kentucky, Lexington, KY 40535, USA
[3] Department of Biomedical Engineering, University of Virginia, Charlottesville, VA 22903, USA
xf4j@virginia.edu

Abstract. Accurate segmentation of Head and Neck (H&N) tumor has important clinical relevance in disease characterization, thereby holding a strong potential for better cancer treatment planning and optimized patient care. In recent times, the development in deep learning-based models has been able to effectively and accurately perform medical images segmentation task that eliminates the problems associated with manual annotation of region of interest (ROI) such as significant human efforts and inter-observer variability. For H&N tumors, FDG-PET and CT carry complementary information of metabolic and structural details of tumor and the fusion of those modalities were explored in this study, which led to significant enhancement in performance level with combined data. Furthermore, deep supervision technique was applied to the segmentation network, where the computation of loss occurs at multiple layers that allows for gradients to be injected deeper into the network and facilitates the training. Our proposed segmentation methods yield promising result, with a Dice Similarity Coefficient (DSC) of 0.731 in our cross-validation experiment. Finally, we developed a linear model for progression-free survival prediction using extracted imaging and non-imaging features.

Keywords: Head and neck tumor segmentation · Deep learning · Survival prediction · Deep supervision · 3D UNet

1 Introduction

Disease characterization in the context of Head and Neck (H&N) tumor has the potential to improve treatment planning from a range of options such as surgery, radiation, and chemotherapy. Radiomics is the field of medical study that aims to extract quantitative features from medical images using algorithms. The development of deep learning models in recent times have been widely used in radiomics, which ultimately has the potential to improve disease characterization and optimized patient care. Some recent studies have

© Springer Nature Switzerland AG 2022
V. Andrearczyk et al. (Eds.): HECKTOR 2021, LNCS 13209, pp. 141–149, 2022.
https://doi.org/10.1007/978-3-030-98253-9_13

adopted those deep learning-based methods using convolutional neural network (CNN) in medical image segmentation applications and leveraged PET-CT in automatic segmentation of lung cancer [4–6] and brain tumor [7]. Automatic segmentation of H&N tumor eliminates the issues of manual annotation process such as long annotation time and inter-observer variability.

To compare and evaluate different automatic-segmentation and survival prediction algorithms, Medical Image Computing and Computer Assisted Intervention (MICCAI) 2021 HECKTOR Challenge was organized using FDG-PET and CT scans of head and neck (H&N) primary tumors along with clinical data [1, 2]. More specifically, the dataset used in this challenge includes multiple-institutional clinically acquired bimodal scans (FDG-PET and CT), focusing on oropharyngeal cancers and clinical data of the patients. Since training deep learning model on 3D images pose memory issue, patch-based approach was used in this study. FDG-PET and CT carry complementary information of metabolic and structural details of tumor, and so the fusion of those modalities was explored in this study. Finally, we developed a linear model for survival prediction using extracted imaging and non-imaging features, which, despite the simplicity, can effectively reduce overfitting and regression errors.

2 Methods

For the automatic segmentation of head and neck tumor task, the steps in our proposed method include image pre-processing, patch extraction, training multiple models using a generic 3D U-Net [3] structure with different hyper-parameters, deployment of each model for prediction, and image post-processing. For the survival task, the steps include feature extraction, model fitting, and deployment. Details are described in different sections as follows.

2.1 Image Pre-processing

Positron Emission Tomography (PET) measures metabolic activity of the cells of the body tissues. Computed Tomography, on the other hand, focuses on the morphological tissue properties. Fusing the information from the two modalities, PET and CT, was one of the major tasks during image pre-processing. CT scans are expressed in Hounsfield units (HU), and since CT scan images in our dataset include head and neck region, − 1024 HU to 1024 HU was selected as an appropriate window. Each 3D CT image was normalized to 0 to 1 by dividing the HU values by the window range. Since bounding boxes were provided in the dataset, each 3D CT and 3D PET images were cropped using bounding boxes coordinates provided in patient reference. To do so, bounding boxes coordinates were first translated into their respective index values. Isotropic resampling was performed to each 3D images to avoid training a network that must learn from potentially large-scale invariant. Then, for each subject, 3D CT and 3D PET images were fused along the last dimension so that the input image has two channels. To reduce the effects from different contrasts and different subjects, each 3D image was normalized by subtracting their global mean values and divided by the pixel intensity global standard deviation values.

2.2 Patch Extraction

In a 3D U-Net model architecture, using large input image poses several challenges. Some of the challenges include memory issue, long training time and class imbalance. Since the memory of a moderate GPU is about 11 Gb or 12 Gb, the network needs to greatly reduce the number of features and/or the layers to fit the model into the GPU. Reducing the number of features and/or the layers, oftentimes, leads to reduced network performance. Similarly, the training time will increase significantly as more voxels contribute to calculation of the gradients at each step and the number of steps cannot be proportionally reduced during optimization. Since the portion of the image with tumor is usually relatively smaller compared to the whole image, there will be a class imbalance problem. The class imbalance, therefore, will cause the model to focus on non-tumor voxels if trained with uniform loss, or will be prone to false positives if trained with weighted loss that favors the tumor voxels. Therefore, to more effectively utilize the training data, patch-based segmentation approach was applied where smaller patches were extracted from each subject. Patch-based approach also provides a natural way to do data augmentation since we can extract patches differently. The patch size was as a tunable hyper parameter during training. While larger patch size helps extract more global information from medical images, it also comes with the cost of memory issue and longer training time. For implementation, a random patch was extracted from each subject using 90% probabilities of selecting a patch with tumor (foreground) during each epoch. Since the head and neck region is anatomically symmetrical along the left-right direction, a random left-right flip with a 50% probability of being flipped after the patch selection step. Similarly, further data augmentation was applied to improve the robustness of the model by forcing the model to learn from different scale variant, orientation, contrast, brightness, and resolution. Spatial augmentation (including scaling and rotation), gaussian noise, gaussian blur, contrast, brightness, and down-sampling techniques were used. The scaling range used was 0.7 to 1.4, and the rotation range is $+30°$ to $-30°$. To avoid border artifact from data augmentation, the initial selected patch was slightly larger than the required patch. After performing data augmentation, center crop was applied on the augmented images to match the patch size required by the network for training.

2.3 Network Structure and Training

A 3D U-Net network structure with 5 encoding and 5 decoding blocks was used, as shown in Fig. 1. For each encoding block, a VGG like network with two consecutive 3D convolutional layers was used with kernel size of 3. It was followed by the rectified linear unit (ReLU) activation function. Finally, instance normalization layers were used. A large number of features were used in the first encoding block to improve the expressiveness of the network. Sum of cross entropy and dice was used as the loss function. Similar to convention U-Net structure, the spatial dimension was halved whereas the number of features was doubled during each encoding block. For the decoding blocks, symmetric blocks were used with skip-connections from corresponding encoding blocks. Features were concatenated to the de-convolution outputs, and the segmentation map of the input

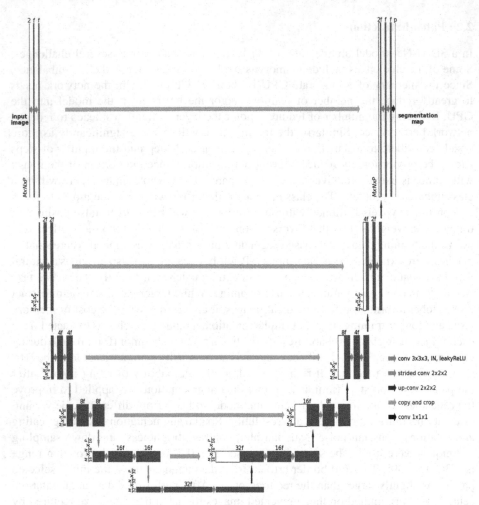

Fig. 1. 3D U-Net structure with 5 encoding and 5 decoding blocks.

patch was expanded to the binary class (tumor and non-tumor/background) ground truth labels.

Deep supervision technique was applied to the network, where the computation of loss occurs at each decoding block (except for the bottleneck layer and first decoding block). This approach allows for gradients to be injected deeper into the network and facilitates the training at each layer [8]. The final loss is the sum of the weighted loss computed at each decoding block. For the optimizer, stochastic gradient descent (SGD) with polynomial learning rate decay was used. The initial learning rate used was 0.01 and the ending learning rate was decayed to 0. While larger patch size could significantly increase the training time, it also has the potential to significantly improve the model performance and thus patch size of $128 \times 128 \times 128$ was used for training to balance the goal of large patch size selection and memory management.

During preliminary experimental phase, data from HECKTOR 2020 Challenge was used (201 training data). Cross-validation was applied during hyperparameter tuning, where 80% of the training data (160 cases) was used as training set and the rest 20% (41 cases) was used as validation set. The training schedule used was 3000 epochs and batch size 1. Hyperparameter tuning was performed for model selection, where multiple models were created using tunable parameters: 1) patch size; 2) different modalities; 3) type of convolution. While larger patch size could significantly increase the training time, it also has the potential to significantly improve the model performance. CT focuses on morphological tissue properties and PET focuses on metabolic activity of the cells of the tissues, and thus combining the information from both modalities can help improve the model performance. Similarly, using dilated convolution allows the network to have a larger receptive field and could lead to extracting better global information. Table 1 shows the parameters used to train six different models, where N denotes the input size, Modality denotes the image modality used during training and Convolution denotes whether conventional convolution or dilated convolution was used in the network.

Table 1. Detailed parameters for all 6 3D U-Net models.

Model #	N	Modality	Convolution
1	64, 96, 96	CT + PET	Conventional
2	64, 96, 96	CT + PET	Dilated
3	128, 128, 128	CT	Conventional
4	128, 128, 128	PET	Conventional
5	128, 128, 128	CT + PET	Conventional
6	128, 128, 128	CT + PET	Dilated

The training time was about 72 h per model and the training was performed on a Nvidia Tesla V100 SXM2 GPU with 16 Gb memory. TensorFlow framework was used to create and train the models.

2.4 Prediction

As discussed in the earlier section, using a large input size or whole images poses challenge of memory issue during training and so the patch-based training approach was implemented. The same challenge was faced during deployment, where the whole image cannot fit into the GPU memory. To combat the problem, sliding window approach was used to get the output for each testing subject. The sliding window approach, however, creates an issue of unstable predictions when sliding the window across the whole image without overlaps. When overlapping compared to the one without overlapping, the prediction time increases 8 times since it has to overlap in each dimension of the voxel. In deployment, stride size of 4 was used and the output probability was averaged. For each window, the original image and left-right flipped image were both predicted, and the average probability after flipping back the output of the flipped input was used as the output after thresholding the probability.

2.5 Image Post-processing

The last step before final submission was image post-processing. During post-processing, each final segmentation label (binary mask) was resampled using correct pixel spacing and origin based on the original CT 3D images after being cropped using the bounding box coordinates. Pixel spacing and origin were the main components used for resampling, since images were already cropped using bounding boxes during image pre-processing step. The post-processing was performed because of the requirement for the submission in the Challenge, where the output needs to be in the original CT resolution and cropped using respective bounding boxes.

2.6 Survival Prediction

To predict the progression-free survival time measured in days, extracted images features and non-image features were used to construct a linear regression model. 3 image-based features were used, where the features were calculated from ground-truth label during training and predicted label during validation. Two of the image-based features were tumor volume (V), tumor surface area (S). The volume was calculated by summing up the foreground voxels and surface area was calculated by summing up the magnitude of the gradients along three directions were obtained, as described in the following equations

$$V_{ROI} = \sum_{i,j,k} s_{i,j,k} \tag{1}$$

$$S_{ROI} = \sum_{i,j,k} s_{i,j,k} \sqrt{\left(\frac{\partial s}{\partial i}\right)^2 + \left(\frac{\partial s}{\partial j}\right)^2 + \left(\frac{\partial s}{\partial k}\right)^2} \tag{2}$$

in which ROI denotes a specific foreground class (tumor) and $s_{i,j,k} = 1$ for voxels that are classified to belong to this ROI and $s_{i,j,k} = 0$ otherwise. The third imaging feature was the image-based classification model output, where each class (6 classes total) denotes a certain range of the progression-free survival days. For this, the progression-free survival score was put into 6 bins first. A separate classification model was then trained, where the input was the whole 3D images, and the target was the class/ bin of progress-free survival. This method allows for the classification model to approximately estimate the progression-free survival time by looking at the images. Therefore, the third imaging feature allows to give additional important information to the survival prediction model.

For non-imaging features, there were both numerical and categorical variables. The numerical variables were age and weight; and the categorical features were alcohol use, center ID, chemotherapy, gender, HPV status, T-stage, N-stage, M-stage, TNM-group and tobacco use. The missing values were imputed with median value for numerical variables and mode value for categorical variables. A linear regression model after normalizing the input features to zero mean and unit standard deviation was fit with the training data. The predicted progression-free survival days was multiplied by -1 to get the anti-concordant value required in Challenge submission.

3 Results

3.1 Head and Neck Tumor Segmentation

Table 2 below shows the results of all 6 3D U-Net models (mean $\pm$ standard deviation). For patch size, the result show that larger patch size (Models 5 and 6) outperformed the models with smaller patch size (Models 1 and 2). For different modalities, we can clearly see that the fused imaging data (Models 5 and 6) outperformed models trained on individual modality (Models 3 and 4). Finally for convolution type, conventional convolution used in Models 1 and 5 outperformed Models 2 and 6 respectively.

Table 2. Results for all 6 3D U-Net models.

Model #	Validation dice (CT)	Validation dice (PET)	Validation dice (CT + PET)
1	-	-	0.720 ± 0.217
2	-	-	0.701 ± 0.220
3	0.620 ± 0.208	-	-
4	-	0.743 ± 0.160	-
5	0.081 ± 0.137	0.509 ± 0.170	0.798 ± 0.145
6	0.073 ± 0.142	0.546 ± 0.199	0.791 ± 0.129

During model selection, Model 5 was selected as the best and final model. The results shown above are results of models trained and tested on HECKTOR 2020 dataset. Model 5 was trained using HECKTOR 2021 dataset (18 added case in training dataset, and 6 cases added in validation dataset), and the validation dice was 0.731. The final model was trained with all 224 cases from HECKTOR 2021 Challenge dataset. Finally, segmentation was performed on the testing data of 101 subjects. The average Dice Similarity Coefficient (DSC) and Hausdorff Distance at 95% (HD95) for the test set were calculated by the MICCAI 2021 (HECKTOR) Challenge organizers after submission. The DSC and HD95 were 0.6851 and 4.1932 respectively.

3.2 Survival Prediction

The same split from head and neck tumor segmentation task was used for cross-validation of survival prediction model. During cross-validation, the mean absolute error was about 500 days (progression-free survival days). While the performance of survival prediction model does not seem to be very promising based on the result on the validation set, a linear model is robust against overfitting. The survival prediction was performed on 101 testing cases, which was evaluated by the MICCAI 2021 (HECKTOR) Challenge organizers after submission. Two separate tasks were in the Challenge for survival prediction: Task 2 for survival prediction without the ground-truth segmentation of the tumor, and Task 3 for the survival task with ground-truth segmentation provided through docker. The Concordance index (C-index) score on our prediction model output was 0.4342 on Task

2 and 0.5089 on Task 3. The results for Task 2 and Task 3 are in line with the expected result that ground truth would help the performance of survival task as opposed to the predicted tumor.

4 Discussion and Conclusions

In this paper we developed an automatic head and neck tumor segmentation method using deeply supervised 3D UNet and a survival prediction method using linear model. The algorithms developed for head and neck tumor has the potential of being transferred to other tumors and/or similar study, especially in respect to using the information from multi-modality images. For segmentation task, the preliminary results showed improvement with larger patch size and proved the benefits from the complementary information from PET and CT scans in tumor segmentation. Furthermore, during testing, overlap in the sliding window with larger number of strides improved the performance during cross-validation. Thus, during deployment, stride of 4 was used for prediction in the testing set. For survival prediction task, linear model was selected specifically to have a robust model against overfitting.

During model selection process of segmentation task, only 6 models were trained with hyperparameter tuning due to limitation in computation time. With further hyperparameter tuning, the performance could be improved. Similarly, ensemble modeling could also further improve the performance. However, ensemble modeling is time-consuming as we need to train multiple models and so we refrained from using that approach. For the survival prediction task, experimenting with different combinations of imaging and non-imaging features could have yield better performance.

Acknowledgements. This project has been funded in whole or in part with Federal funds from the National Cancer Institute, National Institutes of Health, Department of Health and Human Services, under Contract No. 75N91020C00048 and with Research Grant from Varian Inc.

References

1. Andrearczyk, V., et al.: Overview of the HECKTOR challenge at MICCAI 2021: automatic head and neck tumor segmentation and outcome prediction in PET/CT images. In: Andrearczyk, V., Oreiller, V., Hatt, M., Depeursinge, A. (eds.) HECKTOR 2021. LNCS, vol. 13209, pp. 1–37. Springer, Cham (2022)
2. Oreiller, V., et al.: Head and neck tumor segmentation in PET/CT: the HECKTOR challenge. Med. Image Anal. **77**, 102336 (2021)
3. Ronneberger, O., Fischer, P., Brox, T.: U-Net: convolutional networks for biomedical image segmentation. In: Navab, N., Hornegger, J., Wells, W.M., Frangi, A.F. (eds.) MICCAI 2015. LNCS, vol. 9351, pp. 234–241. Springer, Cham (2015). https://doi.org/10.1007/978-3-319-24574-4_28
4. Zhao, X., Li, L., Lu, W., Tan, S.: Tumor co-segmentation in PET/CT using multi-modality fully convolutional neural network. Phys. Med. Biol. **64**(1), 015011 (2018). https://doi.org/10.1088/1361-6560/aaf44b

5. Kumar, A., Fulham, M., Feng, D., Kim, J.: Co-learning feature fusion maps from PET-CT images of lung cancer. IEEE Trans. Med. Imaging **39**(1), 204–217 (2020). https://doi.org/10. 1109/TMI.2019.2923601

6. Li, L., Zhao, X., Lu, W., Tan, S.: Deep learning for variational multimodality tumor segmentation in PET/CT. Neurocomputing **392**, 277–295 (2020). https://doi.org/10.1016/j.neucom. 2018.10.099

7. Blanc-Durand, P., Van Der Gucht, A., Schaefer, N., Itti, E., Prior, J.O.: Automatic lesion detection and segmentation of 18F-FET PET in gliomas: a full 3D U-Net convolutional neural network study. PLoS ONE **13**(4), e0195798 (2018). https://doi.org/10.1371/journal.pone.019 5798

8. Isensee, F., Jaeger, P.F., Kohl, S.A., Petersen, J., Maier-Hein, K.H.: nnU-Net: a self-configuring method for deep learning-based biomedical image segmentation. Nat. Meth. **18**, 203–211 (2020)

Deep Learning Based GTV Delineation and Progression Free Survival Risk Score Prediction for Head and Neck Cancer Patients

Daniel M. Lang[1,2,3]($\boxtimes$) (iD), Jan C. Peeken[1,2,4] (iD), Stephanie E. Combs[1,2,4] (iD), Jan J. Wilkens[2,3] (iD), and Stefan Bartzsch[1,2] (iD)

[1] Institute of Radiation Medicine, Helmholtz Zentrum München, Munich, Germany
[2] Department of Radiation Oncology, School of Medicine and Klinikum rechts der Isar, Technical University of Munich (TUM), Munich, Germany
[3] Physics Department, Technical University of Munich, Garching, Germany
daniel.lang@tum.de
[4] Deutsches Konsortium für Translationale Krebsforschung (DKTK), Partner Site Munich, Munich, Germany

Abstract. Head and neck cancer patients can experience significant side effects from therapy. Accurate risk stratification allows for proper determination of therapeutic dose and minimization of therapy induced damage to healthy tissue. Radiomics models have proven their power for detection of useful tumors characteristics that can be used for patient prognosis. We studied the ability of deep learning models for segmentation of gross tumor volumes (GTV) and prediction of a risk score for progression free survival based on positron emission tomography/computed tomography (PET/CT) images. A 3D Unet-like architecture was trained for segmentation and achieved a Dice similarity score of 0.705 on the test set. A transfer learning approach based on video clip data, allowing for full utilization of 3 dimensional information in medical imaging data was used for prediction of a tumor progression free survival score. Our approach was able to predict progression risk with a concordance index of 0.668 on the test data. For clinical application further studies involving a larger patient cohort are needed.

Keywords: Deep learning · Survival analysis · Head neck

1 Introduction

Head and neck tumors represent a very diverse group of cancers with a heterogeneous range of subtypes. Multimodality therapy options involve surgery, chemotherapy and radiotherapy. But up to 40% of the patients experience locoregional failure during the first two years after treatment [3], highlighting the need

Team name: *DMLang*

© Springer Nature Switzerland AG 2022
V. Andrearczyk et al. (Eds.): HECKTOR 2021, LNCS 13209, pp. 150–159, 2022.
https://doi.org/10.1007/978-3-030-98253-9_14

for more effective therapies and clinically relevant patient stratification in order to improve treatment outcome.

Testing for an infection with the human papillomavirus (HPV) represents an established biomarker for oropharyngeal cancer (OPC), a subtype of head and neck squamous cell carcinomas (HNSCC), with HPV positive cases having a 74% reduced risk of cancer related death [8]. This lead the American Joint Committee of Cancer (AJCC) in their 8th edition staging manual to define HPV positive and negative OPC cases as separate entities with independent survival risk stratification [1]. However, *Haider et al.* [10] showed that further improvement in patient stratification in terms of overall survival (OS) and progression free survival (PFS) can be achieved by application of positron emission tomography/computed tomography (PET/CT) based radiomics models.

Task of the MICCAI 2021 HEad and neCK TumOR (HECKTOR) segmentation and outcome prediction challenge [2, 18] was the development of models for segmentation of the gross tumor volume (GTV) and computation of a risk score for PFS based on 3D PET/CT imaging and clinical data for OPC patients. We studied the ability of deep learning models for this task. A Unet-like architecture was trained for segmentation. For prediction of the PFS score we have chosen to follow the video clip based pretraining approach developed previously [14], due to the relative small size of the training set. The video clip classification network C3D [21] was used as a baseline model. Video data is publicly available in large data set sizes and features a three dimensional structure. Therefore, pretraining on video data allows for full exploitation of three dimensional information in the downstream task. C3D processes its input data by 3D convolutional layers, i.e. handling all three dimensions in the same manner, which makes it a perfect fit as baseline model used for feature extraction. In order to be capable of survival data modeling we followed the discrete-time survival model approach developed by *Gensheimer and Narasimhan* [6].

2   Materials and Methods

The HECKTOR 2021 dataset included 224 training cases coming from 5 centers and 101 test cases from two centers. One of the centers involved in the test set was also present in the training data, the other one was completely independent. Each case involved a PET/CT image, a bounding box of size $144\,\text{mm} \times 144\,\text{mm} \times 144\,\text{mm}$, specifying the rough region of interest in the images, and clinical patient data including: center, age, gender, TNM edition and staging, tobacco and alcohol consumption, performance status, HPV status and treatment procedure (radiotherapy only vs. chemoradiotherapy). For the training cases a segmentation map of the primary gross tumor volume (GTV) and progression free survival data in terms of time and event status was present. The CT imaging data was given in hounsfield units (HU) and the PET images in units of standardized uptake value (SUV).

2.1   Preprocessing

We performed a stratified split based on patients HPV status, survival event, center and treatment procedure to separate the training cohort into a train set of 190 cases and a validation cohort of 34 cases.

PET/CT images were resampled to an isotropic voxel size of $1\,\text{mm}^3$, using 3rd order b-spline-interpolation. Segmentation maps were resampled using nearest neighbor interpolation. CT voxel values were clipped at -250 and 250 HU and PET voxel values at 0 and 100 SUV and then rescaled to range from 0 to 255. For the segmentation task all images were cropped to patches of smaller size using the bounding boxes provided. For PFS risk score prediction, bounding boxes generated from the GTV were used for cropping. In order to secure a minimal image size of $100\,\text{mm} \times 100\,\text{mm} \times 100\,\text{mm}$, GTVs with smaller size were center cropped using those dimensions. The region of interest was selected based on the ground truth GTV maps provided for the training cases and the GTV maps generated in the segmentation task for the test set.

2.2   Segmentation

A 3D Unet-like model [4] was used for the segmentation task, Fig. 1. However, instead of forwarding the feature maps obtained right before the spatial dimension reduction step via skip connections from the encoder to the decoder path we have chosen to pass the feature maps right after dimension reduction. This reduces the number of network parameters due to the smaller size of those feature maps.

The main building block of the model consisted of a convolutional layer with a stride of one followed by a convolutional layer with a stride of larger size for downsampling and a transpose convolution for upsampling. Strided convolutional layers preserved the number of feature maps, stride one layers extended/reduced the number of feature maps by a factor of two. Both convolutional layers were followed by a dropout layer.

The best performing model consisted of two building blocks with a kernel size of $5 \times 5 \times 5$ and a stride of $3 \times 3 \times 3$, followed by three building blocks with a kernel size of $3 \times 3 \times 3$ and a stride of $2 \times 2 \times 2$ and respective feature map sizes of 16, 32, 64 and 128 in the downsampling part. The upsampling part used the same parameters in inverted order and the bottleneck layer had a feature map size of 256. Dropout rate was given by 0.25 for all layers. In the final layer sigmoid activation was used, all other layers used ReLU activation.

Augmentation techniques used included: flipping on the coronal and the sagittal plane, rotation by a multiple of 90°, addition of Gaussian noise with zero mean and standard deviation of 2, a constant global voxel value offset between -2 and 2 and application of a random elastic deformation grid sampled from a normal distribution with standard deviation between zero and one and voxel size of $3 \times 3 \times 3$.

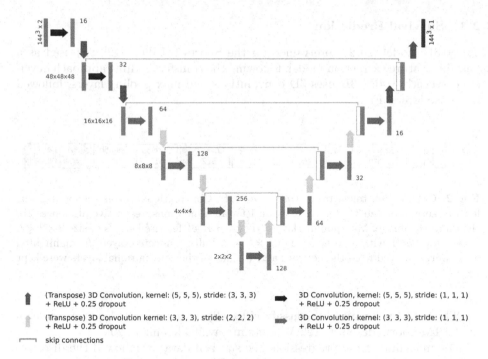

Fig. 1. The segmentation network. The architecture followed the basic Unet structure [4]. In contrast to the conventional architecture we have chosen to forward the feature maps obtained right after spatial dimension reduction instead of the last feature maps before this step from the encoder to the decoder path via skip connections. This reduces the number of network parameters. Convolutional layers with a stride larger than one were chosen for spatial dimension reduction, the number of feature maps were kept fix for those layers. Whereas, convolutional layers with a stride of one extended/reduced the number of feature maps by a factor of two.

The Adam optimizer was used for optimization of the Dice loss. Learning rate was given by 10^{-4} and batch size by 4. The PET/CT images were fed to the model using multi-modality input with a dimension of $144 \times 144 \times 144 \times 2$.

The model was trained for 150 epochs, the final model was chosen based on the epoch with minimal loss.

As post-processing steps we have chosen to threshold the voxel values by a value of 0.50, keep only the largest connected voxel area and apply binary dilation with a kernel size of $2 \times 2 \times 1$.

2.3   Survival Prediction

The C3D model [20,21] pretrained on the Sports-1M data [12] was used as a baseline feature extraction model, following the transfer learning approach developed recently [14]. C3D uses 3D convolutions and max pooling layers, followed by dense layers, Fig. 2.

Fig. 2. C3D model, taken from *Tran et al.* [21]. Convolutional layers, denoted *Conv*, feature kernels of size $3 \times 3 \times 3$ and stride 1 for all dimensions, respective filter sizes are shown in the image. Max-pooling layers, denoted *Pool*, feature kernels of size $2 \times 2 \times 2$, except for *Pool1* with a kernel size of $1 \times 2 \times 2$. Fully connected layers *fc*, highlighted gray, were removed from the network and weights of the convolutional layers were kept fix.

The model was trained to classify video clips into one of 487 sports activities of the 1M-Sports data set [12]. Weights are available online [20].

For modeling of the progression free survival data we followed the discrete-time survival approach of *Gensheimer and Narasimhan* [6]. Each output neuron of the deep neural network represented the conditional probability of surviving a discrete time interval. Loss for time interval j was given by the negative log likelihood function

$$\sum_{i=1}^{d_j} \ln(h_j^i) + \sum_{i=d_j+1}^{r_j} \ln(1 - h_j^i), \tag{1}$$

with h^i the hazard probability for individual i during j and r_j the number of individuals having not experienced failure or censoring before the interval, with d_j of them suffering failure during the interval. The overall loss was given by the sum of losses for all time intervals. The approach allows for a time-varying baseline hazard rate and non-proportional hazards.

All dense layers were removed from the C3D model and weights of the convolutional layers were kept fix during training. Randomly initialized dense layers were added after the last convolutional layer of the C3D model. Clinical features were added to the features obtained after the first dense layer. All intermediate dense layers were followed by a ReLU activation and a dropout layer with a rate of 0.5. Sigmoid activation was applied after the last layer. The best performing model involved layers of size 512 and 256 and an output size of 15 corresponding to time intervals covering a maximum of 10 years of survival with the first 5 years split into intervals of half a year and all subsequent intervals with a width of one year. The model can be seen in Fig. 3.

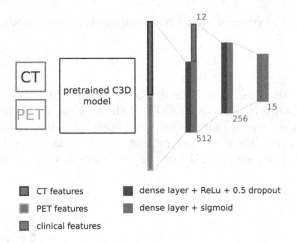

CT features ▪ dense layer + ReLu + 0.5 dropout

PET features ▪ dense layer + sigmoid

clinical features

Fig. 3. The survival model. The pretrained C3D model was used to generate feature vectors for the CT and PET images which were concatenated and feed to a dense layer. Features of this dense layer were then concatenated with the clinical features and feed to further dense layers, ending in an output representing a survival probability for consecutive intervals.

We have chosen to use a $96 \times 96 \times 96$ dimensional input for the network. Image patches generated from GTV bounding boxes in the preprocessing step were cropped to the required input size by random cropping during training and center cropping for validation and testing. Images were then rearranged to fulfill the input requirements of the pretrained model, i.e. 3 consecutive layers were fed to the color channels. A batch size of 16 was applied during training and the Adam optimizer with a learning rate of 5×10^{-5} was used to minimize the negative log likelihood [6]. Manual hyperparameter tuning as performed for model optimization. The model was trained for 75 epochs, the respective final model was chosen based on the best performing c-index score.

The same flipping, rotation and elastic deformation augmentation as for the segmentation task was applied during training. Gaussian noise was sampled with a variance between zero and one.

For the final prediction, the expected progression free survival time T_{PFS}^{j} per patient j was then computed using

$$T_{\text{PFS}}^{j} = \sum_{k=0}^{15} h_{k+1}^{j} \prod_{l=0}^{k} (1 - h_{l}^{j}) \, T_{l}^{\max}, \tag{2}$$

with h and $T^{\max}$ the hazard probability and upper time limit per output interval l and the last hazard probability set to $h_{k>15} = 1$. We then used min-max scaling to normalize the survival times between zero and one and computed a risk score R^{j} by

$$R^{j} = 1 - \tilde{T}^{j}, \tag{3}$$

with $\tilde{T}$ the normalized survival times.

The code for the survival model has been made publicly available[1].

3   Results

For the segmentation task the Dice similarity score (DSC) was given by 0.733 (0.096−0.912) and 0.762 (0.418−0.885) for the training and validation set and 0.71 for the test set. Error margins were obtained by computation 5% and 95% percentiles from the dice scores per patient. Figure 4 shows the Dice similarity score for the training set, with 10 of the 190 cases having a score lower than 0.10.

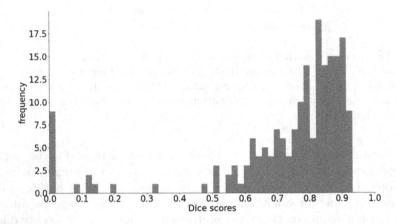

Fig. 4. Segmentation train set results. The histogram shows Dice score results for the training set, with 16 of the 190 cases having a Dice score smaller than 0.50.

The survival model achieved a c-index of 0.899 (0.865 − 0.931) and 0.833 (0.728−0.930) on the training and validation set, for the test set the c-index score was given by 0.668. Error margins were computed using bootstrap resampling with 10,000 samples and computation of 5% and 95% percentiles.

4   Discussion

We trained a model that was able to segment head and neck gross tumor volumes with a Dice similarity score of 0.705. For some of the cases the model completely failed to identify the GTV volume, which we accredit to two factors. First of all, head and neck cancer CT images are prone to artifacts that impair the performance of machine learning models [16]. Second, different pathophysiological mechanisms can lead to uncommon uptake of fluorine-18-fluorodeoxy-D-glucose (FDG) in PET images [19], which also affects the performance of the algorithm.

---

[1] https://github.com/LangDaniel/HECKTOR2021.

However, for general verification further statistical assessment on larger patient cohorts is needed.

In *Gudi et al.* [9] the interobserver DSC between three experienced radiation oncologists for PET/CT based GTV segmentation of head and neck cancer patients was found to be 0.69. Even though, the cohort used in this study differs from our data we conclude that our model is able to sufficiently detect the GTV in the majority of the cases and could be used to annotate tumor regions in a standardized way.

For the task of progression free survival prediction, we achieved a c-index score of 0.668. *Leijenaar et al.* developed a CT based radiomics model for prediction of overall survival in oropharyngeal cancer patients which was able to stratify patients with a c-index of 0.628. *Feliciani et al.* [5] employed PET/CT based radiomics for prediction of PFS in head and neck cancer patients and achieved an c-index of 0.76. However, only data from one institution was used in this study with no external validation. For the HECKTOR data set 53 of the 101 test set cases came from a center not involved in training. We accredit the difference in the train/validation and testing scores to this partial external testing. Radiomic features are known to be affected by imaging scanners and protocols [7,15,17,22] and convolutional neural networks are sensitive to such domain shifts [11,13].

In a previous study [14] we employed the video clip pretrained model for prediction of head and neck cancer patients HPV status based on CT images. Transfer learning helped the model to cope with the relative small data set size of 675 training set cases. With the task at hand involving less cases we conclude a even more pronounced benefit from the procedure, even though direct comparison for verification is necessary.

A model capable of survival prediction has the power to function as a input factor for improvement of determination of therapeutic dose. Patients with a good prognosis may profit from therapy deintensification and can therefore be prevented from therapy induced side effects. However, further model development and testing on a larger patient cohort is necessary for such application.

5   Conclusion

We have demonstrated the ability of video clip transfer learning for prediction of PFS in oropharynx cancer patients. For general verification further studies involving larger patient cohorts are needed.

References

1. Amin, M.B., et al.: The eighth edition AJCC cancer staging manual: continuing to build a bridge from a population-based to a more "personalized" approach to cancer staging. CA: Can. J. Clin. **67**(2), 93–99 (2017)

2. Andrearczyk, V., et al.: Overview of the HECKTOR challenge at MICCAI 2021: automatic head and neck tumor segmentation and outcome prediction in PET/CT images. In: Andrearczyk, V., Oreiller, V., Hatt, M., Depeursinge, A. (eds.) HECKTOR 2021. LNCS, vol. 13209, pp. 1–37. Springer, Cham (2022)
3. Chajon, E., et al.: Salivary gland-sparing other than parotid-sparing in definitive head-and-neck intensity-modulated radiotherapy does not seem to jeopardize local control. Radiat. Oncol. **8**(1), 1–9 (2013)
4. Çiçek, Ö., Abdulkadir, A., Lienkamp, S.S., Brox, T., Ronneberger, O.: 3D u-net: learning dense volumetric segmentation from sparse annotation. In: Ourselin, S., Joskowicz, L., Sabuncu, M.R., Unal, G., Wells, W. (eds.) MICCAI 2016. LNCS, vol. 9901, pp. 424–432. Springer, Cham (2016). https://doi.org/10.1007/978-3-319-46723-8_49
5. Feliciani, G., et al.: Radiomic profiling of head and neck Cancer: 18F-FDG PET texture analysis as predictor of patient survival. Contrast media & molecular imaging 2018 (2018)
6. Gensheimer, M.F., Narasimhan, B.: A scalable discrete-time survival model for neural networks. PeerJ **7**, e6257 (2019)
7. Ger, R.B., et al.: Comprehensive investigation on controlling for CT imaging variabilities in radiomics studies. Sci. Rep. **8**(1), 1–14 (2018)
8. Gillison, M.L., et al.: Evidence for a causal association between human papillomavirus and a subset of head and neck cancers. J. Natl Cancer Inst. **92**(9), 709–720 (2000)
9. Gudi, S., et al.: Interobserver variability in the delineation of gross tumour volume and specified organs-at-risk during IMRT for head and neck cancers and the impact of FDG-PET/CT on such variability at the primary site. J. Med. Imaging Radiat. Sci. **48**(2), 184–192 (2017)
10. Haider, S.P., et al.: Potential added value of PET/CT radiomics for survival prognostication beyond AJCC 8th edition staging in oropharyngeal squamous cell carcinoma. Cancers **12**(7), 1778 (2020)
11. Hendrycks, D., Dietterich, T.: Benchmarking neural network robustness to common corruptions and perturbations (2019)
12. Karpathy, A., Toderici, G., Shetty, S., Leung, T., Sukthankar, R., Fei-Fei, L.: Large-scale video classification with convolutional neural networks. In: Proceedings of the IEEE conference on Computer Vision and Pattern Recognition, pp. 1725–1732 (2014)
13. Kurakin, A., Goodfellow, I., Bengio, S.: Adversarial examples in the physical world (2016)
14. Lang, D.M., Peeken, J.C., Combs, S.E., Wilkens, J.J., Bartzsch, S.: Deep learning based HPV status prediction for oropharyngeal cancer patients. Cancers **13**(4), 786 (2021)
15. Larue, R.T., et al.: Influence of gray level discretization on radiomic feature stability for different CT scanners, tube currents and slice thicknesses: a comprehensive phantom study. Acta Oncologica **56**(11), 1544–1553 (2017)
16. Leijenaar, R.T., et al.: External validation of a prognostic CT-based radiomic signature in oropharyngeal squamous cell carcinoma. Acta Oncologica **54**(9), 1423–1429 (2015)
17. Lu, L., Ehmke, R.C., Schwartz, L.H., Zhao, B.: Assessing agreement between radiomic features computed for multiple CT imaging settings. PLoS ONE **11**(12), e0166550 (2016)
18. Oreiller, V., et al.: Head and neck tumor segmentation in PET/CT: the HECKTOR challenge. Med. Image Anal. **77**, 102336 (2021)

19. Purohit, B.S., Ailianou, A., Dulguerov, N., Becker, C.D., Ratib, O., Becker, M.: FDG-PET/CT pitfalls in oncological head and neck imaging. Insights into Imaging **5**(5), 585–602 (2014). https://doi.org/10.1007/s13244-014-0349-x
20. Tran, D., Bourdev, L., Fergus, R., Torresani, L., Paluri, M.: C3D: generic features for video analysis (2014). http://vlg.cs.dartmouth.edu/c3d/
21. Tran, D., Bourdev, L., Fergus, R., Torresani, L., Paluri, M.: Learning spatiotemporal features with 3d convolutional networks. In: Proceedings of the IEEE International Conference on Computer Vision, pp. 4489–4497 (2015)
22. Zhao, B., Tan, Y., Tsai, W.Y., Schwartz, L.H., Lu, L.: Exploring variability in CT characterization of tumors: a preliminary phantom study. Transl. Oncol. **7**(1), 88–93 (2014)

Multi-task Deep Learning for Joint Tumor Segmentation and Outcome Prediction in Head and Neck Cancer

Mingyuan Meng⬤, Yige Peng⬤, Lei Bi(✉)⬤, and Jinman Kim⬤

School of Computer Science, The University of Sydney, Sydney, Australia
`lei.bi@sydney.edu.au`

Abstract. Head and Neck (H&N) cancers are among the most common cancers worldwide. Early outcome prediction is particularly important for H&N cancers in clinical practice. If prognostic information can be provided to treatment planning at the earliest stage, the patient's 5-year survival rate can be significantly improved. However, traditional radiomics methods for outcome prediction are limited to a prior skillset in hand-crafting image features and also limited by its reliance on manual segmentation of primary tumors, which is intractable and error-prone. Multi-task learning is a potential approach to realize outcome prediction and tumor segmentation in a unified model. In this study, we propose a multi-task deep model for joint tumor segmentation and outcome prediction in H&N cancer using positron emission tomography/computed tomography (PET/CT) images, in the context of MICCAI 2021 HEad and neCK TumOR (HECKTOR) segmentation and outcome prediction challenge. Our model is a multi-task neural network that simultaneously predicts the risk of disease progression and delineates the primary tumors using PET/CT images in an end-to-end manner. Our model was evaluated for outcome prediction and tumor segmentation in the cross-validation of training set (C-index: 0.742; DSC: 0.728) and on the testing set (C-index: 0.671; DSC: 0.745).

Keywords: Outcome prediction · Tumor segmentation · Multi-task deep learning

1 Introduction

Head and Neck (H&N) cancers are among the most common cancers worldwide [1]. Outcome prediction is a major concern for patients with H&N cancer, as it provides early prognostic information that is needed for treatment planning. Outcome prediction is a regression task that predicts the survival outcomes of patients, e.g., progression-free survival (PFS). However, compared to typical regression tasks such as clinical score prediction, outcome prediction is more challenging because survival data is often right-censored, which means the time of events occurring is unclear for some patients (i.e., censored patients). Recently, several radiomics studies based on positron emission tomography/computed tomography (PET/CT) imaging have been proposed for outcome prediction in patients with H&N cancer [2-4]. However, establishment and validation of

© Springer Nature Switzerland AG 2022
V. Andrearczyk et al. (Eds.): HECKTOR 2021, LNCS 13209, pp. 160–167, 2022.
https://doi.org/10.1007/978-3-030-98253-9_15

radiomics models require manual segmentation of primary tumors and nodal metastases for every patient, which is intractable and error-prone. Therefore, an outcome prediction model that do not require the manual segmentation is highly desirable to support clinical practice.

The MICCAI 2021 HEad and neCK TumOR (HECKTOR) segmentation and outcome prediction challenge [5, 6] aims to develop algorithms for tumor segmentation and outcome prediction in H&N cancer using PET/CT images. A total of three tasks were proposed in this challenge: (1) Task 1: the automatic segmentation of H&N primary tumors in PET/CT images; (2) Task 2: the prediction of patient outcomes (i.e., PFS) from PET/CT images; (3) Task 3: the prediction of PFS from PET/CT images (same as Task 2), except that the ground truth segmentation of primary tumors will be made available as inputs. In this study, we focused on Task 1 and Task 2, and aimed to propose a unified model that directly predicted the patient outcomes from PET/CT images (without the need for manual segmentation) and automatically delineated H&N primary tumors in PET/CT images.

To address the Task 1 and Task 2 in the HECKTOR challenge, we introduced the concept of multi-task leaning into deep models. Multi-task learning is a subfield of machine learning in which multiple tasks are simultaneously learned by a shared model [7]. This approach improves data efficiency and reduces overfitting by leveraging the shared features learned for multiple tasks [8]. Recently, we identified that tumor segmentation has been used as an auxiliary task in deep multi-task learning and benefited tumor-related tasks such as glioma diagnosis [9], genotype prediction [10], and treatment response prediction [11]. This is likely because the tumor segmentation task can guide the model to extract the features related to tumor regions [11]. Moreover, deep segmentation models can work with small training data (even less than 50 images) [12], which suggests that segmentation losses have high efficiency in leveraging small data. Therefore, combining tumor segmentation and outcome prediction is promising to achieve better performance, especially when using small PET/CT data.

In this study, we propose a 3D end-to-end deep multi-task model for joint outcome prediction and tumor segmentation. Our model can jointly learn outcome prediction and tumor segmentation from PET/CT images. Tumor segmentation, as an auxiliary task, enhances our model to effectively learn from PET/CT data. For survival prediction, our model can predict the risk of disease progression, which is a regression task where a higher predicted risk value indicates a patient with shorter PFS.

2 Materials and Methods

2.1 Patients

The organizer of HECKTOR challenge provided a training set including 224 patients from 5 centers, CHGJ ($n = 55$), CHMR ($n = 18$), CHUM ($n = 56$), CHUP ($n = 23$), and CHUS ($n = 72$). A testing set including 101 patients was used for evaluation and the ground truth of the testing set was not released to the public. A part of the testing set ($n = 48$) is from a center present in the training set (CHUP), while another part ($n = 53$) from a different center (CHUV). We did not use any external dataset to build our model. All patients were histologically confirmed with H&N cancer in the oropharynx. They all underwent FDG-PET/CT before treatment, and the primary gross tumor volume (GTVt) for these patients were annotated by experts. In addition, all patients had clinical reports including T stage, N stage, M stage, TNM group, Human Papilloma Virus (HPV) status, performance status, tobacco consumption, and alcohol consumption. For clinical factors, we performed univariate and multivariate analyses using Cox proportional hazard regression, in which the factors with $P < 0.05$ in univariate analysis were then taken into multivariate analysis. Only HPV status, performance status, and M stage showed significant relevance to PFS in both univariate and multivariate analyses. Therefore, only these three clinical factors were used for building our models.

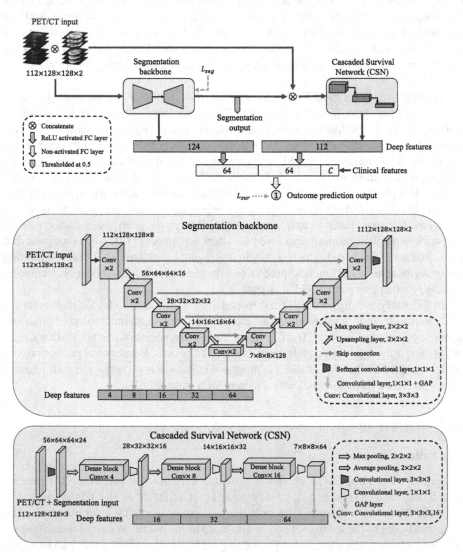

Fig. 1. The workflow of our multi-task deep model for joint survival outcome prediction and primary tumor segmentation.

2.2 Multi-task Deep Model

Our model is a multi-task deep neural network that can simultaneously predict the risk of disease progression and delineate the primary tumors using PET/CT images. This model is based on our recently proposed DeepMTS [13], a deep multi-task model for survival prediction in patients with advanced nasopharyngeal carcinoma. Figure 1 shows the workflow of our model. Our model is composed of two main components: a U-net [12] based segmentation backbone and a DenseNet [14] based cascaded survival network (CSN). Deep features are derived from these two components and fed into Fully Connected (FC) layers for outcome prediction. The whole architecture was trained in an end-to-end manner to minimize the combined loss of segmentation loss L_{seg} and survival prediction loss L_{sur}.

Specifically, the PET and CT images are first concatenated and fed into the segmentation backbone. Then, the segmentation backbone produces a tumor probability map. The tumor probability map and the PET/CT images are further concatenated and fed into the CSN. For the tumor segmentation task, the tumor probability map can be thresholded at 0.5 to obtain a binary tumor segmentation mask. For the outcome prediction task, 124 deep features are derived from the downsampling branch of the segmentation backbone and fed into a Rectified Linear Unit (ReLU) activated FC layer (FC-1); 112 deep features are derived from the CSN and also fed into a ReLU activated FC layer (FC-2). The FC-1, FC-2, and three prognostic clinical factors (HPV status, performance status, and M stage) are concatenated and fed into a non-activated (linear) FC layer with a single neuron (FC-3). The output of FC-3 is the predicted risk of disease progression. Moreover, dropout with 0.5 probability and L2 regularization with 0.1 coefficient are used in all FC layers for regularization.

The segmentation backbone is a modified 3D U-net [12] and its detailed architecture is shown in Fig. 1. Batch Normalization (BN) and ReLU activation are used after each convolutional layer of the segmentation backbone. The CSN is an extended version of 3D DenseNet [14] with 3 dense blocks. The output of each dense block was fed into a $1 \times 1 \times 1$ convolutional layer and a Global Average Pooling (GAP) layer to obtain deep features. The detailed architecture of CSN is shown in Fig. 1 as well. In the dense blocks, BN and ReLU activation are employed before the convolutional layers, while dropout with 0.05 probability is utilized afterwards.

2.3 Loss Function

Our model was trained to minimize a combined loss L as follows:

$$L = L_{seg} + L_{sur} \tag{1}$$

where L_{seg} (Eq. 2) is the loss function of the tumor segmentation task and L_{sur} (Eq. 3) is the loss function of the outcome prediction task.

For the tumor segmentation task, the L_{seg} is a Dice loss [15] as follows:

$$L_{seg} = -\frac{2 \sum_i^N p_i g_i}{\sum_i^N p_i^2 + \sum_i^N g_i^2} \tag{2}$$

where $p_i \in [0, 1]$ is voxels of the predicted tumor probability map, $g_i \in \{0, 1\}$ is the voxels of the ground truth tumor segmentation mask (label), and the sums run over all N voxels of the segmentation space.

For the outcome prediction task, a Cox negative logarithm partial likelihood loss [16] was used as the L_{sur} to handle right-censored survival data:

$$L_{sur} = -\frac{1}{N_{E=1}} \sum_{i:E_i=1} \left(h_i - \log \sum_{j \in \mathcal{H}(T_i)} e^{h_j} \right) \tag{3}$$

where h is the predicted risk, E is an event indicator (0 indicates a censored patient and 1 indicates a patient with disease progression), T is the time of PFS (for $E = 1$) or the time of patient censored (for $E = 0$), $N_{E=1}$ is the number of patients with disease progression, and $\mathcal{H}(T_i)$ is a set of patients whose T is no less than T_i.

2.4 Data Preprocessing

Both PET and CT images were first resampled into isotropic voxels of unit dimension to ensure comparability, where 1 voxel corresponds to 1 mm^3, in which trilinear interpolation and nearest neighbor interpolation were used for the PET/CT images and segmentation masks respectively. Then, the challenge organizers offered bounding boxes to roughly locate oropharynx. Through the bounding boxes, each PET/CT image was cropped into a patch of $144 \times 144 \times 144$ voxels. To compress the usage of GPU memory, we further cropped the central part of each patch into a central patch of $112 \times 128 \times 128$ voxels. The central patches can include most of the oropharynx region, which are available for outcome prediction and tumor segmentation of H&N cancer in the oropharynx. Finally, the PET patches were standardized individually to zero mean and unit variance (Z-score normalization), while the CT patches were clipped to $[-1024, 1024]$ and mapped to $[-1, 1]$.

2.5 Training Procedure

Our model was implemented using Keras with a Tensorflow backend on two 12GB GeForce GTX Titan X GPUs. We used the Adam optimizer with a batch size of 8 to train the model for a total of 10000 iterations. The learning rate was set as 1e-4 initially and then decreased to 5e$-$5, 1e$-$5 at the 2500th, 5000th training iteration. During the training, data augmentation was applied to the input PET/CT patches in real-time to avoid overfitting. The used data augmentation techniques included random translations (up to ± 10 pixels), random rotations (up to $\pm 5°$), and random flipping in the sagittal axis. Furthermore, we sampled an equal number of censored and uncensored samples during the data augmentation process to minimize the problem introduced from the unbalanced dataset.

2.6 Ensemble

Our results on the testing set were obtained using an ensemble of ten models trained and validated on different splits of the training set. Five models were built using leave-one-center-out cross-validation, i.e., the data from four centers was used for training and the data from the fifth center was held out for validation. Other five models were built using 5-fold cross-validation. For outcome prediction, the testing results derived from the ten models were first standardized by Z-score normalization and then averaged together to obtain the final testing results. For tumor segmentation, the testing results derived from the ten models were directly averaged together and then thresholded with 0.5 to obtain the final testing results.

2.7 Evaluation Metrics

The survival prediction task was evaluated using concordance index (C-index) [17]. The C-index metric measures the consistency between the predicted risk and the survival outcomes. The tumor segmentation task was evaluated using Dice Similarity Coefficient (DSC). The DSC metric measures the similarity between the predicted and ground-truth segmentation masks.

3 Results and Discussion

Our validation and testing results are summarized in Table 1. For outcome prediction, our model achieved C-index results of 0.753 and 0.742 on the 5-fold cross-validation and leave-one-center-out cross-validation, respectively. However, we identified a gap between the testing result and validation results: our model achieved a lower C-index of 0.671 on the testing set. This is likely attributed to the fact that the patients from CHUV are only present in the testing set and are therefore poorly represented in the training set. For tumor segmentation, our model achieved DSC of 0.732 and 0.728 on the 5-fold cross-validation and leave-one-center-out cross-validation, respectively. By averaging all ten models, we achieved a DSC of 0.745 on the testing set.

We used the central patches ($112 \times 128 \times 128$) of bounding boxes to build our model, while the tumor regions out of the central patches were discarded. In the training set, only 4 (out of 224) patients have tumors out of the central patches, where the discarded tumor regions are less than 5% of the whole tumor regions. Through this approach, we sacrificed some non-critical information to better leverage the limited GPU memory. During paper revision time, Andrearczyk et al. [18] also proposed a multi-task deep model for joint segmentation and outcome predciton in H&N cancer. However, our method differs in that we introduced a novel cascaded survival network (CSN) on top of the standard multi-task deep learning model for improved survival prediction.

Table 1. The C-index and DSC results on the 5-fold cross-validation, leave-one-center-out cross-validation, and testing set.

Evaluation data	C-index	DSC
Fold1 ($n = 45$)	0.711	0.721
Fold2 ($n = 45$)	0.740	0.769
Fold3 ($n = 45$)	0.844	0.735
Fold4 ($n = 45$)	0.768	0.698
Fold5 ($n = 45$)	0.704	0.739
Average of 5 folds	0.753	0.732
CHGJ ($n = 55$)	0.799	0.784
CHMR ($n = 18$)	0.681	0.664
CHUM ($n = 56$)	0.754	0.698
CHUP ($n = 23$)	0.760	0.782
CHUS ($n = 72$)	0.714	0.712
Average of 5 centers	0.742	0.728
Testing set CHUP ($n = 48$) + CHUV ($n = 53$)	0.671	0.745

References

1. Parkin, D.M., Bray, F., Ferlay, J., Pisani, P.: Global cancer statistics, 2002. CA Cancer J. Clin. **55**(2), 74–108 (2005)
2. Vallieres, M., et al.: Radiomics strategies for risk assessment of tumour failure in head-and-neck cancer. Sci. Rep. **7**(1), 10117 (2017)
3. Bogowicz, M., et al.: Comparison of PET and CT radiomics for prediction of local tumor control in head and neck squamous cell carcinoma. Acta Oncol. **56**(11), 1531–1536 (2017)
4. Castelli, J., et al.: A PET-based nomogram for oropharyngeal cancers. Eur. J. Cancer **75**, 222–230 (2017)
5. Andrearczyk, V., et al.: Overview of the HECKTOR challenge at MICCAI 2021: automatic head and neck tumor segmentation and outcome prediction in PET/CT images. In: Andrearczyk, V., Oreiller, V., Hatt, M., Depeursinge, A. (eds.) HECKTOR 2021. LNCS, vol. 13209, pp. 1–37. Springer, Cham (2022)
6. Oreiller, V., et al.: Head and neck tumor segmentation in PET/CT: the HECKTOR challenge. Med. Image Anal. **77**, 102336 (2021)
7. Zhang, Y., Yang, Q.: A survey on multi-task learning. IEEE Trans. Knowl. Data Eng. (2021)
8. Crawshaw, M.: Multi-task learning with deep neural networks: a survey. arXiv preprint arXiv: 2009.09796 (2020)
9. Xue, Z., Xin, B., Wang, D., Wang, X.: Radiomics-enhanced multi-task neural network for non-invasive glioma subtyping and segmentation. In: Mohy-ud-Din, H., Rathore, S. (eds.) RNO-AI 2019. LNCS, vol. 11991, pp. 81–90. Springer, Cham (2020). https://doi.org/10.1007/978-3-030-40124-5_9
10. Liu, J., et al.: A cascaded deep convolutional neural network for joint segmentation and genotype prediction of brainstem gliomas. IEEE Trans. Biomed. Eng. **65**(9), 1943–1952 (2018)

11. Jin, C., et al.: Predicting treatment response from longitudinal images using multi-task deep learning. Nat. Commun. **12**(1), 1–11 (2021)
12. Çiçek, Ö., Abdulkadir, A., Lienkamp, S., Brox, T., Ronneberger, O.: 3D U-Net: learning dense volumetric segmentation from sparse annotation. In: Ourselin, S., Joskowicz, L., Sabuncu, M.R., Unal, G., Wells, W. (eds.) MICCAI 2016. LNCS, vol. 9901, pp. 424–432. Springer, Cham (2016). https://doi.org/10.1007/978-3-319-46723-8_49
13. Meng, M., Gu, B., Bi, L., Song, S., Feng, D., Kim, J.: DeepMTS: deep multi-task learning for survival prediction in patients with advanced nasopharyngeal carcinoma using pretreatment PET/CT. arXiv preprint arXiv:2109.07711 (2021)
14. Huang, G., Liu, Z., Van Der Maaten, L., Weinberger, K.Q.: Densely connected convolutional networks. In: Proceedings of the IEEE Conference on Computer Vision and Pattern Recognition (2017)
15. Milletari, F., Navab, N., Ahmadi, S.A.: V-Net: fully convolutional neural networks for volumetric medical image segmentation. In: 2016 Fourth International Conference on 3D Vision (3DV). IEEE (2016)
16. Katzman, J.L., Shaham, U., Cloninger, A., Bates, J., Jiang, T., Kluger, Y.: DeepSurv: personalized treatment recommender system using a Cox proportional hazards deep neural network. BMC Med. Res. Methodol. **18**(1), 24 (2018)
17. Harrell, F.E., Jr., Lee, K.L., Mark, D.B.: Multivariable prognostic models: issues in developing models, evaluating assumptions and adequacy, and measuring and reducing errors. Stat. Med. **15**(4), 361–387 (1996)
18. Andrearczyk, V., et al.: Multi-task deep segmentation and radiomics for automatic prognosis in head and neck cancer. In: Rekik, I., Adeli, E., Park, S.H., Schnabel, J. (eds.) PRIME 2021. LNCS, vol. 12928, pp. 147–156. Springer, Cham (2021). https://doi.org/10.1007/978-3-030-87602-9_14

PET/CT Head and Neck Tumor Segmentation and Progression Free Survival Prediction Using Deep and Machine Learning Techniques

Alfonso Martinez-Larraz[1], Jaime Martí Asenjo[2(✉)], and Beatriz Álvarez Rodríguez[3]

[1] Agroviz Inc., Madrid, Spain
[2] Medical Physics Department, HM Sanchinarro Hospital, c\Oña 10, 28050 Madrid, Spain
jmartiasenjo@hmhospitales.com
[3] Radiation Oncology Department, HM Sanchinarro Hospital, c\Oña 10, 28050 Madrid, Spain

Abstract. Three 2D CNN (Convolutional Neuronal Networks) models, one for each patient-plane (axial, sagittal and coronal) plus two 3D CNN models were ensemble using two 3D models for Head and Neck tumor segmentation in CT and FDG-PET images. A Progression Free Survival (PFS) prediction, based on a Gaussian Process Regression (GPR) model was design on Matlab. Radiomic features such as Haralick textures, geometrical and statistical data were extracted from the automatic segmentation. A 35-feature selection process was performed over 1000 different features. An anti-concordant prediction was made based on the Kaplan-Meier estimator.

1 Introduction

Tumor segmentation is one of the most important challenges faced by medical imaging. Due to this, each year competitions are held in which the participating teams must carry out some task related to this field. In this paper we show the solution carried out by our team to solve two of the tasks posed by the HECKTOR [1, 2] competition: The segmentation of head and neck tumors and the prediction of life time from the predicted segmentation. Due to this, the paper is structured in two parts, one for each of the tasks performed.

2 Task 1: Tumor Segmentation

2.1 Materials and Methods

The goal of this task is the segmentation of Head and Neck (H&N) primary tumors. For that purpose, scans of 224 patients are provided [1, 2]. For each patient there are two different image modalities and a segmentation:

- CT (Computer Tomography)
- FDG-PET.

Convolutional Neural Networks (CNNs) are the selected model for the task, as they have proven the best performance in segmentation tasks and especially in medical image segmentation [4, 5]. Our approach is to combine the results of 2D and 3D networks using a stacking ensembling technique.

Data Preprocessing. The provided original data were preprocessed cropping it by a bounding-box provided by the competition organization per patient, resampling to make all have the same resolution and converting the CT scans from float64 to uint16.

Also the original PET images have a different size and resolution than the CT scans hence, there were resampled to the CT size and resolution using a bilinear interpolation.

Ensembling Strategy. Stacking ensembling was used as ensembling strategy. This technique has two principal stages:

During stage 1 the train data for ensemble model was generated. To carry out this stage it is mandatory to split the original dataset in some different and not-intersected subsets. To balance the computational cost and the accuracy, it was chosen 4 as the number of these subsets, each one with a 25% split of total data. For each one of these subsets a model was trained using all data in the other subsets and used to make predictions over the data in this subset. Therefore, 4 different models will be trained for each dataset. Those models will be called "bucket-models". Finally, a new dataset composed of all inferences was released and used as the input data for the ensembler models (see Fig. 1).

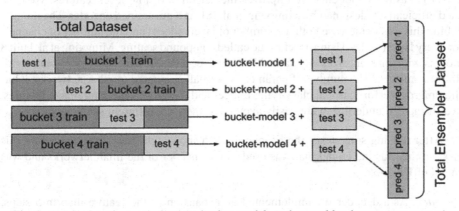

Fig. 1. Detail about how sub-datasets, bucket-models and ensembler dataset were created

During stage 2 a unique model will be trained using the complete dataset to learn using all the information contained in the dataset.

This process was repeated for five times, one time for each-one of the three 2D anatomical axes, one time for the complete 3D images with a model and the last one for the complete 3D images with another different model. The subsets were the same in each repetition. Therefore, there will be five models trained for each, four "bucket-models" and one model trained in the complete dataset. A total of 25 models.

With all those trained models, a dataset to train the ensembler will be created using only the inferences of the "bucket-models" to avoid the overfitting during the ensemblers training process.

Finally, for each patient predictions will be made using all the bucket models plus the all-dataset trained models. For each one of the 2D anatomical axis and the 3D models used five different predictions will be made and averaged to obtain only one prediction per 3D axis and 3D model, a total of five predictions. The ensembler input will be those 5 predictions concatenated using the class axis, and the ensembler output will be the ground truth provided in the dataset (see Fig. 4).

2D Models. For each patient three different datasets are generated, one for each 2D anatomical plane: Axial, Sagittal and Coronal. For each one of the three datasets we have applied a UNet based architecture as it has been proven to show some of the best performance in medical images and organ segmentation [3, 4]. A fully convolutional neural network, based in two parts defines this network: the encoder, which will extract most of the important features from the image; and the decoder, that will expand the featured map until the output can be compared to the ground truth.

U-Nets also have links between these two parts to avoid the so-called gradient vanishing [5]. Our network is based on a U-Net architecture with major changes, both in the encoder and the decoder.

Encoder. As it has been previously reported [6], using a pre-trained network on ImageNet [7] as the U-net encoder improves the detection of low-level features. We have used efficientNet, designed by Mingxing et al. [8], as a pre-trained encoder. The thought behind this was to increase both the number of layers (deep) and the number of channels for every layer (wide). Using a technique called compound scaling, Mingxing et al. built 8 models with different sizes and named EfficientNet-b (0–7). This network presents multiple advantages: The number of parameters is small compared to other networks, it has shown great performance at classifying images and therefore, detecting image features, overfitting and underfitting are easily avoided as there are different sized networks.

After running some tests, the EfficientNet-b5 showed the best performance for this task, so this was the model that was used as the encoder of the final networks and was called effUNet-b5.

Decoder. As a decoder we implemented an expansion of the feature map in 5 steps, doubling the size each time. Each of the expansions is contained within a decoder block with this structure:

$$\text{Conv2D}(3 \times 3) \rightarrow \text{ELU} \rightarrow \text{Upsample2D} \, 2 \times 2 \rightarrow \text{Conv2D}(3 \times 3) \qquad (1)$$

Finally, there is a convolutional layer with a size-1 kernel, that returns a H × W × xC matrix, being H and W the size of the original image and C the number of classes (different organs). We used a softmax activation function to calculate the probability of each class.

Input. As it has been pre-trained on ImageNet, the decoder Input layer was a 3-channel layer. It has been modified to a 3-channel in order to accept sequences consisting of 3 consecutive slices of CT image followed by 3 consecutive slices of FDG-PET image. The slices from CT and FDG-PET were paired. This approach gives some additional spatial information to the model and gives it a broader understanding of the context of the image, instead of just a slice. Therefore increasing its inductive bias.

The 4th, 5th and 6th layer-channels have been initialized with random weights from a Gaussian distribution, with the same mean and standard deviation as the other three pre-trained layers. The prediction was made on the central slide of each of these 3-slides groups (see Fig. 2).

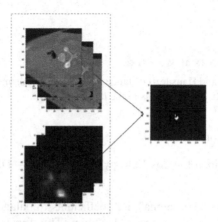

Fig. 2. 6-channels input sequences construction and selected ground truth.

Training. The train dataset was splitted in 4 different and no intersected sub-datasets called buckets. Each one of these buckets has been divided into a train dataset (85%) and a validation dataset (15%) in order to adjust the learning rate and control the underfitting or the underfitting.

Data augmentation has been used in the training set and is based on elastic transformations, rotations, horizontal and vertical flips and zoom-in and zoom-out. For this task the Albumentations library [9] was used.

The final loss function was formed by the combination of three different loss functions:

- Distribution based loss: Cross Entropy Loss like (log likelihood loss) that has shown good performance on multiclass segmentation.
- Region-based loss: DICE loss (Lovasz loss) that improves the convergence.
- Boundary based loss: Hausdorff Distance (HD) loss. The problem of using Hausdorff Distance as loss function is that the Stochastic Gradient Algorithm requires a differentiable loss function and HD is not. To fix this problem we implemented a very

similar and differentiable function, based on "distance transform" function that has been described Davood Karimi et al. [10]. This function was described for binary classification and therefore it had to be adapted to multi-class purposes (Fig. 3).

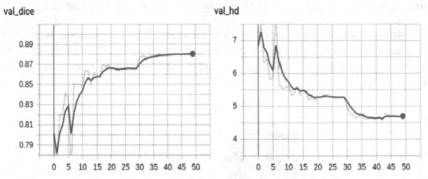

Fig. 3. Example of DICE and Hausdorff distance in validation set using the Soft Dice Loss in the first 30 epochs and the Soft Dice HD Loss in the last 20 epochs.

The final loss function was:

$$SoftDiceHDLoss = alpha\,1 \cdot CrossEntropy + alpha\,2 \cdot (1 - DICE\,) + alpha\,3 \cdot HD \tag{2}$$

Where alpha1, alpha2, and alpha3 are coefficients to balance components so they have the same importance in the loss calculation. This function has a high cost on computation time, but showed better HD and DICE results. The optimizer was Adam [11] with an initial learning rate of (LR) = 1E-04. During all the training there was a Cosine LR-schedule. The batch size was 35 and all models were trained during 300 epochs each one with a softmax activation at the end of the model to calculate the class probabilities.

3D Models. Two different 3D models were used:

UNet3d_ASPP_ds: The 3D network is a standard UNet-3D based on the model proposed by 3D-UNet [12] but modified to a 256 channels central feature map. Also, an Atrous Spatial Pyramid Pooling [13] block was added between the last encoder feature map and the first decoder block. Deep supervision was used to control the loss in all resolutions.

UNet3d_fASPP_SE: This network has the same architecture as *UNet3d_ASPP_ds* but with an Atrous Spatial Pyramid Pooling block between the last encoder feature map and in the middle of all skip-connections. Also, there is no Deep Supervision mechanism. A Concurrent Spatial and Channel Squeeze and Excitation [14] block was added before all ASPP blocks and before all decoder blocks.

Input. The complete CT and FDG-PET 3D images were concatenated and used as input for the model. The ground truth was the complete 3D segmentation provided.

Training. Dataset was divided as it was for the 2D models (85% train, 15% validation) and the same data augmentation was performed. A similar loss function to the 2D model was designed, replacing the Haussdorff Distance component by the Boundary loss [15]. Optimizer was Adam and a cosine learning rate scheduler.

Batch size was only one, due to the GPU memory limitations. The accumulative gradient technique was used in order to avoid an excess on gradient variability [16]. This technique simulates a larger batch size by accumulating several gradients before back propagating. Models were trained for 300 Epochs.

Ensembling. Once the models have been trained, they have been used to calculate predictions for the patients that were not within the segmentation training dataset (Fig. 1). Therefore, a training dataset for the assembler models is achieved without risk of overfitting.

As ensemblers, the same 3D models have been used, an UNet3d_ASPP_ds and an UNet3d_fASPP_SE.

During training, the input of the ensemblers were the concatenated probabilities predictions of the models. There are three 2D models and two 3D models and the ground truth has 2 classes (background and tumor), each ensembler input has $5*2 = 10$ channels.

All details about the ensemblers training are the same as the 3D models training.

Finally, predictions of the two ensemblers were averaged to make the final segmentation.

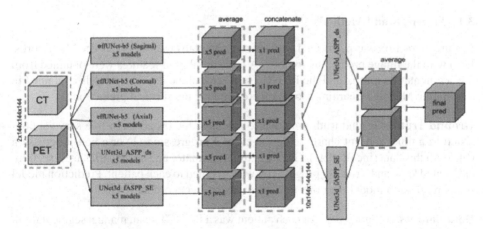

Fig. 4. Final prediction pipeline, for the original images (left) to final prediction (right)

2.2 Results

Final results for the segmentation task in the test set:

Dice Similarity Coefficient (DSC) = 0.740.

All of the results for the train and validation sets are shown in Table 1.

Table 1. Results of the segmentation task per model and set

Set	Model	DICE
Train	EffUNet-b5 (Sagittal)	0,91428
	EffUNet-b5 (Coronal)	0,90076
	EffUNet-b5 (Axial)	0,91226
	UNet3d-ASPP_ds	0,82176
	UNet3d-fASPP_SE	0,83336
Validation	EffUNet-b5 (Sagittal)	0,83386
	EffUNet-b5 (Coronal)	0,83808
	EffUNet-b5 (Axial)	0,82852
	UNet3d-ASPP_ds	0,7908
	UNet3d-fASPP_SE	0,79946
Ensemblers	UNet3d-ASPP_ds (Train)	0,9211
	UNet3d-ASPP_ds (Validation)	0,9201
	UNet3d-fASPP_SE (Train)	0,9245
	UNet3d-fASPP_SE (Validation)	0,9217

3   Task 2: Progression Free Survival (PFS)

3.1   Materials and Methods

Feature extraction was performed with software Matlab (© 1994–2020 The MathWorks, Inc.) with the Image processing toolbox. A large number of features were obtained from the segmentation matrix and the PET/CT scans. Machine learning models contained in the Matlab Machine Learning toolbox were used for the survival prediction.

Ground Truth. Ground truth provided to calculate the progression free survival consisted in a time-to-event chart, being the event a progression (1) or a non-progression (0). With this data (incomplete) a Kaplan-Meier estimator and an actuarial analysis were calculated [17], and a risk for progression was assigned to each patient. Prediction model was a regression model over this risk factor, an anti-concordant result.

Data Sources. Original data for each patient was a PET/CT scan, manual segmentation, and patient info.

In order to study tumor border properties, a new region of interest was created as a boundary region by expanding two pixel outwards and inwards tumor segmentation border. Risk for progression might be related to tumor growth rate. Most mathematical models for tumor growth [18, 19] are related to laplacian and gradient of tumor boundaries, as present a diffusion-like behavior. Mathematical transformations were applied to the segmentation matrix. Laplacian (Eq. 3) and the vector module of the 3D-gradient

(Eq. 4) were obtained for the segmentation matrix. Also, module of 3D gradient multiplied by the distance to the whole tumor centroid (Eq. 5) to take into account the size of the tumor.

$$\Delta F = \frac{\partial^2 F}{\partial x^2} + \frac{\partial^2 F}{\partial y^2} + \frac{\partial^2 F}{\partial z^2} \tag{3}$$

$$|\nabla F| = \sqrt{\left(\frac{\partial F}{\partial x}\right)^2 + \left(\frac{\partial F}{\partial y}\right)^2 + \left(\frac{\partial F}{\partial z}\right)^2} \tag{4}$$

$$|\nabla F| \cdot |r| = |\nabla F| \cdot \left(\sqrt{(x - x_o)^2 + (y - y_o)^2 + (z - z_o)^2}\right) \tag{5}$$

Transformations by Eqs. 3, 4 and 5 were calculated over segmentation mask, being a binary matrix, value 1 inside and value 0 outside the specific region.

Feature Extraction. Features were extracted attending to different characteristics:

Morphological: Matlab function *regionprops3d* was used to obtain the sphere equivalent diameter, extent, surface, volume and principal axis length. Moreover, some features based on biofilm analysis [22], such as porosity, diffusion distance, mean breadth, Euler connectivity number and fractal dimension were also calculated for every class.

Intensity-Based Statistics: Average value, sum, and standard deviation of pixel value within tumor region, and for pixel value after mathematical transformations were collected. Pixel value for every PET and CT image sequence was also considered. Average, entropy, standard deviation, variance, mode, median and kurtosis were collected for every region of interest including the complete tumor and the boundary structure.

Gray Level Co-occurrence: Matlab function *graycoprops* calculates contrast, energy, homogeneity and correlation. Mean values for every class region were added to data collection, including the complete tumor and the ring-boundaries structure. A complete 3D Haralick-feature matrix [20] was calculated centered on tumor centroid [21].

Clinical: Organization provided also patient age, sex, clinical center where the diagnosis, treatment and follow-up were performed and clinical information: tumor staging, patient habits, chemotherapy courses, HPV. Patient habits and HPV were not part of the training as these features were not complete and many values were missing. All tumor staging was downgrade to TNM 7th version for homogeneity.

Feature Selection. After data collection, two different feature selection methods were used: *fsrftest* (ranking for classification using a Fisher F-Test), and *relief* (algorithm based on K-nearest neighbors). Features were then ranked by its importance by adding its rank position for both algorithms. After ranking all features, linear correlation for each feature to all the rest was calculated. Every feature, with an absolute correlation value over 0.5 with a better ranked feature, was deleted.

Training Data and Model. In order to train a robust model, data was extracted from both, the ground truth segmentation and also the segmentation performed by the trained model. Therefore, the number of data entries was twice the number of patients. Final feature number and training model were selected by running tests with a 5-cross validation method. Best ranked features were grouped, starting 5 to 50. Three different models were tested: a Gaussian Process Regression (GPR), an Ensembled Bagged of trees and a Support Vector Machine (SVM).

Best results were obtained by a GPR model trained with 35 best ranked features. Finally, a fine hyperparameter tunning for this model was performed.

3.2   Results

Models were trained over two different ground truths, risk factors based on a Kaplan-Meier estimator and on an Actuarial analysis. Both showed very similar outputs. Results for three different models (GPR, SVM and Emsembled Bagged of trees) were tested on Hecktor platform.

Final result was a concordance index C-Index = 0.628, It was obtained for a GPR model, trained over 35 features and a Kaplan-Meier based risk factor as ground truth.

4   Conclusions

It has been observed that the combination of the 2d and 3d models provides satisfactory results since it allows data processing while taking into consideration different features of the image. The results obtained from this ensembling have shown quality results on the dataset. However, it would be beneficial to find new techniques to regularize the data, to improve the precision when testing, such as label-smoothing or cut-mix, at the data augmentation process. It has also been confirmed that the designed loss functions combination works well on a dataset with these characteristics, with 2 different types of paired images. A way of benefiting from the three channels of the pre-trained backbones of the 2D models has also been created, using images with only one channel. To conclude, it was observed that from the two 3D architectures used, the one that showed slightly better results was the one applied in ASPP blocks in all the skip-connections and the SE blocks; instead of the one that was only applied to the ASPP in the central feature map, even though they had deep-supervision.

The implemented improvements are then part of the pre-processing of the images where it would allow the model to do up-sampling of the PET image instead of doing it through interpolations in the pre-processing. In this manner, introducing linearly inter-polated data to the model will be avoided, considering this process always add noise, and it would be the model the one extracting the characteristics that it would consider necessary of the raw PET data. In addition, it would be of interest to show the models, in some way, the pre-processing crop localization that is being done. Thus, it would not only have the spatial information given by the crop pixels, but also the global position of the crop in the patient's body, which would increase the model's inductive bias.

Clinical prediction models in oncology have shown potential and accuracy in patient outcome predictions. These tools may provide physicians with an experience-based prediction which could be helpful to support clinical decisions.

A model that includes treatment variables, such as chemotherapy or radiotherapy courses information, may improve the accuracy. In case of radiotherapy courses, a dosimetric treatment-dose variables and a dose texture analysis (dosiomic) could be included, as it has shown potential in outcome predictions [23, 24].

References

1. Andrearczyk, V., et al.: Overview of the HECKTOR challenge at MICCAI 2021: automatic head and neck tumor segmentation and outcome prediction in PET/CT images. In: Andrearczyk, V., Oreiller, V., Hatt, M., Depeursinge, A. (eds.) HECKTOR 2021. LNCS, vol. 13209, pp. 1–37. Springer, Cham (2022)
2. Valentin, O., et al.: Head and neck tumor segmentation in PET/CT: the HECKTOR challenge. In: Medical Image Analysis (2021) (under revision)
3. Ronneberger, O., Fischer, P., Brox, T.: U-net: convolutional networks for biomedical image segmentation. In: Navab, N., Hornegger, J., Wells, W., Frangi, A. (eds.) MICCAI 2015. LNCS, vol. 9351, pp. 234–241. Springer, Cham (2015). https://doi.org/10.1007/978-3-319-24574-4_28
4. Milletari, F., Navab, N., Ahmadi, S.-A.: V-Net: Fully Convolutional Neural Networks for Volumetric Medical Image Segmentation. arXiv 2016, arXiv:1606.04797
5. Razvan, P., Tomas, M., Yoshua, B.: On the difficulty of training Recurrent Neural Networks. arXiv: 2013, arXiv:1211.5063
6. Iglovikov, V., Shvets, A.: TernausNet: U-Net with VGG11 Encoder Pre-Trained on ImageNet for Image Segmentation. arXiv 2018, arXiv:1801.05746
7. Deng, J., Dong, W., Socher, R., Li, L.-J., Li, K., Li, F.F.: ImageNet: a large-scale hierarchical image database. In: IEEE Conference on Computer Vision and Pattern Recognition, pp. 248−255 (2009). https://doi.org/10.1109/CVPR.2009.5206848
8. Mingxing, T., Quoc, V.L.: EfficientNet: Rethinking model scaling for convolutional neural networks. In: Kamalika, C., Ruslan, S., (eds.) Proceedings of the 36th International Conference on Machine Learning, volume 97 of Proceedings of Machine Learning Research, 09–15 Jun 2019, pp. 6105–6114. PMLR, Long Beach (2019)
9. Alexander, B., Alex, P., Eugene, K., Vladimir, I.I., Alexandr, A.K.: Albumentations: fast and flexible image augmentations. arXiv: 2018, arXiv:1809.06839
10. Davood, K., Septimiu, E.S.: Reducing the Hausdorff Distance in Medical Image Segmentation with Convolutional Neural Networks. arXiv 2019, arXiv:1904.10030
11. Diederik, P.K., Jimmy, L.B.: Adam: A Method for Stochastic Optimization. arXiv 2014, arXiv:1412.6980
12. Çiçek, Ö., et al.: 3D U-Net: Learning Dense Volumetric Segmentation from Sparse Annotation. https://arxiv.org/abs/1606.06650 (2016)
13. Chen, L., Papandreou, G., Kokkinos, I., Murphy, K., Yuille, A.: Deeplab: Semantic image segmentation with deep convolutional nets, atrous convolution, and fully connected crfs. IEEE Trans. Pattern Anal. Mach. Intell. **40**(04), 834–848 (2018). https://doi.org/10.1109/TPAMI.2017.2699184
14. Roy, A.G., Navab, N., Wachinger, C.: Recalibrating fully convolutional networks with spatial and channel "squeeze and excitation" blocks. IEEE Trans. Med. Imaging **38**(2), 540–549 (2019). https://doi.org/10.1109/tmi.2018.2867261. PMID: 30716024
15. Boundary loss for highly unbalanced segmentation. https://arxiv.org/pdf/1812.07032.pdf

16. Hermans, J., Spanakis, G., Möckel, R.: Accumulated Gradient Normalization (2017)
17. Kaplan, E.L., Meier, P.: Nonparametric estimation from incomplete observations. J. Am. Statist. Assn. **53**, 457–481 (1958)
18. Swanson, K.R., Rostomily, R.C., Alvord, E.C., Jr.: A mathematical modelling tool for predicting survival of individual patients following resection of glioblastoma: a proof of principle. Br. J. Cancer **98**, 113–119 (2008)
19. Swanson, K.R., Bridge, C., Murray, J.D., Ellsworth, C., Alvord, E.C., Jr.: Virtual and real brain tumors: using mathematical modeling to quantify glioma growth and invasion. J. Neurol. Sci. **216**, 1–10 (2003)
20. Haralick, R., et al.: Textural features for image classification. IEEE Trans. Syst. Man Cybern. **3**, 610–621 (1973)
21. Carl, P., Daniel, L.: Cooc3d (2008). https://www.mathworks.com/matlabcentral/fileexchange/ 19058-cooc3d. MATLAB Central File Exchange. Accessed 9 Mar 2020
22. Lewandowski, Z., Beyenal, H.: Fundamentals of Biofilm Research. CRC Press, Florida (2007)
23. Puttanawarut, C., Sirirutbunkajorn, N., Khachonkham, S., Pattaranutaporn, P., Wongsawat, Y.: Biological dosiomic features for the prediction of radiation pneumonitis in esophageal cancer patients. Radiat. Oncol. **16**(1), 220 (2021). https://doi.org/10.1186/s13014-021-019 50-y. PMID: 34775975; PMCID: PMC8591796
24. Murakami, Y., et al.: Dose-based radiomic analysis (dosiomics) for intensity-modulated radiotherapy in patients with prostate cancer: Correlation between planned dose distribution and biochemical failure. Int. J. Radiat. Oncol. Biol. Phys. **S0360–3016**(21), 02627–02634 (2021). https://doi.org/10.1016/j.ijrobp.2021.07.1714. Epub ahead of print. PMID: 34706278

Automatic Head and Neck Tumor Segmentation and Progression Free Survival Analysis on PET/CT Images

Yading Yuan[1(✉)], Saba Adabi[1], and Xuefeng Wang[2]

[1] Department of Radiation Oncology, Icahn School of Medicine at Mount Sinai, New York, NY, USA
yading.yuan@mssm.edu
[2] Department of Biostatistics and Bioinformatics, H. Lee Moffitt Cancer Center and Research Institute, Tampa, FL, USA

Abstract. Automatic segmentation is an essential but challenging step for extracting quantitative imaging bio-markers for characterizing head and neck tumor in tumor detection, diagnosis, prognosis, treatment planning and assessment. The HEad and neCK TumOR Segmentation Challenge 2021 (HECKTOR 2021) provides a common platform for the following three tasks: 1) the automatic segmentation of the primary gross target volume (GTV) in the oropharynx region on FDG-PET and CT images; 2) the prediction of patient outcomes, namely Progression Free Survival (PFS) from the FDG-PET/CT images with automatic segmentation results and the available clinical data; and 3) the prediction of PFS with ground truth annotations. We participated in the first two tasks by further evaluating a fully automatic segmentation network based on encoder-decoder architecture. In order to better integrate information across different scales, we proposed a dynamic scale attention mechanism that incorporates low-level details with high-level semantics from feature maps at different scales. Radiomic features were extracted from the segmented tumors and used for PFS prediction. Our segmentation framework was trained using the 224 challenge training cases provided by HECKTOR 2021, and achieved an average Dice Similarity Coefficient (DSC) of 0.7693 with cross validation. By testing on the 101 testing cases, our model achieved an average DSC of 0.7608 and 95% Hausdorff distance of 3.27 mm. The overall PFS prediction yielded a concordance index (c-index) of 0.53 on the testing dataset (id: deepX).

1 Introduction

Cancers of the head and neck (H&N) are among the most prevalent cancers worldwide (the fifth most common cancer in terms of incidence) [1]. The conventional treatment for people with inoperable H&N cancer is radiotherapy (RT) coupled with chemotherapy [2]. Locoregional failures, on the other hand, have been found in as many as 40% of cases throughout the first two years following

© Springer Nature Switzerland AG 2022
V. Andrearczyk et al. (Eds.): HECKTOR 2021, LNCS 13209, pp. 179–188, 2022.
https://doi.org/10.1007/978-3-030-98253-9_17

treatment [3]. Several radiomic studies have recently been proposed to utilize the vast quantitative features collected from high-dimensional imaging data acquired during diagnosis and therapy in order to identify individuals with a worse prognosis before treatment. With a large patient population, evaluating both PET and CT scan images concurrently to delineate primary tumors and nodal metastases is almost impractical for oncologists. Additionally, to design a treatment plan for radiotherapy, targets and the organs at risk (OARs) need to be delineated manually by radiotherapists which is time-consuming and operator-dependent [4]. Consequently, the automatic segmentation method is very demanding to support clinicians for improved detection, diagnosis, forecast, a treatment plan of and H&N cancer evaluation. The HEad and neCK TumOR segmentation challenge (HECKTOR) segmentation challenge [5–7] provides a large PET/CT dataset for automatic H&N primary tumor segmentation on oropharyngeal cancers including 224 cases for model training and 101 cases for testing from six different institutions (i.e. CHGJ, CHMR, CHUM, CHUS, and CHUP cases for training and cases from CHUP and CHUV for testing. The ground truth was annotated by radiation oncologists, on the CT of the PET/CT study (31% of the patients) or on a different CT scan dedicated to treatment planning (69%) where the planning CT was registered to the PET/CT scans, for training cases [8]. Test cases were annotated on the PET/CT images. Each case includes a co-registered PET-CT set as well as the primary Gross Tumor Volume (GTVt) annotation. Segmentation algorithms performed on the volume of interest (VOI) near GTVt by using the supplied bounding box [9]. These images were resampled to $1 \times 1 \times 1$ mm isotropic resolution and then cropped to a volume size of $144 \times 144 \times 144$.

There are three tasks in this challenge: 1) automatic segmentation of the primary gross target volume (GTV) in the oropharynx region on FDG-PET and CT images; 2) the prediction of patient outcomes, namely Progression Free Survival (PFS) from the FDG-PET/CT images with automatic segmentation results and the available clinical data; and 3) the prediction of PFS with ground truth annotations. The evaluation of automatic segmentation will be based on Dice Similarity Coefficient (DSC), which is computed only within these bounding boxes at the original CT resolution. Concordance index (c-index) will be used to evaluate the performance of PFS prediction for tasks 2 and 3.

2   Related Work

Convolutional neural networks have been effectively used for biomedical image segmentation tasks. However, there is only a few research in the application of Deep convolutional neural networks (DCNNs) for PET/CT automatic tumor segmentation. Moe et al. published a PET-CT segmentation method based on 2D U-Net to outline tumor and metastatic lymph nodes borders with 152 training and 40 testing cases [10]. The expansion of mentioned work is performed by Andrearczyk et al. [5] using V-Net architecture on a publicly available dataset of 202 patients. A multi-modality fully convolutional network (FCN) for tumor co-segmentation in PET-CT images for 84 patients with lung cancer is proposed by Zhao et al. [11]. and Zhong et al. implemented two coupled 3D U-Nets that concurrently co-segmented tumors in PET/CT images for 60 patients with Non-Small Cell Lung Cancer (NSCLC) [12]. The paramount advantage of U-Net [14] and its modifications in automatic PET-CT segmentation is the skip connection design, which allows high resolution features to be used in the coding path as an additional entry to the convolutional layer on the decoding path hence recovers details for image segmentation. The architecture of U-Net corrects the limit of fusion on the same scale if multiple-scale features maps are available on the coding path, i.e. the scale difference of feature maps during fusion become smaller. Studies show features that maps at different scales carry both detailed spatial information (low-level features) and semantic information such as target position (high-level features) [26].

We present a unique encoder-decoder architecture named scale attention networks (SA-Net), to fully exploit multi-scale information where the interconnections between the encoding and decoding pathways are redesigned by using full-scale skip connections instead of scale-wise skip connections in U-Net. SA-Net can now combine low-level fine details with high level semantic information in a single framework. We add the attention mechanism into SA-Net so that as the network learns, the weight on each scale for each feature channel is adaptively modified to accentuate the key scales while suppressing the less important ones [15, 16]. The overall architecture of SA-Net is illustrated in Fig. 1.

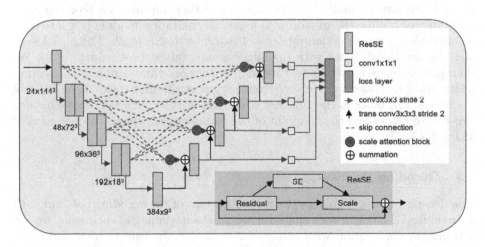

Fig. 1. Architecture of SA-Net. Input is a $2 \times 144 \times 144 \times 144$ tensor followed by one ResSE block with 24 filters. Here ResSE stands for a squeeze-and-excitation block embedded in a residual module [15]. By progressively halving the feature map dimension while doubling the feature width at each scale, the endpoint of the encoding pathway has a dimension of $384 \times 9 \times 9 \times 9$. The output of the encoding pathway has one channel with the same spatial size as the input, i.e., $1 \times 144 \times 144 \times 144$.

3   Methods

3.1   Overall Network Structure

SA-Net uses a standard encoding-decoding architecture, with a asymmetrically larger encoding pathway for learning representative features and a smaller decoding pathway for recovering the segmentation mask in the original resolution. The scale attention blocks (SA-block) combine the outputs of encoding blocks at multiple sizes to learn and pick features with full-scale information. Each patient's PET and CT images are concatenated into a two-channel tensor. $2 \times 144 \times 144 \times 144$ is the dimension of the input to SA-Net. Each voxel value represents the likelihood that the associated voxel belongs to the tumor target, and the network output is a map with a size of $1 \times 144 \times 144 \times 144$.

3.2   Encoding Pathway

The encoding pathway is built upon ResNet [17] blocks, where each block consists of two Convolution-Normalization-ReLU layers followed by additive identity skip connection. We keep the batch size to 1 in our study to allocate more GPU memory resource to the depth and width of the model, therefore, we use instance normalization, i.e., group normalization with one feature channel in each group, which has been demonstrated with better performance than batch normalization when batch size is small   [22]. In order to further improve the representative capability of the model, we add a squeeze-and-excitation module [15] into each residual block with reduction ratio $r = 4$ to form a ResSE block. The initial scale includes one ResSE block with the initial number of features (width) of 24. We then progressively halve the feature map dimension while doubling the feature width using a strided (stride = 2) convolution at the first convolution layer of the first ResSE block in the adjacent scale level. All the remaining scales include two ResSE blocks and the endpoint of the encoding pathway has a dimension of $384 \times 9 \times 9 \times 9$.

3.3   Decoding Pathway

The decoding pathway follows the reverse pattern of the encoding one, but with a single ResSE block in each spatial scale. At the beginning of each scale, we use a transpose convolution with stride of 2 to double the feature map dimension and reduce the feature width by 2. The upsampled feature maps are then added to the output of SA-block. Here, we use summation instead of concatenation for information fusion between the encoding and decoding pathways to reduce GPU memory consumption and facilitate the information flowing. The endpoint of the decoding pathway has the same spatial dimension as the original input tensor and its feature width is reduced to 1 after a $1 \times 1 \times 1$ convolution and a sigmoid function.

In order to regularize the model training and enforce the low- and middle-level blocks to learn discriminative features, we introduce deep supervision at each intermediate scale level of the decoding pathway. Each deep supervision subnet employs a $1 \times 1 \times 1$ convolution for feature width reduction, followed by a trilinear upsampling layer such that they have the same spatial dimension as the output, then applies a sigmoid function to obtain extra dense predictions. These deep supervision subnets are directly connected to the loss function in order to further improve gradient flow propagation. The weights among different deep supervision outputs were fixed as $0.5 : 0.167 : 0.167 : 0.166$ in order to emphasize on the output from the finest scale.

3.4   Scale Attention Block

The proposed scale attention block consists of full-scale skip connections from the encoding pathway to the decoding pathway, where each decoding layer incorporates the output feature maps from all the encoding layers to capture fine-grained details and coarse-grained semantics simultaneously in full scales. As illustrated in Fig. 2, the first stage of the SA-block is to add the input feature maps at different scales from the encoding pathway, represented as $\{S_e, e = 1, ..., N\}$ where N is the number of total scales in the encoding pathway except the last block ($N = 4$ in this work), after transforming them to the feature maps with the same dimensions, i.e., $S_d = \sum f_{ed}(S_e)$. Here e and d are the scale level at the encoding and decoding pathways ($d = 2$ in this example), respectively. The transform function $f_{ed}(S_e)$ is determined as follows. If $e < d$, $f_{ed}(S_e)$ downsamples S_e by $2^{(d-e)}$ times by maxpooling followed by a Conv-Norm-ReLU block; if $e = d$, $f_{ed}(S_e) = S_e$; and if $e > d$, $f_{ed}(S_e)$ upsamples S_e through tri-linear upsamping after a Conv-Norm-ReLU block for channel number adjustment. For S_d, a spatial pooling is used to average each feature to form an information embedding tensor $G_d \in R^{C_d}$, where C_d is the number of feature channels in scale d. Then a $1 - to - N$ Squeeze-Excitation is performed in which the global feature embedding G_d is squeezed to a compact feature $g_d \in R^{C_d/r}$ by passing through a fully connected layer with a reduction ratio of r, then another N fully connected layers with sigmoid function are applied for each scale excitation to recalibrate the feature channels on that scale. Finally, the contribution of each scale in each feature channel is normalized with a softmax function, yielding a scale-specific weight vector for each channel as $w_e \in R^{C_d}$, and the final output of the scale attention block is $\widetilde{S_d} = \sum w_e \cdot f_{ed}(S_e)$.

3.5   Implementation

Our framework was implemented with Python using Pytorch package. All the following steps were performed on the volumes of interest (VOIs) within the given bounding boxes. As for pre-processing, we truncated the CT numbers to [−125, 225] HU to eliminate the irrelevant information, then normalized the CT images with the mean and standard deviation of the HU values within GTVs in the entire training dataset. For PET images, we simply normalized each patient

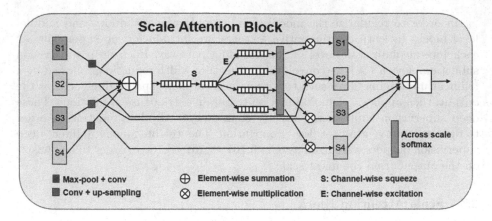

Fig. 2. Scale attention block. Here $S1, S2, S3$ and $S4$ represent the input feature maps at different scales from the encoding pathway.

independently by subtracting the mean and dividing by the standard deviation of the image within the body where the SUVs (Standarized Update Values) are great than zero. The model was trained with a patch size of $144 \times 144 \times 144$ voxels and batch size of 1. We used Jaccard distance, which we developed in our previous studies [18–21], as the loss function in this work. It is defined as:

$$L = 1 - \frac{\sum\limits_{i,j,k} (t_{ijk} \cdot p_{ijk}) + \epsilon}{\sum\limits_{i,j,k} (t_{ijk}^2 + p_{ijk}^2 - t_{ijk} \cdot p_{ijk}) + \epsilon}, \tag{1}$$

where $t_{ijk} \in \{0, 1\}$ is the actual class of a voxel x_{ijk} with $t_{ijk} = 1$ for tumor and $t_{ijk} = 0$ for background, and p_{ijk} is the corresponding output from SA-Net. ϵ is used to ensure the stability of numerical computations.

Training the entire network took 500 iterations from scratch using Adam stochastic optimization method. The initial learning rate was set to 0.003, and learning rate decay and early stopping strategies were utilized when validation loss stopped decreasing. In particular, we kept monitoring the validation loss ($L^{(valid)}$) in each iteration. We kept the learning rate unchanged at the first 250 iterations, but dropped the learning rate by a factor of 0.3 when $L^{(valid)}$ stopped improving within the last 50 iterations. The model that yielded the best $L^{(valid)}$ was recorded for model inference.

In order to reduce overfitting, we first extracted a $244 \times 244 \times 244$ volumes surrounding tumor from the entire image, then randomly cropped $144 \times 144 \times 144$ VOI. We also randomly flipped the input volume in left/right, superior/inferior, and anterior/posterior directions on the fly with a probability of 0.5 for data augmentation. We used 5-fold cross validation to evaluate the performance of our model on the training dataset, in which a few hyper-parameters such as the feature width and input dimension were also experimentally determined. All the experiments were conducted on one Nvidia RTX 3090 GPU with 24 GB memory.

In order to improve the diversity of the models, besides randomly splitting the training set into five folds, we also implemented a leave-one-center-out strategy to training a model with data from four centers while validating the trained model on the remaining center. In addition, we trained models with patch size of $128 \times 128 \times 128$ by resampling image data to $1.25 \times 1.25 \times 1.25$ mm. The segmentation mask was obtained by applying a threshold of 0.5 to the model output.

3.6 PFS Prediction

Survival prediction handling the prediction of the number of days for which patients survive after receiving proper therapy. In this study, we extracted A total of 220 radiomic features from both PET and CT segmented modalities including feature classes, shape, first-order, Grey Level Co-occurrence Matrix features (GLCM) Grey Level Dependence Matrix (GLDM), Grey Level Run Length Matrix (GLRLM), Grey Level Size Zone Matrix (GLSZM), and Neighbouring Grey Tone Difference Matrix Features (NGTDM) using the PyRadiomics package implemented in Python [23]. The features then are manually ranked according to their concordance index ranking and the features with the highest rankings are selected. The selected features were: 1) least axis length, 2) 3d diameter, 3) first order minimum, 4) gldm LowGrayLevelEmphasis, 5) glcm JointEntropy, 6) glszm LowGrayLevelZoneEmphasis from CT and 7) glrlm RunLengthNonUniformity for PET. We added age as our non-imaging and clinical feature. In our method, we performed a less expensive method in term of computational time that applied the concordance index evaluation criteria to each single feature and then compared the effectiveness of all the features using the measures obtained. In this way, each feature can be positioned in a ranking according to the value of the above measure. Then we considered average ranking among the selected features.

4   Results

We trained SA-Net with the training set provided by the HECKTOR 2021 challenge, and evaluated its performance on the training set via different cross validation strategies. Figure 3 shows three examples of automatic segmentation results. Table 1 shows the segmentation results in terms of DSC for each fold.

When applying the trained models on the 101 challenge testing cases, a bagging-type ensemble strategy was implemented to average the outputs of these 15 models to further improve the segmentation performance, achieving an average DSC of 0.7608 and 95% Hausdorff distance of 3.27 mm.

For PFS prediction, We used all 224 training subjects with known progression free survival data in our training process. In the training process we used the average ranking criterion among selected features. The Concordance index for average ranking procedure in training phase was 0.61 while for each individual selected feature it was less than 0.58 (with margin of error of 0.5714 $\pm$0.00844). Table 2 list the validation results for individual features and feature combination.

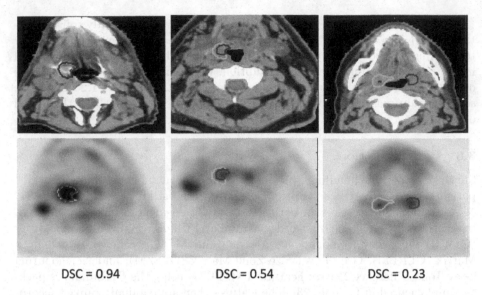

DSC = 0.94   DSC = 0.54   DSC = 0.23

Fig. 3. Three examples of automatic segmentation. The segmentation results (red) and ground truth (green) are displayed on CT (top) and PET (bottom) images respectively. Left: the best result with DSC = 0.94; middle: the median result with DSC = 0.54; right: the worst result with DSC = 0.23 (Color figure online)

Table 1. Segmentation results (DSC) of SA-Net in different cross-validation splits on the training image dataset.

Random splitting	Fold-1 (n = 45)	Fold-2 (n = 45)	Fold-3 (n = 45)	Fold-4 (n = 45)	Fold-5 (n = 44)	Average (n = 224)
Patch size (144 × 144 × 144)	0.765	0.769	0.800	0.745	0.768	0.769
Patch size (128 × 128 × 128)	0.759	0.751	0.785	0.740	0.755	0.758
Leave-one-center-out	CHGJ (n = 55)	CHMR (n = 18)	CHUM (n = 56)	CHUP (n = 23)	CHUS (n = 72)	Average (n = 224)
Patch size (144 × 144 × 144)	0.810	0.700	0.729	0.797	0.757	0.763

Table 2. PFS performance of individual features and the feature combination on the training set.

Features	F-1	F-2	F-3	F-4	F-5	F-6	F-7	Combined
C-index	0.5790	0.5777	0.5769	0.5758	0.5746	0.5718	0.5440	0.61

5   Summary

In this work, we further evaluated the fully automated segmentation model that we previously developed [24,25] for head and neck tumor segmentation from PET and CT images. Our SA-Net replaces the long-range skip connections between the same scale in the vanilla U-Net with full-scale skip connections in order to make maximum use of feature maps in full scales for accurate segmentation. Attention mechanism is introduced to adaptively adjust the weights of each scale features to emphasize the important scales while suppressing the less important ones. Our validation results show that a larger patch size yielded a better segmentation results due to the smaller resampling space. From the leave-one-center-out cross validation, our model demonstrated the best performace on the 55 cases from CHGJ while the poorest performance on the 18 cases from CHMR, indicating that the data distribution of CHMR might be quite different from other centers. The leave-one-center-out results also indicate that the performance variations likely come from the difference of the image acquisition protocol among center instead of the number of cases when applying a trained model to data from a previously unseen center.

Acknowledgment. This work is supported by a research grant from Varian Medical Systems (Palo Alto, CA, USA), UL1TR001433 from the National Center for Advancing Translational Sciences, and R21EB030209 from the National Institute Of Biomedical Imaging And Bioengineering of the National Institutes of Health, National Institutes of Health, USA. The content is solely the responsibility of the authors and does not necessarily represent the official views of the National Institutes of Health.

References

1. Parkin, M., et al.: Global cancer statistics. CA Cancer. J. Clin. **55**(2), 74–108 (2005)
2. Bonner, J., et al.: Radiotherapy plus Cetuximab for localregionally advanced head and neck cancer: 5-year survival data from a phase 3 randomized trial, and relation between Cetuximab-induced rash and survival. Lacent Oncol. **11**(1), 21–28 (2010)
3. Chajon, E., et al.: Salivary gland-sparing other than parotid-sparing in definitive head-and-neck intensity-modulated radiotherapy dose not seem to jeopardize local control. Radiat. Oncol. **8**(1), 132 (2013)
4. Gudi, S., et al.: Interobserver variability in the delineation of gross tumor volume and specified organs-at-risk during IMRT for head and neck cancers and the impact of FDG-PET/CT on such variability at the primary site. J. Med. Imaging Radiat. Sci. **48**(2), 184–192 (2017)
5. Andrearczyk, V., et al.: Automatic segmentation of head and neck tumors and nodal metastases in PET-CT scans. Proc. MIDL **1**–11, 2020 (2020)
6. Andrearczyk, V., et al.: Overview of the HECKTOR challenge at MICCAI 2021: automatic head and neck tumor segmentation and outcome prediction in PET/CT images. In: Andrearczyk, V., Oreiller, V., Hatt, M., Depeursinge, A. (eds.) HECKTOR 2021. LNCS, vol. 13209, pp. 1–37. Springer, Cham (2022)
7. Oreiller, V., et al.: Head and neck tumor segmentation in PET/CT: the HECKTOR challenge. Med. Image Anal. **77**, 102336 (2021)
8. Vallieres, M., et al.: Radiomics strategies for risk assessment of tumour failure in head-and-neck cancer. Sci. Rep. **7**(1), 10117 (2017)
9. Andrearczyk, V., et al. Oropharynx detection in PET-CT for tumor segmentation. In: Irish Machine Vision and Image Processing (2020)

10. Moe, Y.M., et al.: Deep learning for automatic tumour segmentation in PET/CT images of patients with head and neck cancers. In: Proceedings of MIDL (2019)
11. Zhao, X., et al.: Tumor co-segmentation in PET/CT using multi-modality fully convolutional neural network. Phys. Med. Biol. **64**, 015011 (2019)
12. Zhong, Z., et al.: Simultaneous cosegmentation of tumors in PET-CT images using deep fully convolutional networks. Med. Phys. **46**(2), 619–633 (2019)
13. Long, J., et al.: Fully convolutional networks for semantic segmentation. In: CVPR, pp. 3431–3440 (2015)
14. Ronneberger, O., Fischer, P., Brox, T.: U-net: convolutional networks for biomedical image segmentation. In: Navab, N., Hornegger, J., Wells, W.M., Frangi, A.F. (eds.) MICCAI 2015. LNCS, vol. 9351, pp. 234–241. Springer, Cham (2015). https://doi.org/10.1007/978-3-319-24574-4_28
15. Hu, J., et al.: Squeeze-and-excitation networks. In: Proceedings of CVPR 2018, pp. 7132–7141 (2018)
16. Li, X., et al.: Selective kernel networks. In: Proceedings of CVPR 2019, pp. 510–519 (2019)
17. He, K., et al.: Deep residual learning for image recognition. In: Proceedings of CVPR 2016, pp. 770–778 (2016)
18. Yuan, Y., et al.: Automatic skin lesion segmentation using deep fully convolutional networks with Jaccard distance. IEEE Trans. Med. Imaging **36**(9), 1876–1886 (2017)
19. Yuan, Y. Hierachical convolutional-deconvolutional neural networks for automatic liver and tumor segmentation. arXiv preprint arXiv:1710.04540 (2017)
20. Yuan, Y.: Automatic skin lesion segmentation with fully convolutional-deconvolutional networks. arXiv preprint arXiv:1703.05154 (2017)
21. Yuan, Y., et al.: Improving dermoscopic image segmentation with enhanced convolutional-deconvolutional networks. IEEE J. Biomed. Health Informat. **23**(2), 519–526 (2019)
22. Wu, Y., He, K.: Group Normalization. In: Ferrari, V., Hebert, M., Sminchisescu, C., Weiss, Y. (eds.) ECCV 2018. LNCS, vol. 11217, pp. 3–19. Springer, Cham (2018). https://doi.org/10.1007/978-3-030-01261-8_1
23. Griethuysen, J., et al.: Computational radiomics system to decode the radiographic phenotype. Cancer Res. **77**(21), e104–e107 (2017)
24. Yuan, Y.: Automatic head and neck tumor segmentation in PET/CT with scale attention network. In: Andrearczyk, V., Oreiller, V., Depeursinge, A. (eds.) HECKTOR 2020. LNCS, vol. 12603, pp. 44–52. Springer, Cham (2021). https://doi.org/10.1007/978-3-030-67194-5_5
25. Yuan, Y.: Automatic brain tumor segmentation with scale attention network. In: Crimi, A., Bakas, S. (eds.) BrainLes 2020. LNCS, vol. 12658, pp. 285–294. Springer, Cham (2021). https://doi.org/10.1007/978-3-030-72084-1_26
26. Zhou, Z., et al.: UNet++: Redesigning skip connections to exploit multiscale features in image segmentation. IEEE Trans. Med. Imaging **39**(6), 1856–1867 (2020)

Multimodal PET/CT Tumour Segmentation and Prediction of Progression-Free Survival Using a Full-Scale UNet with Attention

Emmanuelle Bourigault[1]([✉]) [iD], Daniel R. McGowan[2] [iD],
Abolfazl Mehranian[3] [iD], and Bartłomiej W. Papież[4,5] [iD]

[1] Department of Engineering, Doctoral Training Centre, University of Oxford,
Oxford, UK
emmanuelle.bourigault@dtc.ox.ac.uk
[2] Department of Oncology, University of Oxford, Oxford, UK
[3] GE Healthcare, Oxford, UK
[4] Big Data Institute, Li Ka Shing Centre for Health Information and Discovery,
University of Oxford, Oxford, UK
[5] Nuffield Department of Population Health, University of Oxford, Oxford, UK

Abstract. Segmentation of head and neck (H&N) tumours and prediction of patient outcome are crucial for patient's disease diagnosis and treatment monitoring. Current developments of robust deep learning models are hindered by the lack of large multi-centre, multi-modal data with quality annotations. The MICCAI 2021 HEad and neCK TumOR (HECKTOR) segmentation and outcome prediction challenge creates a platform for comparing segmentation methods of the primary gross target volume on fluoro-deoxyglucose (FDG)-PET and Computed Tomography images and prediction of progression-free survival in H&N oropharyngeal cancer. For the segmentation task, we proposed a new network based on an encoder-decoder architecture with full inter- and intra-skip connections to take advantage of low-level and high-level semantics at full scales. Additionally, we used Conditional Random Fields as a post-processing step to refine the predicted segmentation maps. We trained multiple neural networks for tumor volume segmentation, and these segmentations were ensembled achieving an average Dice Similarity Coefficient of 0.75 in cross-validation, and 0.76 on the challenge testing data set. For prediction of patient progression free survival task, we propose a Cox proportional hazard regression combining clinical, radiomic, and deep learning features. Our survival prediction model achieved a concordance index of 0.82 in cross-validation, and 0.62 on the challenge testing data set.

Keywords: Medical image segmentation · Head and neck segmentation · Multimodal image segmentation · UNet · Progression free survival

© Springer Nature Switzerland AG 2022
V. Andrearczyk et al. (Eds.): HECKTOR 2021, LNCS 13209, pp. 189–201, 2022.
https://doi.org/10.1007/978-3-030-98253-9_18

1   Introduction

Head and Neck (H&N) tumour is the fifth most prevalent cancer worldwide. Improving the accuracy and efficiency of disease diagnosis and treatment is the rationale behind the developments of computer-aided systems in medical imaging [1,2]. However, obtaining manual segmentations, which can be used for diagnosing and treatment purposes, is time consuming and suffers from intra- and inter-observer biases. Furthermore, segmentation of H&N tumours is a challenging task compared to other parts of the body as the tumour displays similar intensity values to the adjacent tissues making it non distinguishable to the human eye in Computed Tomography (CT) images. Previous attempts at developing deep learning models to segment head and neck tumours suffered from a relatively high number of false positives [3,4]. Currently, in the normal clinical pathway, a combination of Positron Emission Tomography (PET) and CT images plays a key role in the diagnosis of H&N tumors. This multi-modal approach has dual benefits: the metabolic information is provided by PET and anatomical information is available in CT. Furthermore, accurate segmentation of H&N tumors could also be used in automating pipelines for extraction of quantitative imaging features (e.g. radiomics) in prediction of patient survival.

The 3D UNet [5] is one of the most widely employed encoder-decoder architecture for medical segmentation inspired by Fully Convolutional Networks [6]. Promising results have been obtained using 3D UNet based architecture and attention mechanisms with early fusion of PET/CT images [7]. While the performance of the model proposed in [8] was significantly improved when compared to a baseline 3D UNet, a number of false positives was reported where the model was not only segmenting the primary tumour but also other isolated areas such as the soft palate due to tracer overactivity in that region.

In this paper, we propose to segment 3D H&N tumor volume from multimodal PET/CT imaging using a full scale 3D UNet3+ architecture [9] with attention mechanism. Our model, NormResSE-UNet3+, is trained with a hybrid loss function of Log Cosh Dice [10] and Focal loss [11]. The segmentation maps, predicted by our model, are further refined by Conditional Random Fields postprocessing [12,13] to reduce number of false positives and to improve tumour boundary segmentation. For the progression free survival prediction task, we propose a Cox proportional hazard regression model using a combination of clinical, radiomic, and deep learning features from PET/CT images.

The paper is organised as follows. Section 2 outlines the data set and preprocessing steps (Sect. 2.1), the methods used for 3D H&N tumor segmentation and for prediction of progression free survival tasks (Sect. 2.2), and the evaluation criteria (Sect. 2.3). The experimental set-up and results are described in Sect. 3. Finally, the discussion and conclusion are found in Sect. 4.

2   Methods and Data

2.1   Data

Data for Segmentation. PET and CT images used in this challenge were provided by the organisers of the HECKTOR challenge at the 24th International Conference on Medical Image Computing and Computer Assisted Intervention (MICCAI) [1]. The total number of training cases is 224 from 5 centers: CHGJ, CHMR, CHUS, CHUP, and CHUM. The ground-truth annotations are provided by expert clinicians for primary gross tumor volume. For testing, additional 101 cases from two centers, namely CHUP and CHUV, are provided. However, no expert annotations are available to the participants.

Data Preprocessing for Segmentation Task. For segmentation task, we used trilinear interpolation to resample PET and CT images. Bounding boxes of 144 × 144 × 144 voxels were provided by the organizers and used for patch extraction. PET intensities (given in standard uptake value) were normalised with Z-score, while CT intensities (given in Hounsfield unit) were clipped to the range $[-1, 1]$.

Data for Progression Free Survival. Patient clinical data are provided for prediction of progression free survival in days. The covariates (a combination of categorical and continuous variables) are as follows: center ID, age, gender, TNM 7/8th edition staging and clinical stage, tobacco and alcohol consumption, performance status, HPV status, treatment (radiotherapy only, or chemoradiotherapy). Dummy variables were used to encode the categorical variables i.e. the ones mentioned above except for age (continuous). Among the 224 patients for training, the median age was 63 years (range: 34–90 years) with progression event occurred in 56 patients and an average progression survival of 1218 days (range: 160–3067 days). The testing cohort comprised 129 patients with a median age of 61 years (range: 40–84 years).

Data Preprocessing for Progression Free Survival Task. For the prediction of progression free survival, we used multiple imputation for missing values. The multiple imputation models each feature, which contains missing values, as a function of the other features. Then, it uses this estimate in a round-robin fashion for imputing the missing values. At each iteration, a feature is designated as output y and the other features are treated as inputs X. A regressor is fit on (X, y) on the known y used to predict the missing values of y. This process is repeated for each feature and for ten imputation rounds. Feature selection was performed using Lasso regression with 5-fold cross-validation using a combination of features i.e. clinical features, radiomic features, and features extracted from 3D UNet (see Fig. 1). The correlation between those features was evaluated with Spearman's correlation coefficient in order to assess potential redundancy.

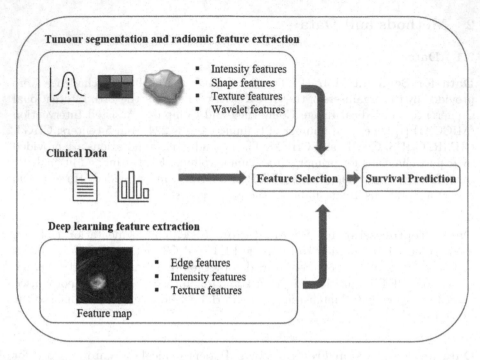

Fig. 1. The pipeline used for the Progression Free Survival task. It consists of three steps following clinical data collection, images acquisition, and pre-processing, then feature extraction, feature selection, and statistical survival prediction.

A threshold of 0.80 was set to filter out highly correlated features. The feature selection process reduced the number of features from 275 to 70 most relevant features. In particular, we kept 7 clinical features (i.e. Age, Chemotherapy, Tstage, Nstage, TNMgroup, TNM edition, and HPV status), 14 radiomics (5 intensity based histogram features, 3 shape features, and 6 texture features from metrics: Gray Level Co-occurrence Matrix (GLCM), Gray Level Run Length Matrix (GLRLM), Gray Level Size Zone (GLSZM)), and 49 deep learning features. Radiomic features were extracted from PET and CT images using the pyradiomics package. Convolutional Neural Networks (CNNs) by using stacks of filtering layers together with pooling and activation layers, are becoming increasingly popular in the field of radiomics [14]. This can be explained by the fact that CNNs do not require prior knowledge and kernels are learned automatically as opposed to hand-crafted features. In this study,deep learning features were extracted at the 5th convolutional layer of our model by averaging feature maps. A vector was created for each feature map, then concatenated to form a single vector of deep learning features. The power of CNNs is their ability to automatically learn multiple filters in parallel thus extracting low and high-level features such as edges, intensity but also texture. Each filter captures different characteristics of the image ultimately enabling CNNs to capture relevant edge, intensity and texture features [15] (see a schematic overview of the survival pipeline in Fig. 1).

2.2   Models Description

Models for Segmentation Task. The UNet model [5], an encoder-decoder architecture, is one of the most widely employed segmentation models in medical imaging. Skip connections are used to couple high-level feature maps obtained by the decoder and corresponding low-level feature maps by the encoder. UNet++ is an extension of UNet that introduce nested and dense skip connections to reduce the merging of dissimilar features from plain skip connections in UNet [16]. However, since UNet++ still fails to capture relevant information from full scales and to recover lost information in down- and up-sampling, [9] proposed UNet3+ to take full advantage of multi-scale features. The design of inter- and intra-connection between the encoder and decoder pathways at full scale enables us to explore both fine and coarse level details (see a schematic overview in Fig. 2). Low level details contain information about the spatial and boundary information of the tumour, while high-level details encode information about the location of tumour. The integration of deep supervision in the decoder pathway is used to reduce false positives. To further reduce false positives and improve segmentation, we make use of attention mechanisms achieving state-of-the-art segmentation results [17]. In particular, we use 3D normalised squeeze-and-excitation residual blocks proposed by [8] and evaluated on PET/CT H&N dataset from MICCAI 2020 challenge [7].

Loss Function for Segmentation Task. The Dice Coefficient is a widely used loss function for segmentation tasks, and is defined as follows:

$$DiceLoss(y,\hat{p}) = 1 - \frac{2\sum_i^N y_i\hat{p}_i + 1}{\sum_i^N y_i + \sum_i^N \hat{p}_i + 1} \tag{1}$$

In addition, 1 is added in the numerator and denominator to ensure that the function is not undefined in cases when $y = \hat{p} = 0$, i.e. the tumour is not present.

The Focal loss [11] is a variation of the Cross-Entropy loss. The Focal loss is well-suited for imbalance problems as it down-weights easy examples to focus on hard ones, and is defined as follows:

$$FocalLoss(y,\hat{p}) = -\frac{1}{N}\sum_i^N y_i(1-\hat{p}_i)^\gamma ln(\hat{p}_i) \tag{2}$$

where γ in the modulating factor is optimised at 2. The Log-Cosh is also popular for smoothing the curve in regression problems [10].

For the data in the HECKTOR challenge, we tested the abovementioned loss functions and their combinations, and in our best performing model, we used a hybrid loss function, the Cosh Log Dice loss combined with the Focal loss defined as follows:

$$LogCoshDiceFocalLoss = log(cosh(DiceLoss)) + FocalLoss \tag{3}$$

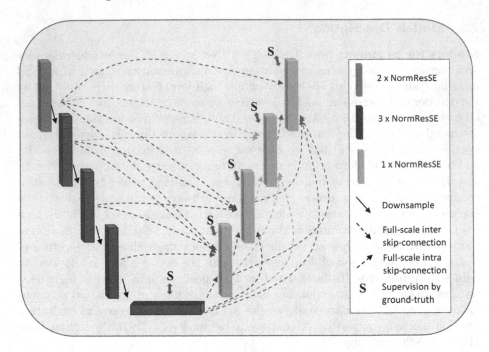

Fig. 2. Architecture of NormResSE-UNet3+ inspired from [8,9]. The input is a $2 \times 144 \times 144 \times 144$ tensor consisting of concatenated PET and CT images. The encoder is made of residual squeeze-and-excitation blocks whose first block has 24 filters following [17]. The output of the encoder has dimension $384 \times 3 \times 3 \times 3$. The decoder path contains full-scale inter- and intra-skip connections as well as a supervision module by ground-truth. The decoder outputs one channel with same size as the input ($1 \times 144 \times 144 \times 144$).

Refining Segmentation Maps. We used 3D Conditional Random Fields (CRF) to refine segmentation maps [12,13]. The segmentation output produced by CNNs tend to be too smooth because of neighbouring voxels sharing spatial information. CRF is a graphical model that captures contextual, shape and region connectivity information thus becoming a popular refinement procedure to improve segmentation performance, for example, [18] used CRF to refine the segmentation outputs as a post-processing step.

Models for Progression Free Survival Task. Cox proportional hazard (CoxPH) regression model is the most commonly used hazard model in the medical field because it effectively deals with censoring. Random Survival Forest is also a popular model for survival time prediction working better for big sample sizes [19,20]. It builds an ensemble of trees on different bootstrap samples of the training data before aggregating the predictions. DeepSurv [21] showed improvements over traditional CoxPH model as it better captures the complex relationship between a patient's features and effectiveness of different treatments.

DeepSurv is a Cox proportional hazards deep neural network, which estimates the individuals' effect based on parametrized weights of the neural network. The architecture is a multi-layer perceptron configurable with the number of hidden layers. In this study, we used 32 hidden layers which are fully-connected non-linear activation layers. Dropout layers are added to reduce over-fitting. The output layer of DeepSurv has a single node with linear activation function to give estimates of log-risk hazard. Compared to traditional Cox regression, which is optimized with the Cox partial likelihood, DeepSurv uses the negative log partial likelihood with the addition of a regularization term. DeepSurv achieved state-of-the-art results for cancer prognosis prediction with concordance index close or higher than 0.8 [22,23].

2.3 Evaluation Metrics

Evaluation Metrics for Segmentation Task. The Dice Similarity Coefficient (DSC) is a region-based measure to evaluate the overlap between the prediction (P) and the ground truth (G). DSC is given as follows:

$$DSC(P,G) = \frac{2|P \cap G|}{|P| + |G|} \tag{4}$$

The DSC ranges between 0 and 1, with a larger DSC denoting better performance.

The average Hausdorff distance (HD) between the voxel sets of ground truth and segmentation is defined as:

$$AverageHD = \frac{1}{2}(\frac{GtoP}{G} + \frac{PtoG}{P}) \tag{5}$$

where GtoP is the directed average HD from the ground-truth to the segmentation, PtoG is the directed average HD from the segmentation to the ground truth, G is the number of voxels in the ground truth, and P is the number of voxels in the segmentation. The 95th percentile of the distances between voxel sets of ground truth and segmentation (HD95) is used in this work to reduce the impact of outliers.

Evaluation Metrics for Survival Task. Harrell's concordance index (C-index) is the most widely used measure of goodness-of-fit in survival models. It is defined as the ratio of correctly ordered (concordant) pairs divided by the total number of possible evaluation pairs. The C-index is used in this study to evaluate survival prediction outcome as it takes into account censoring. The C-index quantifies how well an estimated risk score is able to discriminate among subjects who develop an event from those who do not. In this work, the event of interest is progression. The C-index ranges between 0 and 1 with 1 denoting perfect predicted risk.

Code available at: https://github.com/EmmanuelleB985/Head-and-Neck-Tumour-Segmentation-and-Prediction-of-Patient-Survival

3   Results

3.1   Segmentation Task

The model was trained on 2 NVIDIA A100 GPUs for 1000 epochs. The optimizer used is Adam (0.9,0.999). The scheduler is cosine annealing with warm restarts with the input learning rate value of 10^{-2}, and reducing the learning rate every 25 epochs. A batch size of 2 was used for training and validation. Data augmentation, namely random flipping and random rotation, is used during training to reduce over-fitting. Lifelines and Pycox packages were used for all statistical analyses.

We trained the 3D NormResSE-Unet3+ on a leave-out one center, and we performed model ensembling by averaging the predictions on the test set of the 5 models trained (see Fig. 3 and Table 1). An example of a good quality segmentation map predicted by our model is shown in Fig. 3 (the first row). An example of

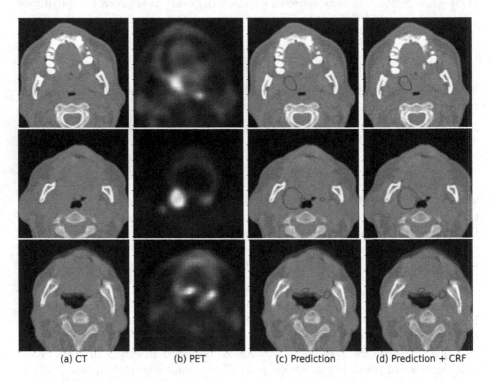

| (a) CT | (b) PET | (c) Prediction | (d) Prediction + CRF |

Fig. 3. Examples of segmentation predictions. From left to right, the axial view of (a) CT image, (b) PET image, and (c) CT overlaid with a segmentation produced by our model (blue) and ground-truth tumour annotation (red), and (d) CT overlaid with segmentation refined by the CRF (green), and ground-truth tumour annotation (red). Two top rows show the examples of the good quality segmentations, while the third row shows an example of the partially correct segmentation, where false positives were not removed by the post-processing. (Color figure online)

the predicted segmentation map, which benefited from the CRF post-processing to reduce false positives is shown in Fig. 3 (the second row). An example of failure of our pipeline to discard false positives from true primary tumour is shown in Fig. 3 (the third row). The quantitative results are summarised in Table 1. For each fold, the segmentation results are presented in terms of DSC. We obtained an average DSC of 0.753 and an average Hausdorff Distance at 95% (HD95) of 3.28 with post-processing and ensembling techniques. On the test set provided by HECKTOR2021, our model achieved an average DSC of 0.7595 and HD95 of 3.27, showing good generalisability.

Table 1. Quantitative results for Segmentation Task. The multiple neural networks for H&N tumor segmentation were trained, and these segmentations were ensembled achieving an average DSC of 0.75 and an average HD95 of 3.28 in cross-validation.

Cross-validation fold	NormResSE-UNet3+		NormResSE-UNet3+ + CRF	
	DSC	HD	DSC	HD95
Fold 1	0.792	3.18	0.822	3.11
Fold 2	0.693	3.43	0.702	3.41
Fold 3	0.728	3.32	0.749	3.29
Fold 4	0.736	3.31	0.738	3.30
Fold 5	0.742	3.29	0.756	3.28
Ensemble	0.738	3.30	0.753	3.28

3.2 Survival Task

We trained three models: CoxPH regression, Random Survival Forest, and Deep-Surv on 5-fold cross-validation splits. Each of the above models were trained with different configurations of clinical, PET/CT radiomic and deep learning features. CoxPH regression was trained with a combination of clinical, CT radiomics, and deep learning features, and achieved the best significant c-index of 0.82 (p-value < 0.05) using a corrected paired two-tailed t-test at the 5% significance level to compare each pair of models [24] (see Table 2). The second best c-index of 0.75 was obtained with CoxPH regression trained with clinical and deep learning features (see Table 2).

The performance on the test set provided by HECKTOR 2021 with the CoxPH regression using clinical, CT radiomics and deep learning features was significantly lower i.e. 0.62 suggesting over-fitting issues.

Table 2. Survival prediction results. The multiple models for survival prediction were trained, and the best model **CoxPH** using clinical, CT radiomics, and deep learning features achieved a C-index of 0.82 in cross-validation.

Survival models	C-index
CoxPH Regression (clinical)	0.70
CoxPH Regression (clinical + PET radiomics)	0.67
CoxPH Regression (clinical + CT radiomics)	0.68
CoxPH Regression (clinical + PET/CT radiomics)	0.72
CoxPH Regression (clinical + deep learning features)	0.76
CoxPH Regression (clinical + CT radiomics + deep learning features)	**0.82**
Random Survival Regression (clinical)	0.59
Random Survival Regression (clinical + PET radiomics)	0.60
Random Survival Regression (clinical + CT radiomics)	0.61
Random Survival Regression (clinical + PET/CT radiomics)	0.59
Random Survival Regression (clinical + CT radiomics + deep learning features)	0.58
DeepSurv (clinical)	0.60
DeepSurv (clinical + PET radiomics)	0.68
DeepSurv (clinical + CT radiomics)	0.69
DeepSurv (clinical + PET/CT radiomics)	0.73
DeepSurv (clinical + PET/CT radiomics + deep learning features	0.65

4   Conclusion

We proposed a multimodal 3D H&N tumor segmentation model, NormResSE-UNet3+, combining the squeeze-and-excitation layers [8] in a UNet3+ [9] architecture to take advantage of full scale features allowing the model to simultaneously focus on the relevant regions of interest. The combination of both local information and global (e.g. context) information aimed to improve the accuracy of the segmentation. We investigated the proposed neural network architecture with different training schemes and different loss functions (the Lovasz-Softmax loss and the Tversky loss [25]), however they did not significantly improve the overall segmentation performance when compared to the hybrid loss function of Log Cosh Dice [10] and Focal loss [11], which was used in our final model. In turn, a method to post-process the predicted segmentation outputs based on uncertainty using Conditional Random Fields [12,13] to filter out false-positives and refine boundaries improved segmentation accuracy.

Future work will include Bayesian uncertainty measurements followed by a tailored post-processing technique based active-contour algorithms [26] for multimodal PET and CT images. Graph-based methods and volume clustering [27] and multi-task learning [28] have been shown to improve segmentation tasks and will also be considered in future work. Increasing the multi-centre sample size for training and validation is expected to strengthen model inferences in order to demonstrate the robustness of the model and its ability to generalise. A larger sample size would also enable it to make stronger inferences to improve the prediction of progression free survival. Further work is required to reduce over-fitting issues in progression free survival e.g. by adding regularization to the model. The addition of mid-level deep learning features effectively improved progression free survival predictions compared to baseline models with only clinical and radiomic features both on training and test sets. The extraction of relevant features is an active area of research and will be the focus of future work along with work on model architectures and custom loss functions.

Acknowledgment. This work was supported by the EPSRC grant number EP/S024093/1 and the Centre for Doctoral Training in Sustainable Approaches to Biomedical Science: Responsible and Reproducible Research (SABS: R^3) Doctoral Training Centre, University of Oxford. The authors acknowledge the HECKTOR 2021 challenge for the free publicly available PET/CT images and clinical data used in this study [1].

References

1. Oreiller, V., et al.: Head and neck tumor segmentation in PET/CT: the HECKTOR challenge. Med. Image Anal. (2021). (under revision)
2. Andrearczyk, V., et al.: Overview of the HECKTOR challenge at MICCAI 2021: automatic head and neck tumor segmentation and outcome prediction in PET/CT images. In: Andrearczyk, V., Oreiller, V., Hatt, M., Depeursinge, A. (eds.) HECK-TOR 2021. LNCS, vol. 13209, pp. 1–37. Springer, Cham (2022)
3. Huang, B., et al.: Fully automated delineation of gross tumor volume for head and neck cancer on PET-CT using deep learning: a dual-center study. Contrast Media Molec. Imaging **2018**, Article ID 8923028, 12 (2018)
4. Andrearczyk, V., et al.: Automatic segmentation of head and neck tumors and nodal metastases in PET-CT scans. In: Proceedings of the Third Conference on Medical Imaging with Deep Learning, in Proceedings of Machine Learning Research, vol. 121, pp. 33–43 (2020)
5. Ronneberger, O., Fischer, P., Brox, T.: U-Net: convolutional networks for biomedical image segmentation. In: Navab, N., Hornegger, J., Wells, W.M., Frangi, A.F. (eds.) MICCAI 2015. LNCS, vol. 9351, pp. 234–241. Springer, Cham (2015). https://doi.org/10.1007/978-3-319-24574-4_28
6. Long, J., Shelhamer, E., Darrell, T.: Fully convolutional networks for semantic segmentation. In: IEEE Conference on Computer Vision and Pattern Recognition (CVPR), vol. 2015, pp. 3431–3440 (2015). https://doi.org/10.1109/CVPR.2015.7298965
7. Andrearczyk, V., et al.: Overview of the HECKTOR challenge at MICCAI 2020: automatic head and neck tumor segmentation in PET/CT. In: Andrearczyk, V., Oreiller, V., Depeursinge, A. (eds.) HECKTOR 2020. LNCS, vol. 12603, pp. 1–21. Springer, Cham (2021). https://doi.org/10.1007/978-3-030-67194-5_1
8. Iantsen, A., Visvikis, D., Hatt, M.: Squeeze-and-excitation normalization for automated delineation of head and neck primary tumors in combined PET and CT images. In: Andrearczyk, V., Oreiller, V., Depeursinge, A. (eds.) HECKTOR 2020. LNCS, vol. 12603, pp. 37–43. Springer, Cham (2021). https://doi.org/10.1007/978-3-030-67194-5_4

9. Huang, H., et al.: UNet 3+: a full-scale connected UNet for medical image segmentation. In: Proceedings of the ICASSP 2020–2020 IEEE International Conference on Acoustics, Speech and Signal Processing (ICASSP), Barcelona, Spain, 4–8 May 2020, pp. 1055–1059 (2020)

10. Jadon, S.: A survey of loss functions for semantic segmentation. In: IEEE Conference on Computational Intelligence in Bioinformatics and Computational Biology (CIBCB) **2020**, pp. 1–7 (2020)

11. Lin, T.Y., Goyal, P., Girshick, R., He, K., Dollar, P.: Focal loss for dense object detection. arxiv 2017. arXiv preprint arXiv:1708.02002 (2002)

12. Boykov, Y., Kolmogorov, V.: An experimental comparison of min-cut/max-flow algorithms for energy minimization in vision. IEEE TPAMI **26**, 1124–1137 (2004)

13. Kamnitsas, K., et al.: Efficient multi-scale 3D CNN with fully connected CRF for accurate brain lesion segmentation (2017). https://doi.org/10.17863/CAM.6936

14. Baek, S., He, Y., Allen, B.G., et al.: Deep segmentation networks predict survival of non-small cell lung cancer. Sci. Rep. **9**(1), 17286 (2019). Accessed 21 Nov 2019

15. Afshar, P., Mohammadi, A., Plataniotis, K.N., Oikonomou, A., Benali, H.: From handcrafted to deep-learning-based cancer radiomics: challenges and opportunities. IEEE Signal Process. Mag. **36**(4), 132–160 (2019)

16. Zhou, Z., Rahman Siddiquee, M.M., Tajbakhsh, N., Liang, J.: UNet++: a nested u-net architecture for medical image segmentation. In: Stoyanov, D., et al. (eds.) DLMIA/ML-CDS -2018. LNCS, vol. 11045, pp. 3–11. Springer, Cham (2018). https://doi.org/10.1007/978-3-030-00889-5_1

17. Hu, J., Shen, L., Sun, G.: Squeeze-and-excitation networks. CoRR, vol. abs/1709.01507 (2017)

18. Chen, L.-C., Papandreou, G., Kokkinos, I., Murphy, K., Yuille, A.L.: Semantic image segmentation with deep convolutional nets and fully connected crfs. In: Proceedings of the International Conference on Learning Representations (ICLR) (2015)

19. Akai, H., et al.: Predicting prognosis of resected hepatocellular carcinoma by radiomics analysis with random survival forest. Diagn. Interv. Imag. **99**(10), 643–651 (2018). Epub 2018 Jun 14 PMID: 29910166

20. Qiu, X., Gao, J., Yang, J., et al.: A comparison study of machine learning (random survival forest) and classic statistic (Cox proportional hazards) for predicting progression in high-grade glioma after proton and carbon ion radiotherapy. Front Oncol. **10**, 551420 (2020). Accessed 30 Oct 2020

21. Katzman, J.L., Shaham, U., Cloninger, A., et al.: DeepSurv: personalized treatment recommender system using a Cox proportional hazards deep neural network. BMC Med. Res. Methodol. **18**, 24 (2018)

22. Kim, D.W., Lee, S., Kwon, S., et al.: Deep learning-based survival prediction of oral cancer patients. Sci. Rep. **9**, 6994 (2019)

23. Kang, S.R., et al.: Survival prediction of non-small cell lung cancer by deep learning model integrating clinical and positron emission tomography data [abstract]. In: Proceedings of the AACR Virtual Special Conference on Artificial Intelligence, Diagnosis, and Imaging, 13–14 January 2021. AACR; Clin. Cancer Res. **27**(5 Suppl), Abstract nr PO-029 (2021)

24. Nadeau, C., Bengio, Y.: Inference for the generalization error. Mach. Learn. **52**, 239–281 (2003)

25. Abraham, N., Khan, N.: A novel focal tversky loss function with improved attention u-net for lesion segmentation. In: 2019 IEEE 16th International Symposium on Biomedical Imaging (ISBI 2019), pp. 683–687 (2019)

26. Swierczynski, P., et al.: A level-set approach to joint image segmentation and registration with application to CT lung imaging. Comput. Med. Imaging Graph. **65**, 58–68 (2018)
27. Irving, B., et al.: Pieces-of-parts for supervoxel segmentation with global context: Application to DCE-MRI tumour delineation. Med. Image Anal. **32**, 69–83 (2016)
28. Zhong, Z., et al.: 3D fully convolutional networks for co-segmentation of tumors on PET-CT images. In: 2018 IEEE 15th International Symposium on Biomedical Imaging (ISBI 2018), pp. 228–231 (2018)

Advanced Automatic Segmentation of Tumors and Survival Prediction in Head and Neck Cancer

Mohammad R. Salmanpour[1,2,3(✉)], Ghasem Hajianfar[3,4], Seyed Masoud Rezaeijo[5], Mohammad Ghaemi[3,6], and Arman Rahmim[1,2]

[1] University of British Columbia, Vancouver, BC, Canada
msalman@bccrc.ca
[2] BC Cancer Research Institute, Vancouver, BC, Canada
[3] Technological Virtual Collaboration (TECVICO Corp.), Vancouver, BC, Canada
[4] Rajaie Cardiovascular Medical and Research Center, Iran University of Medical Science, Tehran, Iran
[5] Department of Medical Physics, School of Medicine, Ahvaz Jundishapur University of Medical Sciences, Ahvaz, Iran
[6] Kerman University of Medical Sciences, Tehran, Iran

Abstract. In this study, 325 subjects were extracted from the HECKTOR-Challenge. 224 subjects were considered in the training procedure, and 101 subjects were employed to validate our models. Positron emission tomography (PET) images were registered to computed tomography (CT) images, enhanced, and cropped. First, 10 fusion techniques were utilized to combine PET and CT information. We also utilized 3D-UNETR (UNET with Transformers) and 3D-UNET to automatically segment head and neck squamous cell carcinoma (HNSCC) tumors and then extracted 215 radiomics features from each region of interest via our Standardized Environment for Radiomics Analysis (SERA) radiomics package. Subsequently, we employed multiple hybrid machine learning systems) HMLS), including 13 dimensionality reduction algorithms and 15 feature selection algorithms linked with 8 survival prediction algorithms, optimized by 5-fold cross-validation, applied to PET only, CT only and 10 fused datasets. We also employed Ensemble Voting for the prediction task. Test dice scores and test c-indices were reported to compare models. For segmentation, the highest dice score of 0.680 was achieved by the Laplacian-pyramid fusion technique linked with 3D-UNET. The highest c-index of 0.680 was obtained via the Ensemble Voting technique for survival prediction. We demonstrated that employing fusion techniques followed by appropriate automatic segmentation technique results in a good performance. Moreover, utilizing the Ensemble Voting technique enabled us to achieve the highest performance.

Keywords: Survival prediction · Head and neck squamous cell carcinoma · Deep leaning · Automatic segmentation · Hybrid machine learning system

V. Andrearczyk et al. (Eds.): HECKTOR 2021, LNCS 13209, pp. 202–210, 2022.
https://doi.org/10.1007/978-3-030-98253-9_19

1   Introduction

Head and neck squamous cell carcinoma (HNSCC) is a malignant tumor of the head and neck that originates from the lips, mouth, paranasal sinuses, larynx, nasopharynx, and other throat cancers. HNSCC cancer, as the sixth deadly cancer, involves over 655000 people yearly worldwide. It is reported that half of these cases result in death [1]. Early diagnosis of brain tumors plays an essential role in improving treatment possibilities. Medical imaging procedures such as Positron Emission Tomography (PET), Single-Photon Emission Computed Tomography (SPECT), Computed Tomography (CT), Magnetic Resonance Spectroscopy (MRS), and Magnetic Resonance Imaging (MRI) are all used to provide helpful information about shape, size, location, and metabolism of brain tumors [2]. PET can discover subtle functional changes at the early stages of a disease process and provide distinct advantages in evaluating brain tumors over anatomical imaging techniques. However, the delineation of these early changes can be operator-dependent. Artificial intelligence and image processing techniques, where segmentation is an essential step, can reduce such bias. Since manual segmentation of brain PET images is individual and time-consuming, many automated procedures may help the acceleration of the segmentation process [3]. The spatial resolution of PET is significantly lower than that of images acquired with CT or MRI. Furthermore, it is challenging to accurately segment brain tumor from PET data because PET has a high intrinsic noise level [4].

Identifying biomarkers employed to predict survival outcomes and response to treatment is an unmet need in oncology [5]. Combining different modalities may improve predictive accuracy by capturing various tumor characteristics to enhance predictive accuracy [6]. Accurate quantification of heterogeneity has excellent potential to identify high-risk patients who may benefit from aggressive treatment and low-risk patients who may be free from toxic side-effects [7]. A study applied multiple fusion techniques on PET and CT images to improve prognosis. Our recent efforts [8–10] showed that employing the most relevant features may enhance the performance of different prediction tasks. Furthermore, most predictor algorithms are not able to work with a large number of input features. Thus, it is necessary to select optimal features to be used as inputs (via feature reduction techniques), as we have demonstrated in a different prediction task.

In this work, we pursue two tasks; namely, i) automatic segmentation of tumors, and ii) survival prediction. For the first task, we utilize deep learning algorithms. Subsequently, we apply multiple hybrid machine learning systems (HMLS), including various features reduction techniques linked with survival prediction algorithms, applied to datasets constructed from clinical and radiomics features to predict Progression Free Survival (PFS).

2   Material and Method

2.1   Data Collection

325 patients with HNSCC cancer were studied in the HECKTOR Challenge [11, 12]. 224 subjects were considered for the training procedure, and 101 subjects were employed to validate our models. Each patient had CT, PET, GTV and clinical information. Table 1 shows some properties of the dataset.

Table 1. List of characteristics of patients

Outcome	Split	# of patients	# of females	# of males	Age range	Age Mean ± STD
PFS	Train	224	57	167	34–90	63 ± 9.5
PFS	Test	101	19	82	40–84	61.22 ± 9.08

2.2 Analysis Procedure

As shown in Fig. 1, we applied different HMLSs [8-10] to predict PFS. The images of the HNSCC dataset for the CT and PET images were in two sizes, 512×512 and 224×224 respectively. Therefore, In the first pre-processing step, all PET images were registered to PET images using rigid registration through an in-house-developed Python code. Image registration was also performed via six degrees of freedom, consisting of three translations and three rotations based on patient-wise registration. Hence, PET images were converted to the size 512 * 512. In addition, to achieve a common image size and reduce the computation time with redundant pixels, image cropping was performed through a bounding box (being equal to $144 \times 144 \times 144$ mm^3). In addition, image normalization was performed via max normalization technique so that the intensity of each slice was normalized between 0 and 1. In this study, resampling was not employed. Moreover, image augmentation techniques were not performed in our study. We also used the loss function as follows:

$$L_{dice} = 1 - \frac{2\sum_{x\in\varphi} pl(x)gl(x)}{\sum_{x\in\varphi} pl^2(x) + \sum_{x\in\varphi} gl^2(x)} \tag{1}$$

Where pl (x) indicates that the probability of belonging to the pixel x for class l and gl (x) is a vector of the ground-truth label, where it is one for the true class and zero for other classes. This formula of dice loss helps solve the problem of unbalanced training data. Thus, we have no weighting parameters between different classes (the background and the vascular tree) during the training process. Hence, the loss Function properly works for the binary segmentation tasks.

First, we employed 10 fusion techniques (FT) to combine PET and CT information and utilized 3D UNETR (UNET with Transformers) and 3D UNET algorithms to automatically segment HNSCC tumor on PET, CT and new images generated from 10 fusion techniques (see Fig. 2). We employed 10 fusions, as shown in Table 2. The image fusion process is defined to convert all the vital information from multiple images to a single image. This single image is more informative and accurate than any single source image, and it consists of all the necessary information. No GTV mask was applied to the fusion techniques. An example of CT, PET, and fused images for a patient is shown in Fig. 3.

We considered 224 patients for the training process and 101 patients for the external test for the segmentation task. 90% out of 224 patients (training dataset) were employed to train the algorithms, and the remaining patients were used to validate the training process. In addition, we applied new images generated from 10 fusion techniques to segmentation algorithms. Subsequently, 215 Radiomics Features (RFs) [13, 14] were extracted from each region of interest in PET, CT and 10 fusion techniques through our standard-ized SERA package [15]. In addition, the extracted RFs were combined with clinical fea-tures. Finally, we employed multiple HMLSs: 13 dimensionality reduction algorithms (DRA) or 15 feature selection algorithms (FSA) linked with Eight survival prediction algorithms (SPA) applied to PET only, CT only, and fused images (PET-CT). We also employed the Ensemble Voting technique for the prediction task. All datasets were nor-malized by the Z-score technique and optimized via 5-fold cross-validation on the train-ing dataset. Dice scores and c-indices were reported to compare models. Table 2 shows a list of the algorithms employed.

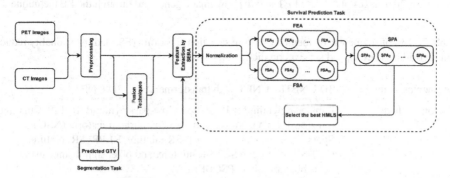

Fig. 1. Graphical view of our study procedure in the survival prediction task

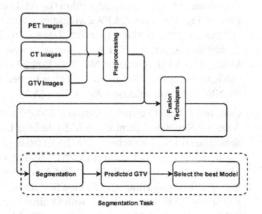

Fig. 2. Graphical view of our study procedure in the segmentation task

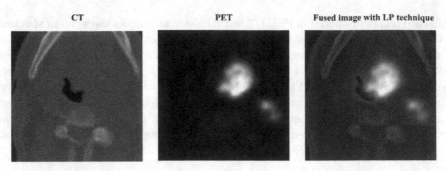

Fig. 3. 2D-Images of CT and PET and the new image generated through the LP technique

Table 2. List of segmentation methods, feature selection algorithm (FSA), dimensionality reduction algorithms (DRA), and survival prediction algorithm (SPA).

Segmentation methods	3D UNETR (UNET with transformers) and 3D UNET
Fusion methods	Laplacian pyramid (LP), Ratio of low-pass pyramid (RLPP), Curvelet transform (CVT), Nonsubsampled contourlet transform (NSCT), Sparse representation (SR), LP-SR mixture, RLPP- SR mixture, CVT-SR mixture, NSCT-SR mixture, and pixel significance using cross bilateral filter (PSCBF)
DRA	Principal Component Analysis (PCA), Kernel_PCA, t-distributed Stochastic Neighbor Embedding (t-SNE), Sammon Mapping Algorithm (SMA), Isomap Algorithm (IsoA), Locally Linear Embedding Algorithm (LLEA), Multidimensional Scaling Algorithm (MDSA), Diffusion Map Algorithm (DMA), Stochastic Proximity Embedding Algorithm (SPEA), Gaussian Process Latent Variable Model (GPLVM), Stochastic Neighbor Embedding Algorithm (SNEA), Symmetric Stochastic Neighbor Embedding Algorithm (SSNEA) and Autoencoders Algorithms (AA)
FSA	Correlation-based Feature Selection (CFS), Feature Selection with Adaptive Structure Learning (FSASL), Infinite Latent Feature Selection (ILFS), Laplacian Score (LS), Lasso, Local Learning based Clustering Feature Selection (LLCFS), Minimum Redundancy Maximum Relevance (MRMR), Relief Algorithm (ReliefA), Unsupervised Discriminative Feature Selection (UDFS), Unsupervised Feature Selection with Ordinal Locality (UFSOL), Select features based on c-index (CindexFS), Mutual Information (MI), Variable hunting (VH), Variable hunting variable importance (VH.VIMP) and Minimal Depth (MD)
SPA	Conditional Survival Forest (CSF), Cox's proportional hazard (CoxPH), Fast Survival SVM (FSSVM), CoxPH model by likelihood-based boosting (CoxBoost), Lasso and Elastic-Net Regularized Generalized Linear Models (GlmNet), Gradient Boosting with Component-wise Linear Models (GlmBoost), Gradient Boosting Machines (GBM), Random Survival Forest (RSF)

3 Results

3.1 Segmentation Results

We applied 2 deep learning techniques on CT, PET, and 10 fused datasets for segmentation. The result obtained from 3D UNET overcame the performance resulting from 3D UNETR. Training parameters of the models are displayed in Table 3. In this paper, we ignored to report the result provided from 3D UNETR because of poor performance compared to 3D UNET. The highest dice score of 0.68 was achieved by the LP fusion technique linked with 3D UNET (as a segmentation technique) on the test dataset and 0.81 on the validation dataset. 3D UNET algorithm had more layers compared to 3D UNETR with a similar number of parameters. It can be why the algorithm resulted in a higher dice score. The performances of segmentation are listed in Table 4.

Table 3. Training parameters of the 3D UNET and 3D UNETR (UNET with transformers) algorithms

Parameters	Models
Batch size	2
Learning rate	0.0001
Adaptive learning rate method	Adam
	$\beta_1 = 0.5, \beta_2 = 0.9$
The number of epochs	1000
Validation set	23
Early stopping	No
Learning rate decay scheduling	No

Table 4. Performance of different datasets in segmentation (sorted by external test dice score).

Modality/fusion	Validation dice score	Test dice score
LP	0.81	0.680
RLPP-SR	0.81	0.649
PET	0.79	0.635
RLPP-SR	0.78	0.635
LP	0.77	0.635

3.2 Survival Prediction Results

For the prediction of PFS, we applied multiple HMLSs, including FSAs/DRAs linked with SPAs on CT, PET and 4 datasets obtained from fusion techniques. The highest c-index of 0.68 was obtained via the Ensemble Voting technique. Table 5 shows some best performances resulting from HMLSs. For Ensemble Voting, we employ all predicted outcomes resulting from all HMLSs. In addition, we employed two filtering methods, such as Median and Mean filters, to smooth the outcome range.

Table 5. Performance of different datasets on survival prediction (sorted by test C-index).

Modality/fusion technique	FSA/DRA	SPA	Validation c-index (Mean ± STD)	Test c-index
–	–	Ensemble voting (mean filter)	0.750 ± 0.030	0.680
LP-SR	MI	CoxPH	0.710 ± 0.020	0.677
–	–	Ensemble voting (median filter)	0.710 ± 0.040	0.676
PET	MI	CoxPH	0.690 ± 0.030	0.668
LP-SR	CindexFS	GBM	0.690 ± 0.040	0.644

4   Discussion

In this work, we utilized 2 deep learning algorithms linked with multiple fusion techniques to segment HNSCC cancer. We applied multiple HMLS (FSAs/DRAs and SPAs) on radiomics feature extracted from each ROI to predict PFS. As shown, fusion techniques add a significant value to segmentation performance compared to original images. Our emphasis on studying multiple survival prediction algorithms is an important new direction. As shown in Table 3, a best performance of ~0.68 was obtained for prediction of PFS. Overall, our work shows that employing HMLSs linked with fusion techniques added a value to enhancement of performance. In short, this work is unique as it studies multiple fusion techniques, two automatic segmentation techniques, multiple SPAs (accompanied with Ensemble Voting technique), DRAs and FSAs, to significantly improve segmentation of tumor as well as prediction of outcome in HNSCC cancer patients. In a previous study [9], we implemented multiple HMLSs, including DRAs/FSAs, SPAs, and 6 datasets constructed from clinical features and radiomics features as extracted from the fused images, CT or PET-only.

5   Conclusion

This study demonstrated enhanced task performance for automatic segmentation of tumors and survival prediction in HNSCC patients. We showed that using fusion techniques followed by appropriate automatic segmentation techniques results in a good performance. We arrived at our highest test performance ~0.68 for the segmentation task through 3D UNET linked with LP, outperforming 3D UNETR. 3D UNET has more layers with a similar number of parameters, which may be a reason for the higher dice score performance. Meanwhile, we obtained our highest test performance ~0.68 for survival prediction through an Ensemble Voting technique.

6 Code Availability

https://github.com/Tecvico/Advanced-Survival-Prediction-and-Automatic-Segmentation-in-Head-and-Neck-Cancer.

Acknowledgement. This study was supported by the Natural Sciences and Engineering Research Council of Canada (NSERC) Discovery Grant RGPIN-2019-06467.

Conflict of Interest. The authors have no relevant conflicts of interest to disclose.

References

1. Wu, Z.H., Zhong, Y., et al.: miRNA biomarkers for predicting overall survival outcomes for head and neck squamous cell carcinoma. Genomics **113**(1), 135–141 (2021)
2. Butowski, N.A.: Epidemiology and diagnosis of brain tumors. CONTINUUM Lifelong Learn. Neurol. **21**, 301–313 (2015)
3. Kumari, N., Saxena, S.: Review of brain tumor segmentation and classification. In: 2018 International Conference on Current Trends towards Converging Technologies (ICCTCT), pp. 1–6. IEEE (2018)
4. Rahmim, A., Zaidi, H.: PET versus SPECT: strengths, limitations and challenges. Nucl. Med. Commun. **29**, 193–207 (2008)
5. Fitzgerald, C.W., Valero, C., et al.: Positron emission tomography–computed tomography imaging, genomic profile, and survival in patients with head and neck cancer receiving immunotherapy. JAMA Otolaryngol. Head Neck Surg. **147**, 1119 (2021)
6. Martens, R.M., Koopman, T., et al.: Multiparametric functional MRI and 18 F-FDG-PET for survival prediction in patients with head and neck squamous cell carcinoma treated with (chemo) radiation. Eur. Radiol. **31**(2), 616–628 (2021)
7. Marur, S., Forastiere, A.A.: Head and neck squamous cell carcinoma: update on epidemiology, diagnosis, and treatment. Mayo Clin. Proc. **91**(3), 386–396 (2016)
8. Salmanpour, M., Shamsaei, M., et al.: Optimized machine learning methods for prediction of cognitive outcome in Parkinson's disease. Comput. Biol. Med. **111**, 1–8 (2019)
9. Salmanpour, M., Shamsaei, M., et al.: Machine learning methods for optimal prediction of motor outcome in Parkinson's disease. Physica Medica **69**, 233–240 (2020)
10. Salmanpour, M., Shamsaei, M., Rahmim, A.: Feature selection and machine learning methods for optimal identification and prediction of subtypes in Parkinson's disease. Comput. Methods Prog. Biomed. **206**, 1–12 (2021)
11. Andrearczyk, V., et al.: Overview of the HECKTOR challenge at MICCAI 2021: automatic head and neck tumor segmentation and outcome prediction in PET/CT images. In: Andrearczyk, V., Oreiller, V., Hatt, M., Depeursinge, A. (eds.) HECKTOR 2021. LNCS, vol. 13209, pp. 1–37. Springer, Cham (2022)
12. Valentin, O., et al.: Head and neck tumor segmentation in PET/CT: the HECKTOR challenge. In: Medical Image Analysis (2021) (under revision)

13. Masoud, R.S., Abedi-Firouzjah, R., Ghorvei, M., Sarnameh, S.: Screening of COVID-19 based on the extracted radiomics features from chest CT images. J. X-ray Sci. Technol. **29**, 1–5 (2021)
14. Rezaeijo, S.M., Ghorvei, M., Alaei, M.: A machine learning method based on lesion segmentation for quantitative analysis of CT radiomics to detect covid-19. In: 2020 6th Iranian Conference on Signal Processing and Intelligent Systems (ICSPIS), pp. 1–5. IEEE (2020)
15. Ashrafinia, S.: Quantitative Nuclear Medicine Imaging using Advanced Image Reconstruction and Radiomics. Ph.D. Dissertation, Johns Hopkins University (2019)

Fusion-Based Head and Neck Tumor Segmentation and Survival Prediction Using Robust Deep Learning Techniques and Advanced Hybrid Machine Learning Systems

Mehdi Fatan[1], Mahdi Hosseinzadeh[1,2], Dariush Askari[1,3], Hossein Sheikhi[1], Seyed Masoud Rezaeijo[4], and Mohammad R. Salmanpour[1,5,6(✉)]

[1] Technological Virtual Collaboration (TECVICO Corp.), Vancouver, BC, Canada
m.salman@bccrc.ca
[2] Tarbiat Modares University, Tehran, Iran
[3] Department of Radiology Technology, Shahid Beheshti University of Medical, Tehran, Iran
[4] Department of Medical Physics, School of Medicine, Ahvaz Jundishapur University of Medical Sciences, Ahvaz, Iran
[5] University of British Columbia, Vancouver, BC, Canada
[6] BC Cancer Research Institute, Vancouver, BC, Canada

Abstract. Multi-level multi-modality fusion radiomics is a promising technique with potential for improved prognostication and segmentation of cancer. This study aims to employ advanced fusion techniques, deep learning segmentation methods, and survival analysis to automatically segment tumor and predict survival outcome in head-and-neck-squamous-cell-carcinoma (HNSCC) cancer. 325 patients with HNSCC cancer were extracted from HECTOR Challenge. 224 patients were used for training and 101 patients were employed to finally validate models. 5 fusion techniques were utilized to combine PET and CT information. The rigid registration technique was employed to register PET images to their CT image. We employed 3D-UNet architecture and SegResNet (segmentation using autoencoder regularization) to improve segmentation performance. Radiomics features were extracted from each region of interest (ROI) via the standardized SERA package, applying to Hybrid Machine Learning Systems (HMLS) including 7 dimensionality reduction algorithms followed by 5 survival prediction algorithms. Dice score and c-Index were reported to compare models in segmentation and prediction tasks respectively. For segmentation task, we achieved dice score around 0.63 using LP-SR Mixture fusion technique (the mixture of Laplacian Pyramid (LP) and Sparse Representation (SR) fusion techniques) followed by 3D-UNET. Next that, employing LP-SR Mixture linked with GlmBoost (Gradient Boosting with Component-wise Linear Models) technique enables an improvement of c-Index ~0.66. This effort indicates that employing appropriate fusion techniques and deep learning techniques results in the highest performance in segmentation task. In addition, the usage of fusion techniques effectively improves survival prediction performance.

© Springer Nature Switzerland AG 2022
V. Andrearczyk et al. (Eds.): HECKTOR 2021, LNCS 13209, pp. 211–223, 2022.
https://doi.org/10.1007/978-3-030-98253-9_20

1   Introduction

Early diagnosis of head and neck tumors plays an essential role in improving treatment possibilities. The medical imaging procedures such as Positron Emission Tomography (PET), Single-Photon Emission Computed Tomography (SPECT), Computed Tomography (CT), Magnetic Resonance Spectroscopy (MRS), and Magnetic Resonance Imaging (MRI) are all used to provide helpful information about shape, size, location, and metabolism of brain tumors [1]. PET can discover subtle functional changes at the early stages of a disease process and provide distinct advantages in evaluating head and neck tumors over anatomical imaging techniques. However, the delineation of these early changes can be operator dependent. Artificial intelligence and image processing techniques, where segmentation is an essential step, can reduce such bias. The spatial resolution of PET is significantly lower than that of images acquired with CT or MRI. Furthermore, it is challenging to segment brain tumors accurately from PET data because PET has a high intrinsic noise level [3].

Squamous cell carcinoma of the head and neck (HNSCC) is a malignant tumor of the head and neck that originates from the lips, mouth, sinuses, larynx, nasopharynx, and other cancers of the larynx. HNSCC cancer, the sixth fatal cancer, affects over 655,000 people worldwide each year. It is reported that half of these patients resulted in death [4]. The identification of biomarkers that predict survival outcomes and response to treatment is an unmet need in oncology [5]. The combination of different modalities can improve prediction accuracy by capturing a variety of tumor features to improve prediction accuracy [6]. Accurate quantification of heterogeneity has great potential to identify high-risk patients who may benefit from aggressive treatment and low-risk patients who may be free from toxic side effects [7]. In one study, several fusion techniques were applied to PET and CT images to improve prognosis. Recent efforts of ours [8-10] have shown that using the most relevant functions can improve the performance of various predictive tasks. In addition, most predictor algorithms are incapable of working with a large number of input features, and therefore it is necessary to select optimal features to use as inputs (via dimension reduction techniques), as we have shown in another prediction task. In this effort, we apply several HMLSs, including several dimensionality reduction and survival prediction algorithms that are applied to datasets constructed from clinical and radiomics features to predict progression free survival. To better understand brain tumor segmentation, specialists look for reliable techniques to do this automatically in less time. This effort uses deep neural techniques to automatically segment the tumor. Further, it aims to employ fusion techniques to improvement performances. Finally, multiple HMLSs apply to dataset with clinical features and radiomics features extracted from the segmented tumor.

2   Material and Methods

2.1   Dataset, PET/CT Acquisition

Our experiments were done on the relevant data obtained from HECKTOR challenge dataset (MICCAI 2021). The images of the HNSCC dataset for the CT and PET images were in two sizes, 512 × 512 and 224 × 224, respectively. Therefore, In the first pre-processing step, all PET images were registered to CT images using rigid registration. Image registration is the method of aligning images so that relationship between them can be seen more quickly. The same term is also applied to explain the alignment of images to a computer model or physical space. Image registration was performed using six degrees of freedom, consisting of three translations and three rotations based on patient-wise registration. The fusion techniques literally fuse CT and PET imaging to a single image. Hence, fusion technique enables the exploitation of the powers of all imaging modalities together, reducing or minimizing the faults of every single modality. In addition, in order to achieve a common image size and also to reduce the computation time with redundant pixels, image cropping was performed using a bounding box (being equal to 144 × 144 × 144 mm^3). To ensure comparable voxel intensities across CT and PET images, image normalization was performed via maximum intensity. Of note, the resampling was not performed. Of note, in our study, image augmentation techniques were not performed. 325 patients with HNSCC cancer were employed [11-13]. 224 subjects [age: 63 ± 9.5] were considered for training procedure and 101 [age: 61.22 ± 9.08] subjects were employed to finally validate our models. Moreover, some fusion techniques are used to combine PET and CT information. We extracted 215 radiomics features from each segmented area via SERA package and added those to clinical features (Table 1). Radiomics is an image quantitative analysis procedure widely used in clinical research and early detection, prognosis, and prediction of treatment response. The purpose of radiomics is to find the relationship between quantitative data extracted from the images and clinical information. Based on this hypothesis, phenotypic differences can be obtained by feature extraction of images [19, 20] (Fig. 1).

Table 1. Radiomics features extracted

Features family	# Features	Image biomarker
Morphology	29	Volume, Surface area, Surface to volume ratio, Compactness 1, Compactness 2, Spherical disproportion, Sphericity, Asphericity, Centre of mass shift, Maximum 3D diameter, Major axis length, Minor axis length, Least axis length, Elongation, Flatness, Volume density (AABB), Area density (AABB), Volume density (OMBB), Area density (OMBB), Volume density (AEE), Area density (AEE), Volume density (MVEE), Area density (MVEE), Volume density (convex hull), Area density (convex hull), Integrated intensity, Moran's I index, Geary's C measure

(*continued*)

Table 1. (*continued*)

Features family	# Features	Image biomarker
Local intensity	2	Local intensity peak and Global intensity peak
Statistics	18	Mean, Variance, Skewness, Median, Minimum, 10th percentile, 90th percentile, Maximum, Interquartile range, Range, Mean absolute deviation, Robust mean absolute deviation, Median absolute deviation, Coefficient of variation, Quartile coefficient of dispersion, Energy, Root mean square
Intensity histogram	23	Mean, Variance, Skewness, Kurtosis, Median Minimum, 10th percentile, 90th percentile, Maximum, Mode, Interquartile range, Range, Mean absolute deviation, Robust mean, absolute deviation, Median absolute deviation, Coefficient of variation, Quartile coefficient of dispersion, Entropy, Uniformity, Maximum histogram gradient, Maximum gradient grey level, Minimum histogram gradient, Minimum gradient grey level
Intensity volume histogram	7	Volume fraction at 10% intensity, Volume, fraction at 90% intensity, Intensity at 10% volume, Intensity at 90% volume, Volume fraction difference between 10% and 90% intensity, Intensity difference between 10% and 90% volume, Area under the IVH curve

(*continued*)

Table 1. (*continued*)

Features family	# Features	Image biomarker
Co-occurrence matrix (3D, averaged and merged)	50	Joint maximum, Joint average, Joint variance, Joint entropy, Difference average, Difference variance, Difference entropy, Sum average, Sum variance, Sum entropy, Angular second moment, Contrast, Dissimilarity, Inverse difference, Inverse difference normalized, Inverse difference moment, Inverse difference moment normalized, Inverse variance, Correlation, Autocorrelation, Cluster tendency, Cluster shade, Cluster prominence, Information correlation 1, Information correlation 2, Joint maximum, Joint average, Joint variance, Joint entropy, Difference average, Difference variance, Difference entropy, Sum average, Sum variance, Sum entropy, Angular second moment, Contrast, Dissimilarity, Inverse difference, Inverse difference normalized, Inverse difference moment, Inverse difference moment normalized, Inverse variance, Correlation, Autocorrelation, Cluster tendency, Cluster shade, Cluster prominence, Information correlation 1, Information correlation 2

(*continued*)

Table 1. (*continued*)

Features family	# Features	Image biomarker
Run length matrix (3D, averaged and merged)	32	Short runs emphasis, Long runs emphasis, Low grey level run emphasis, High grey level run emphasis, Short run low grey level emphasis, Short run high grey level emphasis, Long run low grey level emphasis, Long run high grey level emphasis, Grey level non-uniformity, Grey level non-uniformity normalized, Run length non-uniformity, Run length non-uniformity normalized, Run percentage, Grey level variance, Run length variance, Run entropy, Short runs emphasis, Long runs emphasis, Low grey level run emphasis, High grey level run emphasis, Short run low grey level emphasis, Short run high grey level emphasis, Long run low grey level emphasis, Long run high grey level emphasis, Grey level non-uniformity, Grey level non-uniformity normalized, Run length non-uniformity, Run length non-uniformity normalized, Run percentage Grey level variance, Run length variance, Run entropy

(continued)

Table 1. (*continued*)

Features family	# Features	Image biomarker
Size zone matrix (3D) and Distance zone matrix (3D)	32	Small zone emphasis, Large zone emphasis, Low grey level emphasis, High grey level emphasis, Small zone low grey level emphasis, Small zone high grey level emphasis, Large zone low grey level emphasis, Large zone high grey level emphasis, Grey level non-uniformity, Grey level non uniformity normalized, Zone size non-uniformity, Zone size non-uniformity normalized, Zone percentage, Grey level variance, Zone size variance, Zone size entropy, Small distance emphasis, Large distance emphasis, Low grey level emphasis, High grey level emphasis, Small distance low grey level emphasis, Small distance high grey level emphasis, Large distance low grey level emphasis, Large distance high grey level emphasis, Grey level non-uniformity, Grey level non-uniformity normalized, Zone distance non-uniformity, Zone distance non-uniformity normalized, Zone percentage, Grey level variance, Zone distance variance, Zone distance entropy
Neighbourhood grey tone difference matrix (3D)	5	Coarseness, Contrast, Busyness, Complexity, Strength
Neighbouring grey level dependence matrix (3D)	17	Low dependence emphasis, High dependence emphasis, Low grey level count emphasis, High grey level count emphasis, Low dependence low grey level emphasis, Low dependence high grey level emphasis, High dependence low grey level emphasis, High dependence high grey level emphasis, Grey level non-uniformity, Grey level non-uniformity normalized, Dependence count non-uniformity, Dependence count non-uniformity normalized, Dependence count percentage, Grey level variance, Dependence count variance, Dependence count entropy, Dependence count energy
Total	215	

a   b

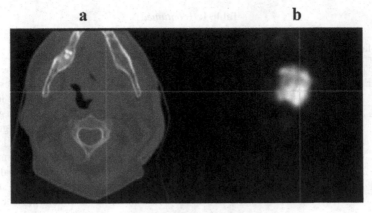

Fig. 1. (a) CT and registered PET (b) images

2.2 Survival Prediction

As shown in Fig. 2 and Fig. 3, we applied different HMLSs [9-11] to predict progression free survival and segment tumor. The used hyperparameter are shown in Table 2. In the pre-processing step, we first registered PET to CT images and then cropped registered images. First, we employed 5 fusion techniques mentioned in Table 2 to combine PET and CT information. Image fusion means combining two images into a single image with the maximum information content without producing details that are non-existent in the given images. We thus utilize 3D-UNet architecture and SegResNet to automatically segment HNSCC tumor on PET, CT and 5 fusion techniques. Subsequently, 215 radiomics features were extracted from each region of interest in PET, CT and 5 fusion techniques via our standardized SERA package and then extracted radiomics features combined with clinical features. Finally, we employed multiple HMLSs: 7 dimensionality reduction algorithms (DRA) linked with 5 survival prediction algorithms (SPA) applied to PET only, CT only and 5 PET-CT datasets generated by image-level fusion strategies. For segmentation and survival prediction task, dice score and c-index were reported to compare models. Table 3 shows a list of the algorithms employed. Of note, in our study, image augmentation techniques were not performed (Fig. 4).

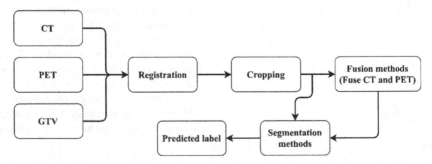

Fig. 2. Graphical view of our prediction algorithm for segmentation task.

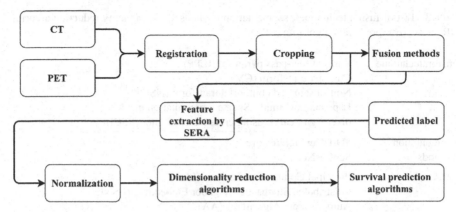

Fig. 3. Graphical view of our prediction algorithm for survival prediction task.

Fig. 4. Applied curvelet transform (CVT) fusion method on CT and PET

Table 2. Indicates the selected hyperparameters

Model	R package	Hyper parameter
Coxph	Survival	feature number: 1–10
glmboost	mboost	mstop: 50–500 feature number: 1–10
GBM	gbm	n.trees: 100, 500, 1000 interaction.depth: 1–5 n.minobsinnode: 3–5 shrinkage: 0.01,0.05,0.1 feature number: 1–10

Table 3. List of fusion techniques, segmentation methods, dimensionality reduction algorithms (DRA), survival prediction algorithm (SPA).

Fusion technique	Ratio of low-pass pyramid (RLPP)
	Curvelet transform (CVT)
	Nonsubsampled contourlet transform (NSCT)
	Laplacian pyramid - Sparse representation mix
	pixel significance using cross bilateral filter (PSCBF)
Segmentation methods	3D-UNet architecture
	SegResNet
DRA	Principal Component Analysis (PCA)
	t-distributed Stochastic Neighbor Embedding (t-SNE)
	Autoencoders Algorithms (AA)
	Lasso
	Relief Algorithm (ReliefA)
	Mutual Information (MI)
	Minimal Depth (MD)
SPA	Cox's proportional hazard (CoxPH)
	Fast Survival SVM (FSSVM)
	CoxPH model by likelihood based boosting (CoxBoost)
	Gradient Boosting with Component-wise Linear Models (GlmBoost)
	Gradient Boosting Machines (GBM)

3   Results

3.1   Segmentation Task

For segmentation, we applied two deep learning algorithms on CT, PET and 5 datasets obtained from fusion techniques. Training parameters of the models are displayed in Table 4. In addition, we used loss function as follows:

$$L_{dice} = 1 - \frac{2\sum_{x\in\varphi} pl(x)gl(x)}{\sum_{x\in\varphi} pl^2(x) + \sum_{x\in\varphi} gl^2(x)} \qquad (1)$$

where pl (x) indicates the probability of belonging to the pixel x for class l and gl (x) is a vector of the ground truth label, where it is one for the true class and zero for other classes. This formula of dice loss helps solve the problem of unbalanced training data. During training, we employed one no weighting parameters between different classes (our case the background and the vascular tree) and the loss function works well for the binary segmentation tasks.

The highest dice score 0.63 of external validation dataset was achieved by Laplacian Pyramid-Sparse representation mixture (LP-SR) fusion technique linked with 3D UNET (as a segmentation technique). In addition, we also received a performance around 0.76 from validation dataset (randomly 10% of dataset as validation). Some of best performances of segmentation task are listed in Table 5.

Table 4. Training parameters of the 3D UNET and 3D UNETR (UNET with Transformers) models

Parameters	Models
Batch size	2
Learning rate	0.0001
Adaptive learning rate method	Adam
	$\beta_1 = 0.5$, $\beta_2 = 0.9$
The number of epochs	1000
Validation set	23
Early stopping	No
Learning rate decay scheduling	No

Table 5. Performance of different datasets in segmentation.

Modality/fusion	Validation dice score	External test dice score	External test Hausdorf distance at 95%
LP-SR	0.76	0.63	5.83
CVT	0.76	0.60	6.35
RLPP	0.75	0.48	6.85
NSCT	0.73	0.40	6.21
PSCBF	0.71	0.40	6.21

3.2 Progression Free Survival Prediction Task

For progression free survival prediction, we applied multiple HMLSs on radiomics features extracted from CT, PET and 5 datasets obtained from fusion techniques. The highest c-index 0.66 was obtained via GlmBoost technique. The performances of progression free survival prediction are listed in Table 6. Of note, the results of this study were finally submitted for the HECKTOR challenge.

Table 6. Survival prediction performance of different datasets.

Modality	DRA	SPA	5-fold cross validation c-index (Mean ± STD)	External test c-index
LP-SR	–	GlmBoost	0.73 ± 0.02	0.66
PET	–	GBM	0.71 ± 0.03	0.65
LP-SR	–	CoxPH	0.73 ± 0.01	0.65
PET	AA	FSSVM	0.72 ± 0.02	0.64
NSCT	MI	CoxPH	0.69 ± 0.03	0.64

4   Conclusion

An automated procedure for the segmentation of head and neck tumor is presented in this study. This procedure was tested on HECKTOR 2021 challenge dataset. The proposed model provides accurate segmentation outcomes which are comparable with manual ground truth annotations. The presented algorithm makes it feasible to produce a patient-specific segmentation of tumor without manual interaction. In summary, 3D UNET linked with LP-SR resulted in the highest segmentation performance ~0.63 on external test and 0.76 on validation datasets compared to other hybrid systems. Furthermore, the highest c-index ~0.66 on external test and 0.73 on validation datasets were achieved via LP-SR + GlmBoost.

5   Code Availability

https://github.com/Tecvico/Fusion-Based-Brain-Tumor-Segmentation-and-Survival-prediction-using-Robust-DL-and-Advanced-HMLS.

Acknowledgement. This study was supported by Technological Virtual Collaboration (TECVICO Corp.) company, Vancouver, BC, Canada. Business number: 762450864RC0001.

Conflict of Interest. The authors have no relevant conflicts of interest to disclose.

References

1. Butowski, N.A.: Epidemiology and diagnosis of brain tumors. CONTINUUM Lifelong Learn. Neurol. **21**, 301–313 (2015)
2. Kumari, N., Saxena, S.: Review of brain tumor segmentation and classification. In: 2018 International Conference on Current Trends towards Converging Technologies (ICCTCT), pp. 1–6. IEEE (2018)
3. Rahmim, A., Zaidi, H.: PET versus SPECT: strengths, limitations and challenges. Nucl. Med. Commun. **29**, 193–207 (2008)

4. Wu, Z.H., Zhong, Y., et al.: miRNA biomarkers for predicting overall survival outcomes for head and neck squamous cell carcinoma. Genomics 113(1), 135–141 (2021)
5. Fitzgerald, C.W., Valero, C., et al.: Positron emission tomography-computed tomography imaging, genomic profile, and survival in patients with head and neck cancer receiving immunotherapy. JAMA Otolaryngol. Head Neck Surg. 147, 1119 (2021)
6. Martens, R.M., Koopman, T., et al.: Multiparametric functional MRI and 18 F-FDG-PET for survival prediction in patients with head and neck squamous cell carcinoma treated with (chemo) radiation. Eur. Radiol. 31(2), 616–628 (2021)
7. Marur, S., Forastiere, A.A.: Head and neck squamous cell carcinoma: update on epidemiology, diagnosis, and treatment. Mayo Clin. Proc. 91(3), 386–396 (2016)
8. Lv, W., Ashrafinia, S., et al.: Multi-level multi-modality fusion radiomics: application to PET and CT imaging for prognostication of head and neck cancer. IEEE J. Biomed. Health Inform. 24(8), 2268–2277 (2016)
9. Salmanpour, M., Shamsaei, M., et al.: Optimized machine learning methods for prediction of cognitive outcome in Parkinson's disease. Comput. Biol. Med. 111, 1–8 (2019)
10. Salmanpour, M., Shamsaei, M., et al.: Machine learning methods for optimal prediction of motor outcome in Parkinson's disease. Physica Medica 69, 233–240 (2020)
11. Salmanpour, M., Shamsaei, M., Rahmim, A.: Feature selection and machine learning methods for optimal identification and prediction of subtypes in Parkinson's disease. Comput. Methods Prog. Biomed. 206, 1–12 (2021)
12. Andrearczyk, V., et al.: Overview of the HECKTOR challenge at MICCAI 2021: automatic head and neck tumor segmentation and outcome prediction in PET/CT images. In: Andrearczyk, V., Oreiller, V., Hatt, M., Depeursinge, A. (eds.) HECKTOR 2021. LNCS, vol. 13209, pp. 1–37. Springer, Cham (2022)
13. Valentin, O., et al.: Head and neck tumor segmentation in PET/CT: the HECKTOR challenge. In: Medical Image Analysis, 2021 (under revision)
14. Jeyavathana, R.B., Balasubramanian, R., Pandian, A.A.: A survey: analysis on pre-processing and segmentation techniques for medical images. Int. J. Res. Sci. Innov. (IJRSI) 3, 11 (2016)
15. Joshi, N., Jain, S., Agarwal, A.: Segmentation based non local means filter for denoising MRI. In: 2017 6th International Conference on Reliability, Infocom Technologies and Optimization (Trends and Future Directions) (ICRITO), pp. 640–644. IEEE (2017)
16. Özgün, Ç., et al.: 3D U-Net: Learning Dense Volumetric Segmentation from Sparse Annotation (2016). arXiv:1606.06650
17. Kohler, R.: A segmentation system based on thresholding. Comput. Graph. Image Process. 15, 319–338 (1981)
18. Zhang, Y.J.: A review of recent evaluation methods for image segmentation. In: Proceedings of the 6th International Symposium on Signal Processing and its Applications (Cat. No. 01EX467), vol. 1, pp. 148–151. IEEE (2001)
19. Masoud, R.S., Abedi-Firouzjah, R., Ghorvei, M., Sarnameh, S.: Screening of COVID-19 based on the extracted radiomics features from chest CT images. J. X-ray Sci. Technol. 29, 1–5 (2021)
20. Rezaeijo, S.M., Ghorvei, M., Alaei, M.: A machine learning method based on lesion segmentation for quantitative analysis of CT radiomics to detect covid-19. In: 2020 6th Iranian Conference on Signal Processing and Intelligent Systems (ICSPIS), 23 Dec 2020, pp. 1–5. IEEE (2020)

Head and Neck Primary Tumor Segmentation Using Deep Neural Networks and Adaptive Ensembling

Gowtham Krishnan Murugesan[ID], Eric Brunner, Diana McCrumb[ID],
Jithendra Kumar[ID], Jeff VanOss[✉][ID], Stephen Moore, Anderson Peck,
and Anthony Chang

BAMF Health, Grand Rapids, MI 49503, USA
jeff.vanoss@bamfhealth.com

Abstract. The ability to accurately diagnose and analyze head and neck (H&N) tumors in head and neck cancer (HNC) is critical in the administration of patient specific radiation therapy treatment and predicting patient survivability outcome using radiomics. An automated segmentation method for H&N tumors would greatly assist in optimizing personalized patient treatment plans and allow for accurate feature extraction, via radiomics or other means, to predict patient prognosis. In this work, a three-dimensional UNET network was trained to segment H&N primary tumors using a framework based on nnUNET. Multimodal positron emission tomography (PET) and computed tomography (CT) data from 224 subjects were used for model training. Survival forest models were applied to patient clinical data features in conjunction with features extracted from the segmentation maps to predict risk scores for time to progression events for every patient. The selected segmentation methods demonstrated excellent performance with an average DSC score of 0.78 and 95% Hausdorff distance of 3.14. The random forest model achieved a C-index of 0.66 for predicting the Progression Free Survival (PFS) endpoint.

Keywords: Head and neck tumor segmentation · Survival prediction · Ensemble · nnUNET · Random survival forest

1 Introduction

Head and Neck cancer (HNC) is the sixth most common cancer worldwide, with 890,000 new cases and 450,000 deaths in 2018 [5,8]. The incidence of HNCs continues to rise and is anticipated to increase by 30% (1.08 million new cases annually) by 2030 [20]. The ability to accurately diagnose and analyze Head and Neck (H&N) tumors in HNCs is a critical component in the development of personalized patient treatment plans. The primary treatment option for HNCs, radiotherapy, requires the careful delineation of the primary gross tumor volume (GTVt) from fluorodeoxyglucose (FDG)-Positron Emission Tomography

© Springer Nature Switzerland AG 2022
V. Andrearczyk et al. (Eds.): HECKTOR 2021, LNCS 13209, pp. 224–235, 2022.
https://doi.org/10.1007/978-3-030-98253-9_21

(PET)/Computed Tomography (CT) scans during diagnosis and treatment. Several studies, in the field of radiomics, have proposed exploiting these already existing PET/CT images to better identify patients with worse prognosis in a non-invasive fashion [4,6,19]. Radiomics has shown tremendous potential in optimizing patient care of H&N tumors by predicting disease characteristics using quantitative image biomarkers extracted from PET/CT [19]. Focusing on metabolic and morphological tissue properties respectively, PET/CT modalities include complementary and synergistic information for cancerous lesion segmentation as well as tumor characteristics that are relevant for predicting patient outcomes [3,17,19]. Both radiotherapy and radiomics rely on current tumor delineation methods that utilize manual annotations which are costly, tedious, and error-prone. An automated segmentation method for H&N tumors would greatly assist in optimizing personalized patient treatment plans via radiomics.

The first HEad and neCK TumOR (HECKTOR) challenge [2], held in 2020, challenged participants to automatically segment of H&N primary tumors in FDG-PET and CT The winning team obtained an average DSC of 0.7591, an average precision of 0.8332, and an average recall of 0.7400 by implementing a 3D-UNET architecture with residual layers and supplemented with Squeeze-and-Excitation Normalization [3,9]. Most of the best performing methods implemented in HECKTOR 2020 used resampled isotropic three-dimensional PET/CT data with UNET like architectures [3,10,17,19]. After the success of the first challenge a second challenge was issued for 2021 with the following tasks: (1) automatic segmentation of H&N primary tumors in FDG-PET and CT images, (2) prediction of patient outcomes, namely Progression Free Survival (PFS) from the FDG-PET/CT images and available clinical data, and (3) prediction of PFS (same as task 2) from the FDG-PET/CT images and available clinical data, except that the ground truth annotations of primary tumors will be made available as inputs to the challengers algorithms through dockers [3].

This paper describes the methods used to complete each task. For task (1) an ensemble of outputs from 10 folds, 5 folds using a regular UNET and five folds using a residual UNET, was proposed using a modified version of the 3D no-newUNET from the MONAI framework [7,10]. The nnUNET is considered a state-of-the-art baseline, among current deep learning methods, for medical segmentation tasks [10]. Residual connections effectively reduce gradient vanishing problems and accelerate the convergence of the model. However, residual connections don't guarantee good performance. Isensee et al. experimented with variations of nnUnet on a variety of medical imaging segmentation challenges [10]. In their testing of residual vs regular unet layers, on average models built without residual connections performed better. Howerver, in a large minority of the cases, nnUnet with residual connections had superior performance. This inisight lead us to ensemble models with residual and non-residual connections in order to achieve and output better than any individual input.

Deep model ensemble results in more robust segmentation by identifying overconfident estimates [14]. Though naive averaging of segmentation models demonstrated significant improvement in segmentation accuracy, it has its trade off as worse performing model may influence the results [16]. Hence, the outputs from the model trained on five-fold cross validation were combined using adaptive ensembling to eliminate false positives. Models were evaluated based on their mean DSC Similarity Coefficient (DSC) and median Hausdorff Distance at 95% (HD95). For tasks (2) and (3), risk scores for individual patient prediction of PFS were generated using the Fast Unified Random Forests for Survival, Regression, and Classification (RF-SRC) in a RPy2 conversion package [18]. Exploratory input features were drawn from the provided EHR data, CT and PET radiomics generated from the segmentation models, Principal Component Analysis (PCA) of the generated radiomics, and in-house metrics related to metastatic tumor volume and number of metastases in the head or body gleaned from the provided PET and CT scans. Predictions were ranked based on the concordance index (C-index) on the test dataset.

2   Materials and Methods

2.1   Dataset Description

Task 1. This task uses the MICCAI 2021 HECKTOR segmentation and outcome prediction challenge dataset which includes a total of 325 patient cases from 6 different centers. The training set comprised of 224 cases from all 5 centers while the test set comprised of 101 cases from only 2 of the centers, one of which is not included in the training set [3,17]. Each case comprised of PET and CT image scans. Only the training set had additional annotated ground truths for the GTVts. Boundary box coordinates at the original CT coordinates were given for all cases to mark the region where the evaluation scores would be computed.

Task 2. This task uses the same dataset used in task (1). Additional patient clinical data is available for all patients in the dataset. Only the training set had censoring and time-to-event (in days) between PET/CT scan and event as the PFS endpoint. Progression was defined based on the RECIST criteria where: either known lesions increased in size (change of T and/or N), or new lesions appeared (change of N and/or M). Disease-specific death was also considered a progression event for patients previously considered stable [3].

Task 3. This task uses the same dataset used in task (2). Additional ground truth contours of the GTVts in the test cases are made available for this task. However, the highest performing models described in this paper, for both tasks (2) and (3), made use of the segmentations generated in Task 1 instead.

2.2   Data Preprocessing

Task 1. The standard preprocessing steps included linearly resampling PET to CT to an isotropic spacing of 1mm and cropping to the bounding box region to reduce image size to 144 × 144 × 144. CT data is windowed to [−250 and 250] HU and then z-score normalized. The FDG-PET data was provided in SUV units based on injection activity and patient mass. However, some patients mass was unknown and estimated. This resulted in SUVs that were not well standardized across the sample population. To address this the PET scans were z-score normalized. The CT and PET arrays were concatenated to create a 2 channel 3D array as input.

Each input was put through data augmentation pipeline as shown in Table 1. Each augmentation corresponds directly to a transform in the MONAI framework.

Table 1. Data augmentation pipeline

Step	Description	Probability
Random Zoom	min_zoom=0.9 max_zoom=1.2	0.15
Random Affine	rotate_range=[30°, 30°, 30°] scale_range=[0.7, 1.4]	0.15
Random 3D Elastic Deformation	sigma_range=(5, 8) magnitude_range=(100, 200) scale_range=(0.15, 0.15, 0.15)	0.10
Random Gaussian Noise	mean=0.0 std=0.1	0.15
Random Gaussian Smooth	sigma_x=(0.5, 1.5) sigma_y=(0.5, 1.5) sigma_z=(0.5, 1.5)	0.10
Random Scale Intensity	factors=(0.7, 1.3)	0.15
Random Shift Intensity	offsets=0.10	0.50
Random Flip	spatial_axis=0	0.10
Random Flip	spatial_axis=1	0.10
Random Flip	spatial_axis=2	0.10
Random Rotate 90°	max_k=3	0.10

Task 2 & 3. Missing values for any model input features in the patient clinical data were imputed with the means of the training set's features values. Ordinal input features were label encoded while the single nominal feature, CenterID, was one-hot encoded retaining all columns. Radiomic features were extracted using the PyRadiomics library [1] and all numeric features were used in combination with the library's available filters.

2.3   Task 1. Modified NnUNET Model

Model Architecture. The nnUNET pipeline has achieved top tier performance in multiple medical imaging segmentation competitions. Analysis of the nnUNET pipeline and model architecture has shown that different variations sometimes perform better than the baseline nnUNET architecture [10]. From this, a standard nnUNET architecture (see Fig. 1) as well as a variant model using residual connections was proposed for training (see Fig. 2).

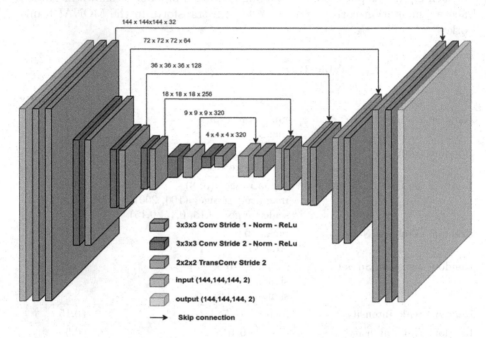

Fig. 1. The layers of the UNET architecture used. The input is a volume of 144 × 144 × 144 with two channels, CT and PT. Input is resampled down five times by convolution blocks with strides of 2. On the decoder side, skip connections are used to concatenate the corresponding encoder layers to preserve spatial information.

The input image size of the 3D nnUNET was 144 × 144 × 144 × 2. Instance normalization and leaky ReLU activation in the network layers was used. This architecture initially used 30 feature maps, which then doubled for each down sampling operation in the encoder (up to 320 feature maps) and then halved for each transposed convolution in the decoder. The end of the decoder has the same spatial size as the input, followed by a 1 × 1 × 1 convolution into 1 channel and a softmax function. Models are trained for five folds with loss function of dice sorensen coefficient (DSC) in combination with weighted cross entropy loss were trained. In order to prevent overfitting augmentation techniques such as random rotations, random scaling, random elastic deformations, gamma correction augmentation, mirroring and elastic deformation, were adopted (Table 2). Each of the five models were trained for 1000 epochs with batch size of eight using SGD optimizer and learning rate of 0.01.

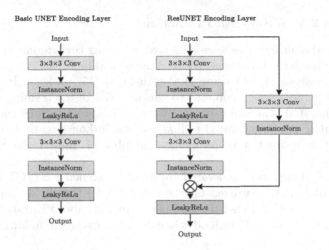

Fig. 2. In one instance of our UNET models, each encoding layer is a series of Convolution, normalization, and activation function repeated twice. In another instance, ResUNET, each encoding layer adds a residual path with convolution and normalization.

Selective Ensemble. For prediction, models used test time data augmentation of each of the 8 permutations of axial flips. These results were averaged to generate a softmax prediction. This prediction was linearly resampled to the original spacing of the CT scan. To get the final prediction a subset of the folds were combined (9 out of 10) and thresholding was used to create the binary prediction mask. The folds selected to ensemble for each sample were independent of other samples. To select the folds to ensemble for a sample, the predictions of all 10 folds were averaged together and a binary mask, P_{avg}, was created by thresholding at 0.1. Thresholding at 0.3 was used to create a mask, P_i, from the prediction from each ith fold. The surface DSC was calculated for each P_i in relation to P_{avg}. The folds were ordered on their surface DSC score and the softmax predictions of the top 9 were averaged together. Thresholding at 0.3 was applied to create the final prediction mask (Table 2).

Table 2. Selective ensembling algorithm

Steps	Selective ensemble for each sample
i	Calculate naive average mask from ten model ensemble, $$P_{(avg)} = \sum_{i=1}^{i=10} P_{(i)}$$
ii	Calculate surface dice between each model's prediction for given sample (P_i) where i = 1..10 and P_{avg}
iii	Remove the model's prediction with lowest surface dice and ensemble other nine models prediction

2.4   Tasks 2 & 3. Survival Prediction

Generated radiomic features were standardized using the training set's features' means and standard deviations before dimensionality reduction with principal component analysis (PCA) using the scikit-learn [15] package. PCA was performed to generate enough columns to capture 95% of the training set variance and each generated principal component was included when constructing a powerset of EHR, PCA of generated radiomics, and in-house scan features used to investigate the optimal combination and number of input features to use for model training.

In-house features were generated from the combined PET/CT scans based off of a calculated standardized uptake value (SUV) threshold specific to each scan. These features included the number of regions above threshold and total volume of these regions. Threshold was calculated using the following equation:

$$SUV_{\text{threshold}} = SUV_{\text{mean}} + 3SUV_{\text{standard deviation}}, \tag{1}$$

Repeated random sub-sampling validation over 100 multiple random splits were tested for each set of input features when searching for the optimal combination and number of independent features. Means and standard deviations of the C-index metric, and individual feature significances outputted using the Variable IMPortance (VIMP) method in the RF-SRC [11–13] package were used to evaluate the performance of every tested combination of input features. Parameters ranging from the survival forest's node depth, number of trees, and variable selection method to various feature stratification schemes and imputations methods were experimented with across the 100 random splits when testing how varying these parameters affected mean model performance across random splits. Table 3 lists the top 12 features that were determined to be significant from most significant to less significant.

Performance status, M-stage, tobacco, HPV, alcohol, TNM, age, and T-stage features were taken directly from the available patient clinical data. The number and volume of regions above threshold came from the in-house generated features. Principal Components 4 and 7 had the 4th and 7th largest retention of variance following PCA.

3   Results

3.1   Task 1 Segmentation Results

The final ensembled predictions on the test set resulted in a mean DSC of 0.78 and a HD95 metric of 3.14. The mean DSC score on the validation set for each fold is shown in Table 4 and the distribution of DSC scores is shown in Fig. 3).

Table 3. Features used to train our highest performing submitted model and their corresponding VIMP significance scores

Feature	Significance
Principal Component 4	0.045
Performance status	0.030
Number of regions in PET scan above threshold	0.021
M-stage	0.018
Volume of regions in PET scan above threshold	0.014
Tobacco	0.012
HPV	0.006
Principal Component 7	0.006
Alcohol	0.005
TNM	0.005
Age	0.003
T-stage	0.003

Table 4. Mean validation DSC score by fold and model architecture

	Fold 0	Fold 1	Fold 2	Fold 3	Fold 4	Average
UNET	0.7639	**0.7181**	0.7296	**0.7642**	0.7188	0.7389
ResUNET	**0.7737**	0.7056	**0.7603**	0.7636	**0.7211**	0.7449

To demonstrate the effect various portions of our method had on the final DSC score, we compared variants of our method against a model trained with the official 3D nnUNET pipeline. In Table 5 we show that the effect of interpolation type and positioning in relation to the ensemble operation. In Table 6 we show the effects of using 10 folds from different architectures vs the baseline nnUNET predictions. Table 7 shows the effect of choosing folds to ensemble globally, or per sample. Table 8 compares our methods result on test data from HECKTOR 2020 and 2021. We began our preperations for the MICCAI HECKTOR 2021 challenge before the data was released. We developed several algorithms and tested them on the MICCAI HECKTOR 2020 dataset. Through combination of algorithm variation, individual fold, and ensemble predictions, we tested a total of 278 times on the MICCAI HECKTOR 2020 test data.

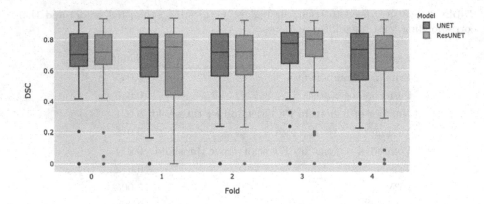

Fig. 3. DSC distribution per fold and model architectures

Table 5. The effect of resampling and ensembling

Pipeline order	HECKTOR 2020 DSC
(baseline) nnUNET 5 fold binary predictions → ensemble → nearest neighbor resample	0.753
nnUNET 5 fold softmax predictions → linear resample → ensemble	0.765

Table 6. The effect ensembling more folds from different architectures

Pipeline order	HECKTOR 2020 DSC
nnUNET 5 fold softmax predictions → linear resample → ensemble	0.765
(ours) UNET + ResUNET 10 fold softmax predictions → linear resample → ensemble	0.780

Table 7. The effect of choosing folds to ensemble globally

Pipeline order	HECKTOR 2020 DSC
(ours) UNET + ResUNET 10 fold softmax predictions → linear resample → ensemble	0.780
(ours) UNET + ResUNET 10 fold softmax predictions → linear resample → selective ensemble	0.782

Table 8. Our submission model performance on HECKTOR 2020 and 2021 test data

Pipeline order	HECKTOR 2020 DSC	HECKTOR 2021 DSC
(ours) UNET + ResUNET 10 fold selective ensemble	0.782	0.780

3.2 Tasks 2 & 3 Progression Prediction Results

Throughout the feature refinement process, mean validation C-index scores across multiple random splits typically ranged between 0.65–0.79. Survival forest scores on the competition's test set faired between 0.63–0.66. We submitted models trained on specific radiomic features and models trained on principal components from PCA analysis of all extracted, numeric radiomic features. The models with specific radiomic features received lower C-indexes on the test set than the models that used the principal component features suggesting that the generally higher C-index scores for models trained on specific features used in development were a result of overfitting.

4 Discussion

4.1 Task 1

In all folds, the UNET and ResUNET performed similarly. However, in three of the five folds, the ResUNET had the higher mean, indicating that neither architecture in inherently superior. All 10 folds had a combined mean DSC of 0.7420 on the validation data.

Table 4 shows that linearly resampling the softmax predictions before ensembling provided an increase in the DSC metric on the baseline model. We were also able to see the effects of ensembling UNET+ResUNET models against the baseline, in Table 5. Adding more folds trained on different architectures increased the accuracy. In Table 6, we show that the selective ensemble method provided a very small increase over a global approach. This was a surprisingly small increase. Our analysis of the training data showed that selective ensemble tended to remove false positives that global ensemble did not, as shown in Fig. 4.

We show in Table 7 that our model's performance in this years and last years test data is very similar. This close clustering indicates a well generalized model. With all these variations we can see that the overwhelming majority of increased performance over baseline comes from ensembling 10 folds from mixed architectures, and resampling softmax values linearly.

4.2 Task 2 & 3

Generally, the C-index scores for the test set fell approximately 0.1 from the scores on the validation portion of the training set. Our highest performing model used features derived from the Task 1 predicted tumors, not the provided

Fig. 4. Axial view of CHUM033. Regular ensemble gives multiple contiguous label (blue). Selective ensemble (red) managed to filter the false positive and also improve the surface boundary to the ground truth (green) (Color figure online)

ground truth tumor mask. Ultimately we permutated over ten of thousands of combinations of features, imputations, stratifications and model parameters and arrived at a strong pool of parameters based on a combination of VIMP significance scores, associated C-index metrics and proportional feature ratios in comparison to the holdout set. Interestingly, our best performing model for Task (3) was based on radiomics generated from our segmentation model's predictions and not the ground truth provided in the task and we are investigating whether there is any significance to this, but currently surmise it was simply random.

5   Conclusion

We have trained regular and residual 3D nnUNET and achieved robust segmentation performance for tumor segmentation (Task 1). We generated risk scores for individual patients using RF-SRC and achieved second rank in the MICCAI 2021 HECKTOR Segmentation and outcome prediction challenge Task 3.

References

1. AIM-Harvard: pyradiomics. https://github.com/AIM-Harvard/pyradiomics
2. Andrearczyk, V., et al.: Overview of the HECKTOR challenge at MICCAI 2020: automatic head and neck tumor segmentation in PET/CT. In: Andrearczyk, V., Oreiller, V., Depeursinge, A. (eds.) HECKTOR 2020. LNCS, vol. 12603, pp. 1–21. Springer, Cham (2021)
3. Andrearczyk, V., et al.: Overview of the HECKTOR challenge at MICCAI 2021: automatic head and neck tumor segmentation and outcome prediction in PET/CT images. In: Andrearczyk, V., Oreiller, V., Hatt, M., Depeursinge, A. (eds.) HECKTOR 2021. LNCS, vol. 13209, pp. 1–37. Springer, Cham (2022)

4. Bogowicz, M., et al.: Comparison of PET and CT radiomics for prediction of local tumor control in head and neck squamous cell carcinoma. Acta Oncologica (Stockholm, Sweden) **56**(11), 1531–1536 (2017)

5. Bray, F., Ferlay, J., Soerjomataram, I., Siegel, R.L., Torre, L.A., Jemal, A.: Global cancer statistics 2018: GLOBOCAN estimates of incidence and mortality worldwide for 36 cancers in 185 countries. CA Cancer J. Clin. **68**(6), 394–424 (2018)

6. Castelli, J., et al.: A pet-based nomogram for oropharyngeal cancers. Eur. J. Cancer (Oxford, England: 1990) **75**, 222–230 (2017)

7. MONAI Consortium: MONAI: Medical Open Network for AI, March 2020. https://github.com/Project-MONAI/MONAI

8. Ferlay, J., et al.: Estimating the global cancer incidence and mortality in 2018: globocan sources and methods. Int. J. Cancer **144**(8), 1941–1953 (2019)

9. Iantsen, A., Visvikis, D., Hatt, M.: Squeeze-and-excitation normalization for automated delineation of head and neck primary tumors in combined PET and CT images. In: Andrearczyk, V., Oreiller, V., Depeursinge, A. (eds.) HECKTOR 2020. LNCS, vol. 12603, pp. 37–43. Springer, Cham (2021). https://doi.org/10.1007/978-3-030-67194-5_4

10. Isensee, F., Jaeger, P.F., Kohl, S.A.A., Petersen, J., Maier-Hein, K.H.: nnU-Net: a self-configuring method for deep learning-based biomedical image segmentation. Nat. Methods **18**(2), 203–211 (2021)

11. Ishwaran, H., Kogalur, U.: Random survival forests for R. R News **7**(2), 25–31 (2007). https://CRAN.R-project.org/doc/Rnews/

12. Ishwaran, H., Kogalur, U.: Fast unified random forests for survival, regression, and classification (RF-SRC) (2021). https://cran.r-project.org/package=randomForestSRC, R package version 2.12.1

13. Ishwaran, H., Kogalur, U., Blackstone, E., Lauer, M.: Random survival forests. Ann. Appl. Statist. **2**(3), 841–860 (2008). https://arXiv.org/abs/0811.1645v1

14. Lan, R., Zou, H., Pang, C., Zhong, Y., Liu, Z., Luo, X.: Image denoising via deep residual convolutional neural networks. SIViP **15**(1), 1–8 (2019). https://doi.org/10.1007/s11760-019-01537-x

15. Scikit learn: scikit learn, September 2021. https://github.com/scikit-learn/scikit-learn

16. Murugesan, G.K., et al.: Multidimensional and multiresolution ensemble networks for brain tumor segmentation. In: Crimi, A., Bakas, S. (eds.) BrainLes 2019. LNCS, vol. 11993, pp. 148–157. Springer, Cham (2020). https://doi.org/10.1007/978-3-030-46643-5_14

17. Oreiller, V., et al.: Head and neck tumor segmentation in PET/CT: the HECKTOR challenge. Med. Image Anal. **77**, 102336 (2021)

18. rpy2: Python R bridge, September 2021. https://github.com/rpy2/rpy2

19. Vallières, M., et al.: Radiomics strategies for risk assessment of tumour failure in head-and-neck cancer. Sci. Rep. **7**(1), 10117 (2017)

20. Wang, X., Li, B.: Deep learning in head and neck tumor multiomics diagnosis and analysis: review of the literature. Front. Genet. **12**, 42 (2021)

Segmentation and Risk Score Prediction of Head and Neck Cancers in PET/CT Volumes with 3D U-Net and Cox Proportional Hazard Neural Networks

Fereshteh Yousefirizi[1]($\boxtimes$) (iD), Ian Janzen[1], Natalia Dubljevic[2], Yueh-En Liu[3], Chloe Hill[4], Calum MacAulay[1,2,5], and Arman Rahmim[1,2,6] (iD)

[1] Department of Integrative Oncology, BC Cancer Research Institute, Vancouver, Canada
frizi@bccrc.ca
[2] Department of Physics and Astronomy, University of British Columbia, Vancouver, Canada
[3] Combined Major in Science, University of British Columbia, Vancouver, BC, Canada
[4] Faculty of Applied Sciences, Simon Fraser University, Burnaby, Canada
[5] Department of Pathology, University of British Columbia, Vancouver, Canada
[6] Department of Radiology, University of British Columbia, Vancouver, Canada

Abstract. We utilized a 3D nnU-Net model with residual layers supplemented by squeeze and excitation (SE) normalization for tumor segmentation from PET/CT images provided by the Head and Neck Tumor segmentation challenge (HECKTOR). Our proposed loss function incorporates the Unified Focal and Mumford-Shah losses to take the advantage of distribution, region, and boundary-based loss functions. The results of leave-one-out-center-cross-validation performed on different centers showed a segmentation performance of 0.82 average Dice score (DSC) and 3.16 median Hausdorff Distance (HD), and our results on the test set achieved 0.77 DSC and 3.01 HD. Following lesion segmentation, we proposed training a case-control proportional hazard Cox model with an MLP neural net backbone to predict the hazard risk score for each discrete lesion. This hazard risk prediction model (CoxCC) was to be trained on a number of PET/CT radiomic features extracted from the segmented lesions, patient and lesion demographics, and encoder features provided from the penultimate layer of a multi-input 2D PET/CT convolutional neural network tasked with predicting time-to-event for each lesion. A 10-fold cross-validated CoxCC model resulted in a c-index validation score of 0.89, and a c-index score of 0.61 on the HECKTOR challenge test dataset.

Keywords: Head and neck cancer · PET-CT · Segmentation · Unified focal loss · Cox regression

1 Introduction

Head and neck (H&N) cancer is the fifth most common cancer diagnosed worldwide and the eighth most common cause of cancer death [1]. While standard therapies combining radiation and chemotherapy are highly effective, they rely heavily on manually generated

© Springer Nature Switzerland AG 2022
V. Andrearczyk et al. (Eds.): HECKTOR 2021, LNCS 13209, pp. 236–247, 2022.
https://doi.org/10.1007/978-3-030-98253-9_22

contours of tumor volumes from medical images. Segmentation is also a crucial bottleneck towards radiomics analysis and prognostication pipelines. This is a labor intensive and time-consuming task that unavoidably suffers from intra- and inter-observer biases [2, 3]. AI approaches to tumor segmentation continue to grow in popularity, and have demonstrated potential for identification of head and neck tumors in PET and CT image tasks. Previous network approaches range from a simple U-net model, to 3D V-nets and 3D deep networks.

Radiomics are hand-engineered statistical features that are predominantly calculated from masked segmentations acquired from medical imaging domains. Moreover, they refer to the use of image analysis to quantify image descriptors, called radiomic features, that are often considered imperceptible to the human eye. Radiomic features typically calculated from PET and CT images are often described as shape, textural, and intensity features. Studies have shown that radiomics help clinicians bridge the gap between quantitative and qualitative image analysis to help them understand the biological processes underlying the image phenomena [4].

Clinically relevant features extracted from PET and CT images are powerful tools for classifying pathological behaviors. Many studies have shown the importance of relevant radiomics features in assessing images and assigning appropriate treatment pathways [2, 4, 18]. Identifying relevant radiomic features requires a robust and algorithmic feature selection process. These methods are required to remove potential evaluation bias while identifying discriminating features that are useful for clinicians to help them quantize lesion characteristics.

The data provided for this study is well suited for a survival analysis and time-to-event prediction task. We explored the use of a number of discrete and continuous time Cox regression models to predict either hazard risk scores or time-to-event predictions for individual head and neck cancers segmented lesions. These models relied on a Multilayer Perception (MLP) neural network backbone, as an extension of the Cox regression model [5], to predict this task. Further explanation of the features that these models were trained on will be elucidated on in the proposed methods.

In the current study, we propose a 3D network for bi-modal PET/CT segmentation based on a 3D nnU-net model with squeeze and excitation (SE) modules and a hybrid loss function (distribution, region and boundary based). Using this network, we constructed an automated radiomic extraction pipeline to generate potentially useful features for a risk prediction score provided by a Cox proportional hazard model.

In the following sections, we first introduce the data provided by the MICCAI 2020 HEad and neCK TumOR segmentation and outcome prediction (HECKTOR) challenge [6, 7]. Our proposed methods and training scheme are additionally explained. The results are then presented, followed by discussion and conclusion.

2 Material and Methods

2.1 Dataset

2.1.1 Description

Operating under the confines of the MICCAI 2021 HECKTOR Data Challenge, we were provided with 224 discrete PET/CT and lesion mask volumes that contained head and neck cancers collected from 5 different sites to train and validate deep learning models. An external cohort of PET/CT and lesion mask volumes (N = 101), collected from two different sites were also provided for these models. This external cohort assessed the performance of each task and gauged its generalizability. For the segmentation task, training and validation splits were executed as a leave-one-out-center-cross-validation to assess model performance on the generalizability on out-of-sample data and random splitting was done. For the risk score prediction task, training and validation splits were done with pseudo-random training and validation splits (90:10 - training: validation) with 10-fold cross-validated. Prior to pseudo-random splitting, we manually selected an "external" validation set of 21 volumes, with an approximate distribution of disease progression and progression free survival of days (15:6 - non event:event). This was done to better assess model ranking ability on out-of-sample data.

2.1.2 Preprocessing

Based on the directives and bounding boxes provided by HECKTOR challenge organizers, the PET and CT images were resampled (by trilinear interpolation), and then cropped, to the resolution of $1 \times 1 \times 1$ mm^3. We clipped the intensities of CT images in the range of $[-1024, 1024]$ Hounsfield Units and then resampled to within a $[-1, 1]$ range. PET images were normalized by Z-score normalization. For the segmentation task, the volumes were resampled back to their original resolutions before verifying their statistical performance. In addition to mirroring (on the axial plane) and rotation (in random directions) for data augmentation, we utilized scaling (with a random factor between 0.8 and 1.2) and elastic deformations to increase the diversity in tumor size and shape.

2.2 Proposed Methods

2.2.1 Segmentation

As the backbone network for medical image segmentation, 3D U-Net [8] and 3D nnU-net [9] have gained much attention among convolutional neural networks (CNNs) due to their good performances. However, the upsampling process involves the recovery of spatial information, which is difficult without taking the global information into consideration [10]. The squeeze & excitation (SE) modules are defined to 'squeeze' along the spatial domain and 'excite' along the channels. SE modules help the model to highlight the meaningful features and suppress the weak ones. CNNs with SE modules frequently achieve top performance, across various challenges (ILSVRC 2017 image classification [10] and Head and Neck Tumor segmentation challenge (HECKTOR 2020) [11]). We utilized the 3D nnU-Net with SE modules after encoder and decoder blocks as the recommended architecture by Roy et al. [12].

Using different loss functions has been shown to affect the performance, robustness and convergence of the segmentation network [13]. Distribution based losses (e.g. cross entropy, Focal loss [14]), region based losses (e.g. Dice), boundary based loss (e.g. Mumford-Shah [15]) or any of their combinations make up the main approaches for loss functions in medical image segmentation tasks. Hybrid loss functions have shown better performance [13, 16] e.g. the sum of cross entropy and Dice similarity coefficient (DSC) proposed by Taghanaki et al. [17] or the Unified Focal loss introduced by Yeung et al. [13] that combined Focal and Dice (the minus of DSC) loss. We used a 3D nnU-Net with SE modules (Fig. 1) and utilized a hybrid loss function that is a combination of the distribution based, region based and boundary based loss functions. The model was trained for 400 epochs using Adam optimizer on two NVIDIA Tesla V100 GPUs 16 GB with a batch size of 2. We used the cosine-annealing schedule to reduce the learning rate from 10^{-3} to 10^{-6} within every 25 epochs.

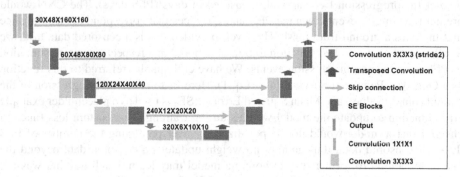

Fig. 1. 3D nnU-Net model for segmentation with SE modules

2.2.2 Radiomic Extraction and Selection

Radiomic features have been shown to improve patient outcomes and assist physicians by quantifying phenotypical behaviors in images as basic statistical inference. We propose utilizing the Pyradiomics Python package [18] to extract radiomic features from both the PET and CT volumes, using provided lesion masks (either ground truth, or from the segmentation model). We extracted approximately 3000 radiomic features from both modalities using this methodology.

To isolate the most discriminating features, we utilized the scikit-learn python package's feature selection algorithms [19]. We performed a robust search grid through the available classification-criterion and feature selection algorithms as an exhaustive method to identify the most discriminating features. Through scikit-learn, we identified four feature algorithms: Sequential Forward Feature Selection (SFS; forwards and backwards), Extra Trees Classifier (ETC), Recursive Feature Elimination (RFE). In addition, we exposed our results to the use of the Maximum Relevance - Minimum Redundancy (MRMR) feature selection algorithm [20], to verify discrete feature discriminating viability. For each feature selection method, we use one of the following criterion models: Linear Discriminant Analysis (LDA), K-Nearest Neighbors (KNN), Bagging methods, Gradient Boosting (GradBoost), eXtreme Gradient Boosting (XGBoost), Support Vector Machines (SVM), Classification and Regression Trees (CART), Quadratic Discriminant Analysis (QDA), and Linear Regression (LR). Each criterion model was provided with the approx. 3000 radiomic features and the ground truth as the binarized indicated progression status of each patient. From there, we searched and logged the top 1, 5, 10, 25,

50, and 100 discriminating features using each combination of feature selection methods and criterion models.

2.2.3 Encoder Feature Extraction

Inspired by the segmentation aspect of the HECKTOR 2021 challenge, we decided that encoded features ought to be implemented in the time-to-event prediction task. We used a multi-input model 2D convolution neural network (CNN) that would be trained to predict the progression free survival, measured in days (PFS days). The CNN would predict this time-to-event given the PET and CT slices that correspond to an axial slice that includes a ground-truth mask. However, provided this is a censored data task, we implemented a custom loss function to train the model that respects an over-estimation of PFS days for non-progression events. We have colloquially referred to this function as a One Way Penalized Survival loss (Eq. 1). Here Pr refers to the progression of the patient (binary value), and Mean Squared Error (MSE) is used as a placeholder example error function to update the model weights during training. This custom loss function ensured that a model would not be penalized for over predicting the number of PFS days for censored data, thus gaining no weight updates on censored data beyond the last observation. To further restrict how the model may learn, batch training was not implemented. This simplistic CNN was trained for 5 epochs, with an Adam optimizer and a learning rate of 0.001. This model achieved a MSE of 483 days between the PFS days and predicted PFS days for the validation set.

$$loss\left(PFS_{pred}, Pr, PFS_{GT}\right) = \begin{cases} 0 \text{ if } Pr = 0 \text{ and } PFS_{pred} > PFS_{GT} \\ 1 \text{ else} \end{cases} \tag{1}$$

As we were provided PET/CT volumes, we extracted a dataset using the same pre-processing methods outlined in section "description". The penultimate layer of 5X fold validation multi-input 2D CNN model trained under this regime was isolated and the features of that layer, averaged per the number of slices where an axial mask was present in the respective volume, were then to be accessed for training our Cox proportional hazard model.

2.2.4 Outcome Prediction

The intention of this experiment is to build on the nested case-control studies conducted by Langholz et al. [21] We extend their work by utilizing the PyTorch framework [22] to construct a case-control Cox proportional hazard model that is able to predict a proportional hazard risk score. We will perform this task by combining the three sources of features listed in section Radiomic Extraction and Selection and Outcome Prediction alongside one-hot-encoded patient demographic features.

These three sources of features are used as the inputs for a case-control Cox Regression model (CoxCC) with a MLP neural network backbone [5] using the pycox python package. This case-control proportional hazard Cox regression model predicts the hazard risk score per patient using a linear combination of patient demographics, selected radiomic features, and encoder features generated from the 2D CNN model. A visualization of the entire feature extraction-to-CoxCC model training block diagram can be seen in Fig. 2. The overall CoxCC model is trained using the loss function described by Kvamme et al. [5] for case-control models, an Adam optimizer, learning rate of 0.0024, a batch size of 32, and an early stopping method that monitored the validation loss (tolerance of 3 epochs) to combat overtraining by saving the optimized model weights.

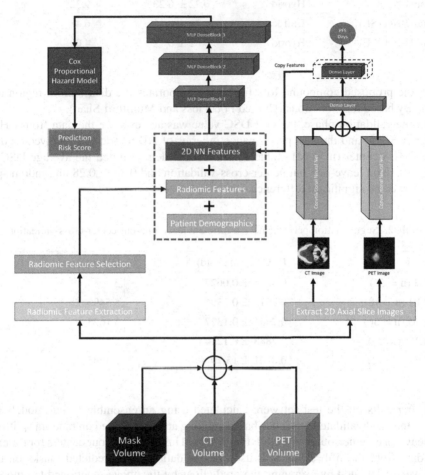

Fig. 2. Overall MICCAI 2021 HECKTOR challenge task 2 data extraction and model training pipeline

3   Results

3.1   Segmentation Results

As Table 1 presents our 3D nnU-Net model with SE modules outperformed 3D nnU-Net using the same loss function for training. The best segmentation performance observed

Table 1. Segmentation performance in different models/losses

Method	Loss	DSC (mean)	HD (median)
3D nnU-Net	Unifocal	0.68 ± 0.19	11.07
3D nnU-Net	Hybrid	0.72 ± 0.23	9.32
3D nnU-Net (SE)	Unifocal	0.79 ± 0.16	6.4
3D nnU-Net (SE)	Hybrid	0.82 ± 0.12	4.09

using the proposed compound loss function incorporates the distribution, region and boundary based loss functions (Unified Focal loss and Mumford-Shah).

In the validation phase, the best DSC value was achieved on the data from CHGJ center (n = 55) and the best performance in terms of HD metric was received on data from CHUS center (n = 72) (Table 2). The model demonstrated the average DSC of 0.82 ± 0.12 on leave-one-out-center-cross-validation and 0.84 ± 0.28 on random split that showed no significant difference.

Table 2. Segmentation performance on different leave-one-out-center-cross-validation

Centers	DSC (mean ± std)	HD (median)
CGHJ (n = 55)	0.8655 ± 0.0627	3
CHMR (n = 18)	0.7919 ± 0.1396	5.5495
CHUM (n = 56)	0.7951 ± 0.1327	3.1623
CHUP (n = 23)	0.7888 ± 0.1214	5
CHUS (n = 72)	0.8101 ± 15.79	2.4495

Our results on the test set were calculated using an ensemble of ten models i.e. five trained and validated on center-based splitting and five trained on random splitting. For leave-one-center-out cross-validation, we used images from four centers for training and data from the fifth center was used for validation. The predicted masks on test data were calculated by averaging the predictions by the above-mentioned ten models (threshold value = 0.5).

3.2 Survival Analysis

We used the concordance index (c-index) to assess survival model performance on this censored dataset. We performed 10-fold cross-validated on these models to verify model efficacy. Thus, when assessing c-index scores on our "external" validation set and the test set, we took the median prediction from these 10 models for each segmented lesion to generate a single prediction risk score from our CoxCC model method.

A grid search involving all criterion models and feature selection methods identified 192 unique combinations of radiomic feature to train a CoxCC model with. We also tested the prediction capability of the CoxCC model against a regular Cox Proportional Hazard (CoxPH) model using the same radiomic features. We identified the boost to predicted risk score capability with the addition of the features described in section "Radiomic Extraction and Selection" to help elucidate final model selection choice. The results of model selection tabulated in Table 3 and an example plot comparing each model against other feature selection methodologies in Fig. 3 and Fig. 4.

Table 3. Identifying the top performing models for Task 2. Acronyms for feature selection method can be found in section "Radiomic Extraction and Selection"

Model type	CoxPH	CoxCC	CoxPH	CoxCC	CoxPH	CoxCC	CoxPH	CoxCC	CoxPH	CoxCC
# of Radiomic features	5	5	10	10	25	25	50	50	100	100
Feature selection method	LDA + ETC	QDA + ETC	LDA + SFS F	SVM + ETC	KNN + ETC	GradBoost + ETC	LR + ETC	GradBoost + SFS F	LR + ETC	GradBoost + SFS F
Training set C-index	0.811	0.883	0.743	0.870	0.774	0.867	0.769	0.891	0.758	0.884
Validation set C-index (10X fold)	0.744	0.871	0.638	0.891	0.646	0.847	0.649	0.886	0.612	0.883
Ext validation set C-index (median)	0.813	0.854	0.865	0.888	0.843	0.888	0.876	**0.910**	0.753	**0.910**
Test set C-index (mean)	–	–	–	–	–	0.576	–	0.604	–	0.612

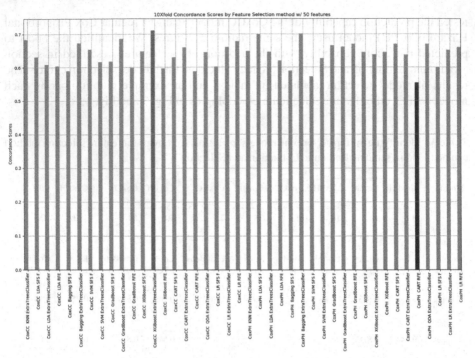

Fig. 3. Close-case Cox (CoxCC) regression method against the Cox Proportional Hazard model (CoxPH) with the validation set c-index score when the features from the custom 2D Neural Network are not introduced during training. All models trained with 50 radiomic features, selected by the routines documented in the x-axis titles and 5 patient demographic features. Compared to Fig. 4, one can see a boost to outcome prediction performance with the addition of the features described in Sect. 2.2.3. [Color: Highest c-index score marked in red, lowest c-index score marked in green]. (Color figure online)

4   Discussion

Our segmentation results show that utilizing SE modules improves the performance of 3D nnU-net model for segmentation. On the other side, a compound loss function composed of distribution, region and boundary based losses also improves the segmentation performance of a 3D nnUnet model with SE modules.

An exhaustive search for the most discriminating radiomic features for predicting hazard risk scores of head and neck cancer patients proved to not have much of a marked effect on these static model architectures. Generally, the features selected were a healthy mix of CT- and PET-based radiomic features as well. Observing Fig. 3, one can conclude that there is little discernible difference in model performance between the CoxCC and CoxPH methods despite additional radiomic features to train with. An average validation set c-index score of 0.62 for the myriad of CoxCC and CoxPH models trained on solely radiomic features and patient demographics does not make for an ideal predictor. This observation is neatly contrasted by the performance of the CoxCC model vs CoxPH model when the 2D NN encoded features are able to be trained with. In Fig. 4

and Table 3, there is a clear boost, approx. +0.09 c-index, to prediction ability on the validation set with the addition of these encoded features during training. Moreover, we see excellent model performance on the external validation set from the CoxCC models with more than 25 radiomic features, seen in Table 3. Particularly noteworthy are the CoxCC models' ability to rank patient risk scores on data that is in the training sets time-to-event distribution.

However, the ranking ability for these models is severely impacted by predicting risk scores for lesions that are out-of-sample data distribution. In this case, prediction on the test set resulted in a drop of approximately 0.30 c-index points from the external validation set to the test set. It is a fair assumption to suppose that these models are over-trained on the 2D NN features from the 5 sites that the training data was collected from. These models showcase a poor ability to generalize currently. Thus, they would require more rigorous regularization techniques, a more exhaustive neural network architecture search, and the introduction of more data samples with which to train.

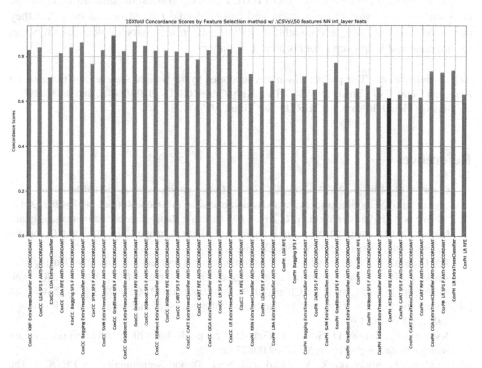

Fig. 4. Showcasing the boost to c-index performance with the close-case Cox (CoxCC) regression method against the Cox Proportional Hazard model (CoxPH) and the validation set c-index score. All models trained with 50 radiomic features, selected by the routines documented in the x-axis tick titles, 5 patient demographic features, and 128 2D Neural Network activation layer features extracted from the penultimate layer. [Color: highest c-index score marked in red, lowest c-index score marked in green]. (Color figure online)

5   Conclusion

The segmentation and survival prediction for head and neck cancers are both difficult tasks. The MICCAI 2021 HECKTOR Challenge provided the groundwork to test our methodologies on the execution of these tasks. SE modules help to improve the performance of nnU-Net models for PET and PET/CT image segmentation. Hybrid loss functions that take into consideration the distribution, region and boundary improve the segmentation performance. The other facet of the HECKTOR challenge, providing an accurate patient risk score, proved to be just as troublesome. Radiomic features were extracted from the PET and CT volumes, and then algorithmically selected. The radiomics were combined with patient demographics and encoded features of a 2D NN trained to predict patient survival in days. This combination of features were trained on by a case-control Cox regression model to estimate an overall patient risk score. While this model achieved impressive results on in-distribution data, it failed to generalize well on out-of-distribution data from other PET/CT collection sites. In summation, head and neck cancers are difficult to segment and can confound patient risk models. Yet they prove to be an intriguing and insightful challenge for researchers, clinical or otherwise, to diagnose and help keep patients alive for longer.

Acknowledgement. This project was in part supported by the Natural Sciences and Engineering Research Council of Canada (NSERC) Discovery Grant RGPIN-2019-06467, and the Canadian Institutes of Health Research (CIHR) Project Grant PJT-173231.

References

1. O'rorke, M., Ellison, M., Murray, L., et al.: Human papillomavirus related head and neck cancer survival: a systematic review and meta-analysis. Oral Oncol. **48**(12), 1191–1201 (2012)
2. Starmans, M.P., van der Voort, S.R., Tovar, J.M.C., et al.: Radiomics: data mining using quantitative medical image features. In: Handbook of Medical Image Computing and Computer Assisted Intervention, pp. 429–456. Elsevier (2020).
3. Jin, D., et al.: Accurate esophageal gross tumor volume segmentation in PET/CT using two-stream chained 3D deep network fusion. In: Shen, D., et al. (eds.) MICCAI 2019. LNCS, vol. 11765, pp. 182–191. Springer, Cham (2019). https://doi.org/10.1007/978-3-030-32245-8_21
4. Tomaszewski, M.R., Gillies, R.J.: The biological meaning of radiomic features. Radiology **298**, 202553 (2021)
5. Kvamme, H., Borgan, Ø., Scheel, I.: Time-to-event prediction with neural networks and Cox regression. arXiv preprint arXiv:1907.00825 (2019)
6. Oreiller, V., Andrearczyk, V.: Head and Neck Tumor Segmentation in PET/CT: The HECKTOR Challenge. Medical Image Analysis (2021). Under revision
7. Andrearczyk, V., et al.: Overview of the HECKTOR challenge at MICCAI 2021: automatic head and neck tumor segmentation and outcome prediction in PET/CT images. In: Andrearczyk, V., Oreiller, V., Hatt, M., Depeursinge, A. (eds.) HECKTOR 2021. LNCS, vol. 13209, pp. 1–37. Springer, Cham (2022)
8. Çiçek, Ö., Abdulkadir, A., Lienkamp, S.S., Brox, T., Ronneberger, O.: 3D U-Net: learning dense volumetric segmentation from sparse annotation. In: Ourselin, S., Joskowicz, L., Sabuncu, M.R., Unal, G., Wells, W. (eds.) MICCAI 2016. LNCS, vol. 9901, pp. 424–432. Springer, Cham (2016). https://doi.org/10.1007/978-3-319-46723-8_49

9. Isensee, F., Petersen, J., Klein, A., et al.: NNU-net: self-adapting framework for U-net-based medical image segmentation. arXiv preprint arXiv:1809.10486 (2018)
10. Hu, J., Shen, L., Sun, G.: Squeeze-and-excitation networks. In: Proceedings of the IEEE Conference on Computer Vision and Pattern Recognition (2018)
11. Iantsen, A., Visvikis, D., Hatt, M.: Squeeze-and-excitation normalization for automated delineation of head and neck primary tumors in combined PET and CT images. In: Andrearczyk, V., Oreiller, V., Depeursinge, A. (eds.) HECKTOR 2020. LNCS, vol. 12603, pp. 37–43. Springer, Cham (2021). https://doi.org/10.1007/978-3-030-67194-5_4
12. Roy, A.G., Navab, N., Wachinger, C.: Recalibrating fully convolutional networks with spatial and channel "squeeze and excitation" blocks. IEEE Trans. Med. Imaging $38(2)$, 540–549 (2018)
13. Yeung, M., Sala, E., Schönlieb, C.-B., et al.: Unified Focal loss: Generalising Dice and cross entropy-based losses to handle class imbalanced medical image segmentation. arXiv preprint arXiv:2102.04525 (2021)
14. Lin, T.-Y., Goyal, P., Girshick, R., et al.: Focal loss for dense object detection. In: Proceedings of the IEEE International Conference on Computer Vision (2017)
15. Kim, B., Ye, J.C.: Mumford-Shah loss functional for image segmentation with deep learning. IEEE Trans. Image Process. 29, 1856–1866 (2019)
16. Zhu, W., Huang, Y., Zeng, L., et al.: AnatomyNet: deep learning for fast and fully automated whole-volume segmentation of head and neck anatomy. Med. Phys. $46(2)$, 576–589 (2019)
17. Taghanaki, S.A., Zheng, Y., Zhou, S.K., et al.: Combo loss: handling input and output imbalance in multi-organ segmentation. Comput. Med. Imaging Graph. 75, 24–33 (2019)
18. Van Griethuysen, J.J., Fedorov, A., Parmar, C., et al.: Computational radiomics system to decode the radiographic phenotype. Can. Res. $77(21)$, e104–e107 (2017)
19. Pedregosa, F., Varoquaux, G., Gramfort, A., et al.: Scikit-learn: machine learning in Python. J. Mach. Learn. Res. 12, 2825–2830 (2011)
20. Peng, H., Long, F., Ding, C.: Feature selection based on mutual information criteria of max-dependency, max-relevance, and min-redundancy. IEEE Trans. Pattern Anal. Mach. Intell. $27(8)$, 1226–1238 (2005)
21. Langholz, B., Goldstein, L.: Risk set sampling in epidemiologic cohort studies. Statist. Sci. 11, 35–53 (1996)
22. Paszke, A., Gross, S., Massa, F., et al.: Pytorch: an imperative style, high-performance deep learning library. Adv. Neural. Inf. Process. Syst. 32, 8026–8037 (2019)

Dual-Path Connected CNN for Tumor Segmentation of Combined PET-CT Images and Application to Survival Risk Prediction

Jiyeon Lee[ID], Jimin Kang[ID], Emily Yunha Shin[ID], Regina E. Y. Kim[ID], and Minho Lee[✉][ID]

Research Institute, NEUROPHET Inc., Seoul 06247, Korea
minho.lee@neurophet.com

Abstract. Automated segmentation methods for image segmentation have the potential to support diagnosis and prognosis using medical images in clinical practice. To achieve the goal of HEad and neCK tumOR (HECKTOR) segmentation and outcome survival prediction in PET/CT images in the MICCAI 2021 challenge, we proposed a novel framework to segment head and neck tumors by leveraging multi-modal imaging using a cross-attention module based on a dual-path and ensemble modeling with majority voting. In addition, we expanded our task to survival analysis using a random forest survival model to predict the prognosis of tumors using clinical information and segmented tumor volume. Our segmentation model achieved a Dice coefficient and Hausdorff distance of 0.7367 and 3.2700, respectively. Our survival model showed a concordance index (C-index) of 0.6459.

Keywords: Deep learning · PET/CT imaging · Tumor segmentation · Survival analysis

1   Introduction

Head and neck tumor accounts for about 4% of all cancers in the United States and is one of the common types of cancer worldwide [20,21]. The long-term survival rate of head and neck tumor has continuously increased and is estimated at around 60% as of 2021 [8,13]. Advancing technology for targeted treatment of head and neck tumor could further improve the prognosis and quality of life in patients. Radiotherapy is one of the standard treatments for this cancer [25], which employs multi-modal imaging of Fluorodeoxyglucose (FDG)-positron emission tomography (PET) and computed tomography (CT) [23]. These imaging modalities provide valuable information for staging and re-staging of the tumors. The exact identification of the cancer properties from those images is thus crucial for the cancer treatment and its planning. While experts, e.g., radiologists, can extract clinical information from these multi-modal images, it is

---

J. Lee, J. Kang and E. Y. Shin—Equal contribution.

© Springer Nature Switzerland AG 2022
V. Andrearczyk et al. (Eds.): HECKTOR 2021, LNCS 13209, pp. 248–256, 2022.
https://doi.org/10.1007/978-3-030-98253-9_23

labor-intensive and could suffer from rater's bias. To address these challenges, several automatic segmentation methods for tumors have been proposed.

In this paper, we describe approaches for both tasks from the HEad and neCK tumOR (HECKTOR) challenge: *task 1)* head and neck tumor segmentation and *task 2)* the prediction of patient outcomes. The segmentation and outcome prediction in PET/CT images in the MICCAI 2021 challenge are to validate the automatic segmentation algorithm [1,14]

Compared to the HECKTOR 2020 challenge, the task is expanded to survival analysis based on segmentation results in 2021, which is also described in this work. To achieve the aim of *task 1*, we propose a novel framework for tumor segmentation, named dual-path-based cross-attention module. The proposed method is designed to take advantage of multi-modal CT and PET images, providing compensatory information to each other for biologically identical tissue. Our network starts with a dual-path encoder, followed by a cross-attention module to maximize the feature information extracted from those multi-modal images. It is hypothesized that our combined design of the dual-path encoder and the cross-attention module could overcome the known issue of dual-path encoders [2]. By introducing the cross-attention module to the dual-path encoder, it is expected that features extracted from the two images will pair well for further training. We refer to this framework as dual-path connected convolution neural network (CNN). For *task 2*, we conducted survival analysis to predict the prognosis based on the size of the tumor from our segmentation approach. We validated our proposed methods using the Dice score, Hausdorff distance (HD), and concordance index (C-index) as evaluation metrics.

2   Materials and Methods

2.1   Dataset

We used the multi-site dataset of the HECKTOR 2021 challenge collected from 5 centers in Canada, Switzerland, and France. All PET images were registered with paired CT images. The ground truth of the primary gross tumor volume (GTVt) was manually annotated in the corresponding CT image resolution. To match the image in the same Euclidean space, we resampled all the images to 1mm spacing with isotropic voxel and cropped the images to the size of 144×144×144 following the ITK convention using the bounding box of tumor regions provided from the challenge. The range of Hounsfield units for CT images is limited from -1024 to 1024 to only focus on the brain structure following the paper [9]. Also, all images were normalized to the same range of −1 to 1. Lastly, contrast stretching is applied to the PET image to emphasize the intensity of the tumor.

2.2   Tumor Segmentation

We proposed a novel framework based on the U-shaped structure [18], including an encoder and decoder for tumor segmentation using the crossed information of the multi-modal images. First, the two images, i.e., CT and PET, were fed into the dual-path encoder with a cross-attention module. Then, a decoder was performed using the cross-attention-based feature map to reconstruct the original space for tumor segmentation. Subsequently, the segmentation network resulted as a binary mask image, i.e., prediction results. Our framework is shown in Fig. 1

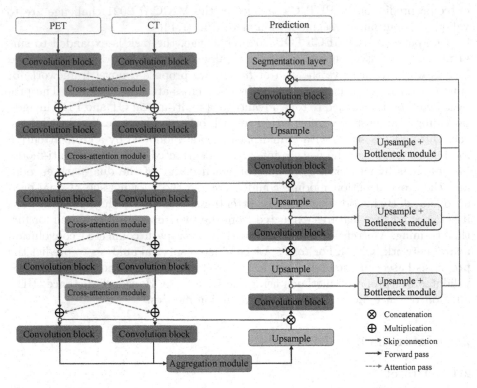

Fig. 1. Overview of our proposed method. The convolution block denotes a module comprised of a convolution kernel, batch normalization, and activation function.

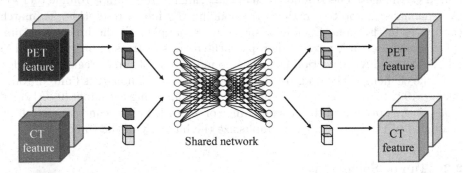

Fig. 2. Our proposed cross-attention module. Each feature is fed into one shared network. The transformed features are multiplied with original feature.

Encoder. The main building block in our encoder is a weight-shared network and cross-attention module. To extract features from the CT and PET images, we designed a dual-path encoder using the Siamese network [3]. For the cross-attention module inspired by [24], the module was constructed based on the channel attention network in Fig. 2. However, unlike the original module, which uses the two representations extracted from one encoder's feature, we used two features extracted from each modality. By doing so, our network performed to share information of the feature corresponding to each modality using a shared network. First, we extracted the image features from each encoder network. Second, we performed a global average pooling using each path's feature. Then, the features were fed into a shared fully connected (FC) layer. Finally, the two features were added and activated using the sigmoid function. The last features from the FC layer were multiplied to input features.

We built the encoder using 4 convolution blocks with cross-attention module. Specifically, the convolution blocks consisted of a $3 \times 3 \times 3$ convolution kernel, batch normalization (BN) [10], and ReLU activation function. As the last stage of the encoder, each feature was merged using an aggregation layer consisting of a $1 \times 1 \times 1$ convolution kernel.

Decoder. In the decoding part, a decoder was performed to upsample the output feature of the encoder corresponding to the original dimension, i.e., a dense prediction map. First, using the upsampling layer, the feature map is twice enlarged. Next, we added a skip connection between the encoder and decoder to obtain fine-grained features for the segmentation. Then, the feature map from the encoder was concatenated with the upsampled features. Finally, the concatenated features were passed through the convolution block. For the decoder network, we used four convolution blocks using a $3 \times 3 \times 3$ kernel, BN, and ReLU activation function with an upsampling layer using a trilinear interpolation.

Ensemble Model. In the decoder, we enabled it to obtain multi-scaled features from each convolution block. We merged the multi-scaled features in the ensemble-manner to improve the segmentation quality. The features were upsampled and then the filter channel size was decreased using the bottleneck module [7]. Finally, the features from the bottleneck network in the decoder were concatenated and fed into the segmentation layer consisting of 2 layers of $1 \times 1 \times 1$ convolution kernels. After the segmentation layer, the prediction map was applied to the softmax activation function.

2.3 Survival Analysis

Clinical information and volumes extracted from each segmented tumor label were used to train a random survival forest model to estimate the survival risk prediction [11]. Age, gender, and other related status information (e.g., tobacco and alcohol usage, performance, and human papillomavirus (HPV) status) were used as demographic features. To correct for bias that differs for each center, Center ID, which indicates centers by number, was included in the clinical information. We also use Tumor-Nodes-Metastasis (TNM) staging and edition features to deliver the specific staging of tumors [5]. Missing values of tobacco, alcohol, and performance status were coded as 0, and missing values for HPV status were coded as -1. No other variables had missing values.

In addition, we selected the volumes of tumors calculated from segmented label maps as imaging features. In the training step, we estimated volumes from the ground truth labels; in the testing step, we estimated volumes from the automated segmented labels from our segmentation model.

To encourage a good fit to our model and prevent outliers from having a strong effect, we removed outliers using Cook's distance [4]. One significantly influential data point was removed. In the evaluation step, the C-index was calculated for this survival model evaluation [6].

2.4   Experimental Settings

To train our proposed network, the learning rate was set to 1e−5, and Adam [12] was used as the optimizer. Cross entropy and generalized Dice loss functions were used [22]. Data augmentation settings were random elastic deformation, random anisotropy, random noise, and random blur. We applied the random noise and random blur methods with a selection probability of 0.7 for data augmentation. The filter size of feature map F was set to (32, 64, 128, 256) in the encoder, and the inverse sorted filter size of the feature map was used in the decoder. To build the cross-attention module, we used two FC layers with F-$(F/2)$-F units. The model and all the data augmentation strategies were implemented using Pytorch [15] and Torchio [16], respectively. We also conducted leave-one-out cross validation to verify our proposed method with evaluation metrics using Dice coefficient and HD.

Survival prediction models and metrics were imported from the scikit-survival library [17]. Outlier detection analysis was conducted using the statsmodels library [19]. In the survival prediction step, we performed 5-fold randomized cross-validation with 100 iterations to evaluate the model with metrics using the C-index.

3   Results

Table 1. Performance of our proposed method with leave-one-out cross validation. Average denotes averaged results of all sites.

	CHGJ	CHUP	CHUM	CHUS	CHMR	Average
Dice score	0.7486	0.7309	0.6117	0.6691	0.6434	0.6808
HD	9.0409	20.0702	13.0182	11.0944	15.5925	13.7633

3.1   Task 1 - Segmentation

The summary of results is shown in Table 1 with leave-one-out cross validation. The CHGJ site had the best Dice score and HD whereas the CHUP site had the second best Dice score but poorest HD. We evaluate our proposed method using a test dataset from the HECKTOR 2021 challenge. The test dataset ($n = 101$) was collected from two-sites, CHUM and CHUV. To leverage the whole information of our proposed framework, the all-trained model using all sites was utilized to obtain the results in an ensemble-manner. We summed up images from each model and conducted a majority voting. Because our outcome is a binary mask about tumor regions, we applied a threshold to our results as ≥ 3. From the ensemble results, we obtained a Dice score and HD of 0.7367 and 3.2700, respectively.

3.2 Task 2 - Survival Prediction

Only one subject (CHMR014) was removed in the outlier removing stage. As a result of cross validation with ground truth labels, the averaged C-index was 0.6508 ± 0.0802. When we applied this model to separated evaluation data provided by the HECKTOR 2021 challenge, the C-index was recorded as 0.6495. This performance was higher than the model without tumor volumes (C-index = 0.6422).

4   Discussion

Head and neck tumor accounts for about 4% of all cancers in the United States and is one of the common types of cancer worldwide. The segmentation of head and neck tumors and predicting a long-term survival rate could improve patients' quality of life. To this end, previous research has been proposed to segment paired CT and PET images and to make prognoses using imaging factors. In our work, we proposed the dual-path connected CNN framework to leverage multi-modal image properties and our experiment showed an ensembled-segmentation result with a reasonable Dice score of 0.7367. Moreover, we demonstrated that combining clinical information with imaging factors, i.e., tumor volume, performs better than without imaging factors.

In task 1, the test dataset shows a better HD and Dice score than the validation results. In particular, the HD of the test dataset (3.2700) was much better than the validation site's HD (13.7633). Based on the results, we infer that combining majority voting and ensemble method have a positive impact on evaluation.

Our model demonstrates the importance of tumor volume as an imaging factor. In addition to task 2 that the outperformed model using segmented tumor volumes, we also compared two survival analysis models with and without outcomes from our segmentation model on the training data. In this additional experiment to represent the effectiveness of the segmentation model, the survival model with tumor volumes showed a significantly higher C-index than without tumor volumes (two dependent sample t-test, $p = 0.013$, 5-fold cross-validation 100 iterations). In this regard, using volumes of segmented tumor regions seems to have a clinical impact.

When dealing with missing values in survival analysis, we coded HPV status as -1 because the variable does not have a structural relationship between other variables and shows different results in a Kaplan-Meier estimator (Fig. 3). Meanwhile, because other missing values were closely related to the TMN edition (the tobacco, alcohol, and performance status are missing only if the TMN edition is 7), they are coded as 0 even if they show different curves in Kaplan-Meier estimators.

In this study, the survival analysis model used only tumor volume as an imaging factor instead of other possible multi-modal variables. Although this design confirms the effectiveness of our segmentation model, it is necessary to consider other imaging factors to predict a more precise prognosis.

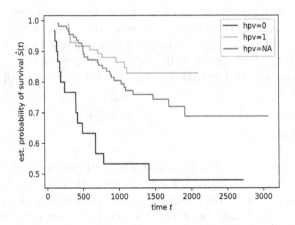

Fig. 3. The Kaplan-Meier estimator of human papillomavirus status. When the value is missing, the probability of survival is greatly reduced. NA, not available.

5   Conclusion

Combining PET/CT images has been widely used in treatment planning and tumor diagnosis. The HECKTOR 2021 challenge is held to validate the effect of treatment and automatic segmentation algorithm. To address the goal of validation, we proposed a novel framework to head and neck tumor segmentation using cross-attention network and ensemble modeling. In addition, we developed a survival risk prediction algorithm using clinical information and image features of tumors. In the experiment, we demonstrated the validity of our proposed method using various evaluation metrics.

Acknowledgements. This research was supported by the Korea Health Technology R&D Project through the Korea Health Industry Development Institute (KHIDI) and Korea Dementia Research Center (KDRC), funded by the Ministry of Health & Welfare and Ministry of Science and ICT, Republic of Korea (grant number: HU20C0315).

References

1. Andrearczyk, V., et al.: Overview of the HECKTOR challenge at MICCAI 2021: automatic head and neck tumor segmentation and outcome prediction in PET/CT images. In: Andrearczyk, V., Oreiller, V., Hatt, M., Depeursinge, A. (eds.) HECK-TOR 2021. LNCS, vol. 13209, pp. 1–37. Springer, Cham (2022)
2. Baltrušaitis, T., Ahuja, C., Morency, L.P.: Multimodal machine learning: a survey and taxonomy. IEEE Trans. Pattern Anal. Mach. Intell. **41**(2), 423–443 (2018)
3. Chopra, S., Hadsell, R., LeCun, Y.: Learning a similarity metric discriminatively, with application to face verification. In: 2005 IEEE Computer Society Conference on Computer Vision and Pattern Recognition (CVPR 2005), vol. 1, pp. 539–546. IEEE (2005)
4. Cook, R.D.: Detection of Influential Observation in Linear Regression. Technometrics **19**(1), 15–18 (1977)
5. Edge, S.B., Byrd, D.R., Carducci, M.A., Compton, C.C., Fritz, A., Greene, F., et al.: AJCC Cancer Staging Manual, vol. 7. Springer, New York (2010)
6. Harrell, F.E., Jr., Lee, K.L., Mark, D.B.: Multivariable prognostic models: Issues in developing models, evaluating assumptions and adequacy, and measuring and reducing errors. Stat. Med. **15**(4), 361–387 (1996)
7. He, K., Zhang, X., Ren, S., Sun, J.: Deep residual learning for image recognition. In: Proceedings of the IEEE Conference on Computer Vision and Pattern Recognition, pp. 770–778 (2016)
8. Howlader, N., et al.: SEER Cancer Statistics Review, 1975–2018. National Cancer Institute (2021)
9. Iantsen, A., Visvikis, D., Hatt, M.: Squeeze-and-excitation normalization for automated delineation of head and neck primary tumors in combined PET and CT images. In: Andrearczyk, V., Oreiller, V., Depeursinge, A. (eds.) HECKTOR 2020. LNCS, vol. 12603, pp. 37–43. Springer, Cham (2021). https://doi.org/10.1007/978-3-030-67194-5_4
10. Ioffe, S., Szegedy, C.: Batch normalization: accelerating deep network training by reducing internal covariate shift. In: International Conference on Machine Learning, pp. 448–456. PMLR (2015)
11. Ishwaran, H., Kogalur, U.B., Blackstone, E.H., Lauer, M.S.: Random survival forests. Ann. Appl. Stat. **2**(3), 841–860 (2008)
12. Kingma, D.P., Ba, J.: Adam: A Method for Stochastic Optimization. arXiv preprint arXiv:1412.6980 (2014)
13. Massa, S.T., Osazuwa-Peters, N., Boakye, E.A., Walker, R.J., Ward, G.M.: Comparison of the Financial Burden of Survivors of Head and Neck Cancer with Other Cancer Survivors. JAMA Otolaryngol. Head Neck Surg. **145**(3), 239–249 (2019)
14. Oreiller, V., et al.: Head and neck tumor segmentation in PET/CT: the HECKTOR challenge. Med. Image Anal., 102336 (2021)
15. Paszke, A., et al.: Pytorch: An Imperative Style, High-Performance Deep Learning Library. Adv. Neural. Inf. Process. Syst. **32**, 8026–8037 (2019)
16. Pérez-García, F., Sparks, R., Ourselin, S.: TorchIO: a Python Library for efficient loading, preprocessing, augmentation and patch-based sampling of medical images in deep learning. Computer Methods and Programs in Biomedicine, p. 106236 (2021)
17. Pölsterl, S.: scikit-survival: a library for time-to-event analysis built on top of scikit-learn. J. Mach. Learn. Res. **21**(212), 1–6 (2020)

18. Ronneberger, O., Fischer, P., Brox, T.: U-Net: convolutional networks for biomedical image segmentation. In: Navab, N., Hornegger, J., Wells, W.M., Frangi, A.F. (eds.) MICCAI 2015. LNCS, vol. 9351, pp. 234–241. Springer, Cham (2015). https://doi.org/10.1007/978-3-319-24574-4_28

19. Seabold, S., Perktold, J.: Statsmodels: econometric and statistical modeling with python. In: Proceedings of the 9th Python in Science Conference, vol. 57, p. 61. Austin, TX (2010)

20. Siegel, R.L., Miller, K.D., Fuchs, H.E., Jemal, A.: Cancer statistics. CA: Cancer J. Clin. **71**(1), 7–33 (2021). https://doi.org/10.3322/caac.21654

21. Siegel, R.L., Miller, K.D., Jemal, A.: Cancer statistics. CA: A Cancer J. Clinicians **66**(1), 7–30 (2016)

22. Sudre, C.H., Li, W., Vercauteren, T., Ourselin, S., Jorge Cardoso, M.: Generalised dice overlap as a deep learning loss function for highly unbalanced segmentations. In: Cardoso, M.J., et al. (eds.) DLMIA/ML-CDS -2017. LNCS, vol. 10553, pp. 240–248. Springer, Cham (2017). https://doi.org/10.1007/978-3-319-67558-9_28

23. Vallieres, M., et al.: Radiomics strategies for risk assessment of tumour failure in head-and-neck cancer. Sci. Rep. **7**(1), 1–14 (2017)

24. Woo, S., Park, J., Lee, J.Y., Kweon, I.S.: Cbam: convolutional block attention module. In: Proceedings of the European Conference on Computer Vision (ECCV), pp. 3–19 (2018)

25. Yeh, S.A.: Radiotherapy for head and neck cancer. In: Seminars in Plastic Surgery. vol. 24, pp. 127–136. Thieme Medical Publishers (2010)

Deep Supervoxel Segmentation for Survival Analysis in Head and Neck Cancer Patients

Ángel Víctor Juanco-Müller[1,2](✉) [iD], João F. C. Mota[2] [iD], Keith Goatman[1] [iD], and Corné Hoogendoorn[1] [iD]

[1] Canon Medical Research Europe Ltd., Edinburgh, UK
victor.juancomuller@mre.medical.canon
[2] Heriot-Watt University, Edinburgh, UK

Abstract. Risk assessment techniques, in particular Survival Analysis, are crucial to provide personalised treatment to Head and Neck (H&N) cancer patients. These techniques usually rely on accurate segmentation of the Gross Tumour Volume (GTV) region in Computed Tomography (CT) and Positron Emission Tomography (PET) images. This is a challenging task due to the low contrast in CT and lack of anatomical information in PET. Recent approaches based on Convolutional Neural Networks (CNNs) have demonstrated automatic 3D segmentation of the GTV, albeit with high memory footprints (≥ 10 GB/epoch). In this work, we propose an efficient solution (~ 3 GB/epoch) for the segmentation task in the HECKTOR 2021 challenge. We achieve this by combining the Simple Linear Iterative Clustering (SLIC) algorithm with Graph Convolution Networks to segment the GTV, resulting in a Dice score of 0.63 on the challenge test set. Furthermore, we demonstrate how shape descriptors of the resulting segmentations are relevant covariates in the Weibull Accelerated Failure Time model, which results in a Concordance Index of 0.59 for task 2 in the HECKTOR 2021 challenge.

1 Introduction

The first two years after therapy of Head and Neck (H&N) cancer are critical, as up to 40% of recurrences occur during that period [1]. Although radiological studies may estimate individual patient outcomes [2,3], and thus help to deliver personalised treatment, they require accurate delineation of the Gross Tumour Volume (GTV), in both Computed Tomography (CT) and Positron Emission Tomography (PET) images. Obtaining such a delineation is a difficult task due to the low contrast of CT and lack of anatomical detail of PET.

Consequently, automatic segmentation algorithms have great potential to scale up prognosis studies to larger populations, thus increasing their statistical power. In 2020, the first edition of the HEad and neCK TumOR Segmentation (HECKTOR) challenge [4] attracted a large number of participants. The edition in 2021 of the challenge [5,6] includes patients from a new centre, and two new tasks related to regression of survival risk scores.

© Springer Nature Switzerland AG 2022
V. Andrearczyk et al. (Eds.): HECKTOR 2021, LNCS 13209, pp. 257–265, 2022.
https://doi.org/10.1007/978-3-030-98253-9_24

The best performing algorithms in the competition are based on Convolutional Neural Networks (CNNs) such as UNet [7], but they have a high-memory footprint, requiring about 10 GB per epoch during training. In this paper, we efficiently approach GTV segmentation as a supervoxel classification task. Our method requires three times less GPU memory, around 3 GB per epoch.

We then use shape descriptors of the produced segmentations, together with patient clinical data, to fit a Weibull Accelerated Failure Time (Weibull AFT) model [8]. We show that the shape features used to fit this model are as relevant as the presence of metastasis for outcome prediction.

2   Methods

This section describes first the Deep Learning approach to segmentation, and then the Weibull AFT model used for risk score regression.

2.1   Deep Supervoxel Segmentation

Here we explain the segmentation pipeline, starting with the supervoxel generation and graph extraction, then the neural network design, and finally the loss function.

Supervoxel and Graph Formation. The input data are Regions of Interest (RoIs), with physical extension of $(144, 144, 144)$ mm, extracted from two registered PET and CT scans of a H&N cancer patient (we give more details on these RoIs in Sect. 3.2). Since the PET modality carries most information about tumour presence [2,3], we process it with the Simple Linear Iterative Clustering (SLIC) algorithm [9] to find supervoxels.

We then build a Region Adjacency Graph (RAG), whose nodes represent supervoxels, and two nodes are connected if the corresponding supervoxels share a wall. Next, a neural network is trained to classify the nodes as tumour or background. Finally, we merge the supervoxels classified as tumour to retrieve the output binary segmentation.

To train the network, we compute target labels for each node. Whenever the ratio of tumour voxels in a given supervoxel is larger than a threshold τ, we assign that supervoxel the label TUMOUR. Note that even if all supervoxels are correctly classified, the resulting mask still differs from the reference mask. This disagreement is quantified by the *achievable accuracy*. In our experiments we found that setting $\tau = 0.7$ resulted in the highest achievable accuracy in a subset of the training data (we explain how divided the data in Sect. 3.1).

Model Architecture. The proposed network, depicted in Fig. 1, has a Multi Layer Perceptron (MLP) encoder and a Graph Convolution Network (GCN) [10] decoder, interconnected by skip connections. The MLP and GCN blocks used throughout the network, depicted in Fig. 2, have residual connections [11], ReLU activations and Batch Normalization [12].

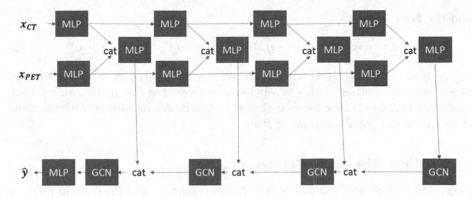

Fig. 1. Supervoxel classification network. x_{CT} and x_{PET} are the input node signals, $\hat{y}$ output probabilities and *cat* refers to concatenation.

The input node features are the voxel values within the cuboid centred at the corresponding supervoxel centroid in the CT and PET RoIs. The encoder network fuses the PET and CT information at four different levels. The decoder has four consecutive residual GCN blocks [10], which aggregate global information. An MLP with final sigmoid activation produces the output probabilities.

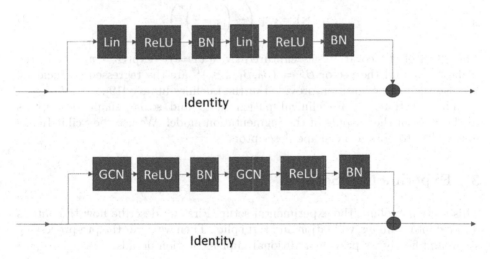

Fig. 2. MLP (top) and GCN (bottom) residual blocks. BN stands for Batch Normalization [12]. If the number of input and output channels differs, the identity is upsampled or downsampled accordingly.

Loss Function. Following [13], we tackle the imbalance of the dataset by minimizing a combination of the Dice Loss [14]

$$\text{Dice} = 1 - \frac{y \cdot \hat{y} + 1}{y \cdot y + \hat{y} \cdot \hat{y} + 1} \tag{1}$$

and the Focal Loss [15]

$$\text{Focal} = \text{mean}\left((1 - \hat{\boldsymbol{y}})^{\gamma} \cdot \boldsymbol{y} \cdot \log(\hat{\boldsymbol{y}}) + \hat{\boldsymbol{y}}^{\gamma} \cdot (1 - \boldsymbol{y}) \cdot \log(1 - \hat{\boldsymbol{y}})\right), \qquad (2)$$

where $\boldsymbol{y}$ and $\hat{\boldsymbol{y}}$ respectively are binary vectors of supervoxel target labels and predicted probabilities, and γ is a hyperparameter that, as in [15], we set to 2. The main benefit of these losses is that they handle the imbalance without using statistics of the population under study.

2.2 Patient Risk Score Regression

According to RECIST guidelines [16], progression can be defined as an increase in tumour diameter greater than 20%. The time elapsed between the scan of a patient and a tumour progression (or another event like new metastases or death related to the tumour) is called *Progression Free Survival* (PFS) [16].

Because a number of patients withdraw the study before progression is recorded, the resulting dataset is said to be *right censored*. This fact limits the applicability of standard regression techniques and requires special methods that deal with the uniqueness of survival data. Here, we fit a Weibull AFT model [8], which defines the survival probability at time t given the covariates $\boldsymbol{x}$ as:

$$S(t; \mathbf{x}) = \exp\left(-\left(\frac{t}{\lambda(\boldsymbol{x})}\right)^{\rho}\right). \qquad (3)$$

The effect of the covariates is captured by $\lambda(\boldsymbol{x}) = \alpha \cdot \exp(\boldsymbol{\beta}^T \cdot \boldsymbol{x})$, where the scalars α, ρ and the vector $\boldsymbol{\beta}^T = (\beta_0, \beta_1, ...\beta_n)^T$ are the regressed coefficients. We use the implementation provided in the Lifelines library [19].

The covariates $\boldsymbol{x}$ are clinical patient data and scalar shape descriptors retrieved from the outputs of the segmentation model. We use the Scikit-Image library [20] to obtain the shape descriptors.

3 Experimental Setup

This section explains the experimental setup. First we describe how the data is divided into training, validation and test splits. Then we cover the pre-processing stage and finally we provide additional implementation details.

3.1 Data Splitting

The top ranked team in the past edition showed that there were no significant distribution shifts among centres [13]. After checking that this condition still applies to the increased dataset, we formed a random split with a 200 training and 24 validation cases, ensuring that each set contained patients from every centre. We accomplished this by first forming sub-splits for each centre and then concatenating them.

3.2   Preprocessing

The challenge organisers provided a list of RoIs, of physical size $(144, 144, 144)$ mm, with voxel level annotations. We use these regions to validate our model. During training, we randomly cropped RoIs of the same physical extension, ensuring they contained tumour voxels.

To augment the dataset, before extracting the RoIs, we applied random rotations across the coordinate axes in the range $[5, 15]°$. To keep the complexity of SLIC and neural networks tractable, we resampled the input RoIs to isotropic volumes of size $(72, 72, 72)$. Then, to process the PET RoI, the SLIC parameters were set to compactness of 1 and supervoxel size of $(8, 8, 8)$ (for a detailed description of these parameters, we refer the reader to [9]).

Finally, following [13], we scaled both the CT and PET data to the $[-1, 1]$ interval for better training stability. In the CT case, we clipped the intensity values to the range $[-1024, 1024]$ Hounsfield Units before rescaling.

3.3   Training

During training, we saved a checkpoint whenever the validation metric, which was the Dice score of the segmentation output, increased. The training was stopped when this metric did not improve for 50 epochs. We used the ADAM [17] optimizer where the learning rate, initially set to 10^{-3}, was modulated with the Cosine Annealing with Warm Re-starts [18] policy.

3.4   Inference

We used the validation split introduced in Sect. 3.1 to evaluate both the segmentation and PFS score prediction. When testing the segmentation model, we used the union of the outputs of the four last checkpoints, which corresponded to the best four measured validation metrics.

4   Results

Here we present the results, first for the segmentation model, and then for the risk score prediction task.

4.1   Segmentation Results

We evaluated the final voxel level output with the Dice score (DSC) and Hausdorff distance in mm at 95th percentile (HD95). Table 1 provides the results in terms of these metrics on the validation and test sets. Following the challenge organisers evaluation criteria, we compute the sample mean of the DSC and median of the HD95 to average over samples.

Figure 3 shows the histogram of Dice score in the validation set and visualizations of the predicted and ground truth segmentations over a merged CT-PET image for the worst, average and best cases in the validation set.

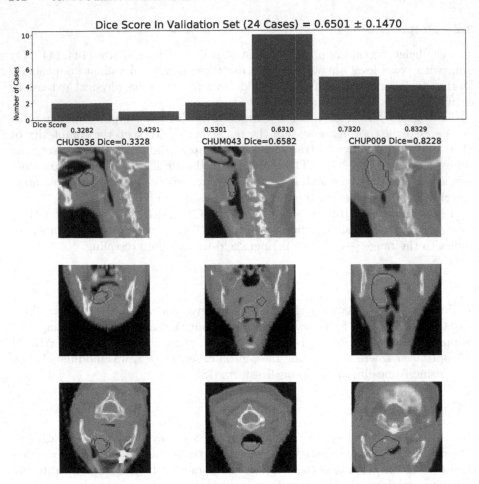

Fig. 3. Top row: histogram of Dice score after running inference on the 24 patients of the validation dataset. Bottom rows: superimposed CT and PET planes with predicted (red) and ground truth (green) segmentations of the worst (left), average (middle) and best (right) cases on the superimposed PET (blue stains) and CT images. (Color figure online)

Table 1. Results of the proposed segmentation method in the validation and test sets.

Set	Dice (↑)	HD95 (mm) (↓)
Validation	0.6545	4.078
Test	0.6331	6.126

4.2 Patient Risk Score Results

The obtained risk scores were evaluated with the Concordance Index, a metric that generalises the Area Under the Curve (AUC) for censored data. A value of 0.5 corresponds to random predictions. Table 2 provides the results for the PFS prediction task on the validation and test sets. Table 3 shows the value of the Weibull AFT regressed coefficients, along with their p-values, for each of the covariate variables.

Table 2. Results of the Progression Free Survival task.

Set	C-Index (↑)
Validation	0.5942
Test	0.5937

Table 3. Summary of covariates in the Weibull AFT model fit. The ones with lowest p-value, in bold, are patient metastasis status (M-Stage) and two descriptors of tumour shape (Euler number and Surface Area).

Covariate	Coefficient (β)	p-value (↓)
Alcohol	−0.24	0.36
Chemotherapy	0.54	0.27
HPV status	0.58	0.04
M-stage	−1.42	≪0.005
N-stage	0.09	0.34
TNM edition	−0.72	0.03
TNM group	−0.18	0.24
Tobacco	0.15	0.61
Euler number	−0.45	≪0.005
Extent	6.23	0.03
Filled area	0.77	0.05
Solidity	−2.32	0.22
Surface Area	−1.35	≪0.005

5   Discussion and Conclusion

We have proposed an innovative segmentation algorithm that yields accurate results and has a smaller memory footprint, of about 3 GB per epoch, compared to CNN methods like U-Net [7], which require 10 GB or higher per epoch.

Furthermore, we have shown that the produced segmentations provide meaningful features for survival regression algorithms. We have demonstrated this fact by fitting a Weibull AFT model and looking at the p-value of coefficients regressed for each covariate, where the surface area and Euler number of the segmented tumour were as relevant as the metastasis status of the patients.

Future work will explore how to improve the results, for example, by applying additional data augmentation following the physics and workflow of the imaging process. We will also explore what type of data poses particularly difficult challenges to our approach.

A challenge of the proposed method is the computation of the SLIC mask, which can be time consuming for larger RoIS. We will examine how to make this stage more efficient. We will also extend the methods to other anatomies and modalities, focusing on how to adapt the SLIC parameters automatically.

References

1. Chajon, E., et al.: Salivary gland-sparing other than parotid-sparing in definitive head-and-neck intensity-modulated radiotherapy does not seem to jeopardize local control. Radiat. Oncol. **8**(1), 1–9 (2013)
2. Castelli, J., et al.: A PET-based nomogram for oropharyngeal cancers. Eur. J. Cancer **75**, 222–230 (2017)
3. Bogowicz, M., et al.: Comparison of PET and CT radiomics for prediction of local tumor control in head and neck squamous cell carcinoma. Acta Oncol. **56**(11), 1531–1536 (2017)
4. Andrearczyk, V., et al.: Overview of the HECKTOR challenge at MICCAI 2020: automatic head and neck tumor segmentation in PET/CT. In: Andrearczyk, V., Oreiller, V., Depeursinge, A. (eds.) HECKTOR 2020. LNCS, vol. 12603, pp. 1–21. Springer, Cham (2021). https://doi.org/10.1007/978-3-030-67194-5_1
5. Andrearczyk, V., et al.: Overview of the HECKTOR challenge at MICCAI 2021: automatic head and neck tumor segmentation and outcome prediction in PET/CT images. In: Andrearczyk, V., Oreiller, V., Hatt, M., Depeursinge, A. (eds.) HECKTOR 2021. LNCS, vol. 13209, pp. 1–37. Springer, Cham (2022)
6. Oreiller, V., et al.: Head and neck tumor segmentation in PET/CT: the HECKTOR challenge. Med. Image Anal. **77**, 102336 (2021)
7. Ronneberger, O., et al.: U-Net: convolutional networks for biomedical image segmentation. In: International Conference on Medical Image Computing and Computer-Assisted Intervention, pp. 234–241 (2015)
8. Kalbfleisch, J.D., Prentice, R.L.: The Statistical Analysis of Failure Time Data. Wiley, New York (1980)
9. Achanta, R., et al.: SLIC superpixels compared to state-of-the-art superpixel methods. IEEE Trans. Pattern Anal. Mach. Intell. **34**(11), 2274–2282 (2012)
10. Kipf, T.N., Welling, M.: Semi-supervised classification with graph convolutional networks. In: 5th International Conference on Learning Representations (2017)
11. He, K., et al.: Deep residual learning for image recognition. In: 29th IEEE Conference on Computer Vision and Pattern Recognition, pp. 770–778 (2016)
12. Ioffe, S., et al.: Batch normalization: accelerating deep network training by reducing internal covariate shift. In: International Conference on Machine Learning, pp. 448–456 (2015)
13. Iantsen, A., Visvikis, D., Hatt, M.: Squeeze-and-excitation normalization for automated delineation of head and neck primary tumors in combined PET and CT images. In: Andrearczyk, V., Oreiller, V., Depeursinge, A. (eds.) HECKTOR 2020. LNCS, vol. 12603, pp. 37–43. Springer, Cham (2021). https://doi.org/10.1007/978-3-030-67194-5_4

14. Milletari, F., et al.: V-Net: fully convolutional neural networks for volumetric medical image segmentation. In: 4th International Conference on 3D Vision, pp. 565–571 (2016)
15. Lin, T., et al.: Focal loss for dense object detection. In: Proceedings of the IEEE International Conference on Computer Vision, pp. 2980–2988. (2017)
16. Eisenhauer, E.A., et al.: New response evaluation criteria in solid tumours: revised RECIST guideline (version 1.1). Eur. J. Cancer **45**(2), 228–247 (2009)
17. Kingma, D.P., Ba, J.: Adam: a method for stochastic optimization. In: 3rd International Conference on Learning Representations (2015)
18. Loshchilov, I., Hutter, F.: SGDR: stochastic gradient descent with warm restarts. In: 5th International Conference on Learning Representations (2017)
19. Davidson-Pilon, C.: Lifelines: survival analysis in Python. J. Open Source Softw. **4**(40), 1317 (2019). https://doi.org/10.21105/joss.01317 (2019)
20. van der Walt, S., et al.: scikit-image: image processing in Python. J. PeerJ **2**, e453 (2014)

A Hybrid Radiomics Approach
to Modeling Progression-Free Survival
in Head and Neck Cancers

Sebastian Starke[1], Dominik Thalmeier[2], Peter Steinbach[1],
and Marie Piraud[2(✉)]

[1] Helmholtz AI, Helmholtz-Zentrum Dresden-Rossendorf, Bautzner Landstrasse 400,
01328 Dresden, Germany
[2] Helmholtz AI, Helmholtz Zentrum München, Ingolstädter Landstraße 1,
85764 Neuherberg, Germany
marie.piraud@helmholtz-muenchen.de

Abstract. We present our contribution to the HECKTOR 2021 challenge. We created a Survival Random Forest model based on clinical features, and a few radiomics features that have been extracted with and without using the given tumor masks, for Task 3 and Task 2 of the challenge, respectively. To decide on which radiomics features to include into the model, we proceeded both to automatic feature selection, using several established methods, and to literature review of radiomics approaches for similar tasks. Our best performing model includes one feature selected from the literature (Metabolic Tumor Volume derived from the FDG-PET image), one via stability selection (Inverse Variance of the Gray Level Co-occurrence Matrix of the CT image), and one selected via permutation-based feature importance (Tumor Sphericity). This hybrid approach turns-out to be more robust to overfitting than models based on automatic feature selection. We also show that simple ROI definition for the radiomics features, derived by thresholding the Standard Uptake Value in the FDG-PET images, outperforms the given expert tumor delineation in our case.

Keywords: Progression-free survival · Survival random forests · Radiomics

1 Introduction

With the advent of precision oncology, Progression-Free Survival (PFS) prediction models, based on non-invasive observations at diagnosis-time, are of great interest, as they could faciliate personalized clinical decision making. Head and neck cancers (HNCs) are among the most common cancers worldwide, and various factors are known to influence their incidence rate, as well as their five-year

S. Starke, D. Thalmeier, P. Steinbach and M. Piraud—Contributed equally to this work.

V. Andrearczyk et al. (Eds.): HECKTOR 2021, LNCS 13209, pp. 266–277, 2022.
https://doi.org/10.1007/978-3-030-98253-9_25

survival rate, such as gender, age and the location of the tumor [17]. Developing a unified model for all stages and all subtypes of HNCs is, however, very challenging, as the type of survival endpoints (lesion growth, metastasis or death) differ between patients, and HNCs are very heterogenous. The HECKTOR 2021 challenge [1,2,13] proposes to develop such a prediction model, based on non-invasive FDG-PET/CT images at diagnosis, together with clinical information and other known biomarkers for cancer progression, such as the HPV status [3]. In the following, we first introduce the solution we developed for Task 3 of the challenge, and therefore make use of the ground truth annotations of primary tumors that are provided. Based on these results, we also submitted a 'quick and dirty' solution to Task 2, replacing the provided masks by thresholding the FDG-PET images, in an attempt to understand the importance of delineation of the tumors by experts.

2 Preliminary Considerations

2.1 Dataset Description

The training cohort of the Hecktor challenge 2021 consisted of 224 patients diagnosed with HNC from five different centers located in Canada, Switzerland and France (1 = CHGJ, 2 = CHUS, 3 = CHMR, 4 = CHUM, 5 = CHUP). For each patient, a pretreatment CT image and an FDG-PET image were available. Both images were already registered and delineated by experts according to the challenge organizers. Additionally, clinical parameters which included treatment center, gender, age, smoking history, drinking history, staging (T, N, M stages and TNM group), HPV status and chemotherapy treatment were available, but included missing values as well. The distribution of the TNM groups and missing data is uneven over the different centers.

The outcome of interest, the PFS, is defined as either a size increase of known lesions (change of T and/or N stage) or an appearance of new lesions (change of N and/or M stage). Disease-specific death is also considered a progression event for patients previously considered stable. Finally, the progression status is also given for the training set as not all patients had progressed at the end of the observation period.

The test set consists of 101 patients, 48 originating from one of the training centers (5 = CHUP) and 53 from a new center (CHUV), with a very heterogeneous distribution of TNM group and missing data among the centers as well.

2.2 Cross-Validation Strategy

In order to build, evaluate and compare machine learning models for the prediction of PFS, we first created patient folds from the training data to measure out-of-sample performance using k-fold cross-validation (CV). Due to the relatively low number of samples we chose to use k = 3 to divide the data into training and validation splits. This enabled us to identify prognostic features and to choose model hyperparameters.

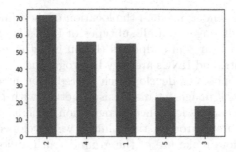

Fig. 1. Distribution of patients among the different centers in the training set.

To select models that generalize well to unseen centers in our validation procedure, we made sure that each split contains one center exclusively in the validation set. Considering the uneven distribution of patients among the centers (see Fig. 1), we decided to use centers $1 = \text{CHGJ}$, $3 = \text{CHMR}$ and $4 = \text{CHUM}$ for this. As $5 = \text{CHUP}$ is also present in the test set, we decided to split it into 3 equally-sized sets (5a, 5b and 5c), stratified by the number of events, and to include one of those sets in each of the validation folds. Finally, to obtain folds of comparable size, we split patients from center $2 = \text{CHUS}$ into two sets 2s and 2l of sizes 34 and 38 respectively, and ensure that, patients from center 2 are present in all training folds. The final CV splits are presented in Table 1.

Table 1. Cross-validation splits

Fold	Training centers	Validation centers
1	1, 2s, 2l, 3, 5b, 5c	4, 5a
2	2s, 2l, 3, 4, 5a, 5c	1, 5b
3	1, 2s, 4, 5a, 5b	2l, 3, 5c

2.3   Evaluation of the Model

The ranking of the challenge will be based on the concordance index (C-index) on the test data. We therefore did not try to estimate the PFS endpoint directly, e.g. in days, but used the risk score as a surrogate. To select the best model, the validation C-index is computed for each split of Table 1 and averaged over the three splits. The model retrained on the whole training set is used for prediction on the test set.

3 Methods

3.1 Feature Engineering

Clinical Feature Pre-processing. We modified the clinical features as follows:

- We create a one-hot encoded version of the T, N and M-stage columns, as well as of the TNM group, ignoring the sub-stages and sub-groups
- We create the new one-hot features $S < X$, with $S \in \{T, N, M\}$ which take value 1 for all patients in S-stage lower than X, and 0 otherwise.
- Since especially the HPV status is known to be a prognostic biomarker for HNC [18] but was unavailable for many patients in the dataset, we included a K-nearest neighbor feature imputation method into our pipeline taken from the 'scikit-learn' python package [15]. We considered this step especially important due to the fact that HPV status was available for all patients from center $5 = \mathrm{CHUP}$ which was going to be part of the test set as well.
- All scalar features were finally standardized to mean of zero and standard deviation of one before being used in the model.

Radiomics Feature Extraction. We extracted radiomics features from both the CT and the PET images using the 'pyradiomics' python package [20]. We extract all default features from the following classes: firstorder (First Order), glcm (Gray Level Co-occurrence Matrix), glszm (Gray Level Size Zone Matrix), glrlm (Gray Level Run Length Matrix) and ngtdm (Neighbouring Gray Tone Difference Matrix), setting the bin-width for discretizing the image to 10 for the CT and 0.2 for the PET images (heuristically adjusted based on manual inspection). We also extracted the default shape 3D features from the available masks. Except for the bin-width we used the default settings of pyradiomics version 3.0.1. We therefore ended up with a total of 172 image features, which are named using the following convention: Modality_class_FeatureName.

Additionally, we extracted the following features:

- the Metabolomic Tumor Volume (PT_MTV) using the method favored in [7], i.e. by thresholding the Standardized Uptake Value (SUV) to 2.5 in the PET, and computing the volume of the largest connected component,
- the Total Lesion Glycolosis (PT_TLG) as the product of the tumor volume (as extracted with 'pyradiomics' from the given masks) and the mean SUV

$$Mask\_shape\_VoxelVolume \times PT\_firstorder\_Mean,$$

following Ref. [14].

3.2 Modeling Approach

Baseline. With our evaluation strategy in place, we first looked into simple Cox proportional hazards (Cox) models using the pre-processed clinical features, estimating the performance by the average concordance index over our three validation folds. We experimented with manual combinations of clinical features selected from apriori knowledge and fitted Cox proportional hazards models using the 'scikit-survival' [16] python package, as reported in Table 2. We used the combined model as our baseline.

We then explored more elaborate models like the random survival forests (RF) [8] from the 'scikit-survival' package [16]. The number of neighbors to use for the missing value imputation, as well as the number of estimators in the random forest were estimated in a grid search in our CV setting, and finally set to K = 20 (nearest neighbors) and n = 300 (number of trees). Table 2 shows the performance of our baseline random forest model, on clinical features, which shows a mean C-index of 0.724 over the three validation folds.

Table 2. Performance of Cox and Random Forest models with clinical features

Features	Model	Validation C-index			
		Fold 1	Fold 2	Fold 3	Mean
Gender	Cox	0.575	0.472	0.435	**0.494**
Age	Cox	0.565	0.491	0.372	**0.476**
T < 4	Cox	0.647	0.608	0.553	**0.603**
HPV	Cox	0.673	0.623	0.544	**0.613**
Tobacco	Cox	0.604	0.650	0.530	**0.595**
Alcohol	Cox	0.382	0.474	0.542	**0.466**
Age, T < 4, HPV, Gender, Alcohol, Tobacco	Cox	0.784	0.593	0.548	**0.642**
Age, T < 4, HPV, Gender, Alcohol, Tobacco	RF	0.790	0.762	0.619	**0.724**

Stability Feature Selection. In order to identify a small number of useful image features in a systematic way we used the stability selection approach [12]. It consists of applying L1-regularized Cox regression on multiple boostrapped datasets for a range of regularization strength parameters. Finally, features are selected that appear with a high rate in these bootstraps (i.e. their coefficients are larger than the median coefficient size for a given fraction of all bootstraps). This gives a robust set of features that are predictive for the regression task. Compared to a vanilla L1-regularization approach, stability selection has the advantage that its selection threshold can be set to a fixed value that generalizes well between tasks. Threshold values that work well in practice range from 0.6 to 0.8. We chose 0.6 because the stricter values resulted in very few stable features.

Permutation-Based Feature Importance. To identify important features for the model, we also implemented permutation-based feature importance, using the 'eli5' python package. The feature importance is derived from the decrease in the model score, when the feature values are randomly shuffled [5]. We then ranked image-features according to their importance for the model.

Sequential Feature Selection. To end up with an optimal number of features, we also proceeded to Sequential Feature Selection from the 'scikit-learn' python package [15]. Given a set of features, and an estimator, this Sequential Feature Selector adds or removes features to form a feature subset in a greedy fashion. At each stage, it chooses the best feature to add or remove based on the cross-validation score of the estimator.

4   Results for Task 3

4.1   Feature Selection

Stability Selection. We applied stability selection on all the radiomics features to the training sets of all three CV splits. It retained between 5 and 7 features for each split, and

– PT_glcm_ClusterShade

was stable in all folds, and the following features:

– CT_glcm_InverseVariance
– CT_glcm_MCC
– PT_firstorder_Kurtosis
– PT_glcm_Imc1

were retained in 2 of our 3 CV splits.

Permutation Importance. We also computed the permutation-based importance for all radiomics features. We then ranked them, and heuristically chose to retain the top 10 features:

– Mask_shape_Sphericity
– CT_glrlm_RunLengthNonUniformity
– CT_glrlm_ShortRunEmphasis
– Mask_shape_Maximum3DDiameter
– CT_glrlm_RunLengthNonUniformityNormalized
– CT_glcm_Idm
– PT_glrlm_ShortRunLowGrayLevelEmphasis
– CT_ngtdm_Complexity
– PT_ngtdm_Complexity
– CT_glszm_GrayLevelNonUniformity

Features from the Literature. Additionally we hand-picked five further radiomics features,

– PT_glszm_SmallAreaLowGrayLevelEmphasis
– CT_firstorder_Energy
– CT_glszm_LargeAreaHighGrayLevelEmphasis
– CT_glszm_SizeZoneNonUniformity

– CT_glrlm_GrayLevelVariance

that have been directly identified as important features in previous related radiomics studies [4, 19], or are correlated to them. We also retain,

– PT_MTV
– PT_TLG

which have been reported as predictive biomarkers for HNC patient outcome in [6, 14], and

– N < 2,

as the spreading to lymph nodes has also been shown to have a prognostic value in [6, 19].

4.2   Submitted Models

Model A: Sequential Feature Selection. Using the clinical features, and all features listed in Sect. 4.1, we proceeded to Sequential Feature Selection. In the backward mode, only four features were retained in the model, namely:

– HPV
– T < 4
– CT_glcm_InverseVariance
– Mask_shape_Sphericity

This gave a mean validation C-Index of **0.779** on our three folds, as reported in Table 3.

Model B: Manually Selected Features. By testing the features from the stability selection with our full pipeline we found that adding PT_firstorder_Kurtosis improved the performance over our baseline considerably, on all three splits. We also found that Mask_shape_Sphericity did improve the performance of our model further, on all three validation splits. We therefore used it as well, and Model B shows a mean C-index of **0.753** over the three validation folds, see Table 3.

Model C: Manually Selected Features. To avoid overfitting on the available dataset, we also submitted a model mainly based on the features found in the literature. Based on our clinical baseline, we added N < 2, PT_MTV and the Mask_shape_Sphericity as well as CT_glcm_InverseVariance, which was retained by the sequential feature selection. This gave a mean validation C-Index of **0.734**.

Table 3. Performance of the models prepared for task 3

Features	Model	Validation C-index				Test C-index
		Fold 1	Fold 2	Fold 3	Mean	
Model A: HPV, T < 4, CT_glcm_InverseVariance, Mask_shape_Sphericity	RF	0.829	0.828	0.682	**0.779**	**0.540**
Model B: Gender, Age, T < 4, Mask_shape_Sphericity, PT_firstorder_Kurtosis, HPV, Tobacco, Alcohol	RF	0.813	0.765	0.682	**0.753**	**0.615**
Model C: Gender, Age, T < 4, HPV, Tobacco, Alcohol, N < 2, PT_MTV, Mask_shape_Sphericity, CT_glcm_InverseVariance	RF	0.781	0.765	0.655	**0.734**	**0.616**

5   Results for Task 2

As the computation of the PT_MTV (see Sect. 3.1 and [7]) also provides a way to define a Region of Interest (ROI) around the tumor area, we decided to compute the radiomics features from masks created by thresholding the SUV to 2.5 in the PET images. Inspired from our results for Task 3, we then submitted the following models (the corresponding results are reported in Table 4):

Model 1, using the same features as Model A.

Model 2, using the same features as Model B.

Model 3, using the same features as Model C.

Model 4, obtained by running the whole pipeline with the newly derived features (stability selection, feature importance and sequential feature selection), in the backward mode, in which 15 features were retained (see Table 4).

Model 5, obtained by running the whole pipeline with the newly derived features (stability selection, feature importance and sequential feature selection), in the forward mode. Three features were retained: PT_glszm_LargeAreaLowGrayLevelEmphasis (PT_glszm_LALGLE), CT_glcm_Imc1, and N < 2.

Table 4. Performance of the models prepared for task 2

Features	Model	Validation C-index				Test C-index
		Fold 1	Fold 2	Fold 3	Mean	
Model 1: HPV, T < 4, CT_glcm_InverseVariance, Mask_shape_Sphericity	RF	0.721	0.622	0.553	**0.631**	**0.624**
Model 2: Gender, Age, T < 4, Mask_shape_Sphericity, PT_firstorder_Kurtosis, HPV, Tobacco, Alcohol	RF	0.833	0.743	0.564	**0.713**	**0.649**
Model 3: Gender, Age, T < 4, HPV, Tobacco, Alcohol, N < 2, PT_MTV, Mask_shape_Sphericity, CT_glcm_InverseVariance	RF	0.647	0.678	0.583	**0.636**	**0.659**
Model 4: PT_glszm_ZoneEntropy, CT_glszm_GrayLevelVariance, CT_glrlm_RunEntropy, CT_glszm_SizeZoneNonUniformity, T < 4, Tobacco, PT_glcm_Correlation, Gender, CT_glcm_Imc1, PT_glcm_Imc1, Chemotherapy, CT_glrlm_GrayLevelVariance, CT_firstorder_Energy, PT_ngtdm_Strength, TNM < 4	RF	0.824	0.822	0.708	**0.784**	**0.606**
Model 5: N < 2, PT_glszm_LALGLE, CT_glcm_Imc1	RF	0.876	0.754	0.720	**0.783**	**0.558**

6   Discussion

Final Models. In order to create robust models, we aimed at simple models, based on clinical features, with a few radiomics features selected via stability selection, feature importance and literature review. We used Random Survival Forests, from the 'scikit-survival' package [16], with $n = 300$ estimators, and handled missing clinical data via K-nearest neighbor feature imputation, with $K = 20$.

Our models for Task 3 performed well in our 3-fold CV setting, as we can see from the mean validation C-indices reported in Table 3. After retraining the models on all available training data, we submitted them to the HECKTOR challenge platform, and got C-indices of **0.540**, **0.615** and **0.616** on the test data for models A, B and C respectively. We therefore see that the fully automatic feature selection models greatly overfit on the training data, and that the models obtained by hand-picking radiomics features from the literature and from the feature selection performed better at test time.

The same applies for Task 2, as models 4 and 5, which fully rely on automatic feature selection, perform the worst at test time. And models 2 and 3, which use the same combinations of hand-picked features than models B and C, perform the best, with test C-indices of **0.649** and **0.659**, respectively. More importantly, we were surprised to see that our submissions for Task 2 systematically overperform the submissions for Task 3 using the same features, indicating that deriving the ROIs for the radiomics features from the PET images with simple thresholding methods is better than using the tumor delineation by experts. This conclusion is in line with previous observations [7]. We therefore submitted Model 3 to Task 3 as well, and this provides our best model for both tasks.

Model Features. First, we note that the sets of radiomics features selected by Permutation Importance and Stability Selection do not overlap. This could be due to large correlations between the features.

In addition to feature selection, we also analyzed the final models, by computing the Shapley values, using the 'shap' python package [11], and random forest permutation importance. We found that Mask_shape_Sphericity and HPV are consistently very predictive features in all models. Note that even if Mask_shape_Sphericity was not selected by the stability selection, it did show up as the most important feature of random forest-based permutation importance when using all extracted radiomics features. Those observations are consistent with previous knowledge, as

- Sphericity is highly correlated to the Spherical Disproportion, which was found to be predictive of local tumor growth in head and neck squamous cell carcinoma in [4],
- the HPV status is known to be a prognostic biomarker for HNC [18]

A new insight from our modeling, is that CT_glcm_InverseVariance shows up as a predictive biomarker in this dataset, and was automatically selected by stability selection and the following sequential feature selection. We did not find this feature reported in the related literature before. The predictive power of Age is less clear, as this feature is quite important in the permutation based approach, but is ranked quite low with Shapley values. Gender is the least important feature in both approaches and Tobacco and Alcohol are ranked low in both approaches as well.

Outlook. We originally looked for a radiomics signature by computing many radiomics features, and using established feature selection techniques to create our model. But models using hand-picked features from the literature, that were not automatically selected, also performed quite well in CV and even better at test time. This might indicate that we did not find the optimal feature selection method, but also that the training dataset is too small for a full-blown radiomics approach. Our hybrid approach, partly driven by domain knowledge and literature review, and partly by automatic feature selection, enabled us to compensate for that and avoid overfitting.

Very interestingly, our quick-and-dirty approach to generate systematic segmentation masks by thresholding the PET images, ended-up creating better models than those using the same radiomics features, extracted from expert tumor delineations. We observed that the MTV volumes obtained with our thresholding method are significantly larger than those of the expert reference, and that our thresholding method retained other structures than the main tumor. This indicates that capturing the surrounding of the tumor and/or further structures like the lymph nodes in the region of interest to compute radiomics features could be more informative than the tumor itself, as was already observed in previous radiomics studies [9,10,21], and deserves further investigation.

References

1. Andrearczyk, V., et al.: Overview of the HECKTOR challenge at MICCAI 2021: automatic head and neck tumor segmentation and outcome prediction in PET/CT images. In: Andrearczyk, V., Oreiller, V., Hatt, M., Depeursinge, A. (eds.) HECKTOR 2021. LNCS, vol. 13209, pp. 1–37. Springer, Cham (2022)
2. Andrearczyk, V., et al.: Automatic segmentation of head and neck tumors and nodal metastases in PET-CT scans. In: Medical Imaging with Deep Learning, pp. 33–43. PMLR (2020)
3. Baumann, M., et al.: Radiation oncology in the era of precision medicine. Nat. Rev. Cancer 16(4), 234–249 (2016)
4. Bogowicz, M., et al.: Comparison of PET and CT radiomics for prediction of local tumor control in head and neck squamous cell carcinoma. Acta Oncol. 56(11), 1531–1536 (2017). pMID: 28820287. https://doi.org/10.1080/0284186X.2017.1346382
5. Breiman, L.: Random forests. Mach. Learn. 45(1), 5–32 (2001)
6. Castelli, J., et al.: A PET-based nomogram for oropharyngeal cancers. Eur. J. Cancer 75, 222–230 (2017)
7. Im, H.J., Bradshaw, T., Solaiyappan, M., Cho, S.Y.: Current methods to define metabolic tumor volume in positron emission tomography: which one is better? Nucl. Med. Mol. Imaging 52(1), 5–15 (2018)
8. Ishwaran, H., Kogalur, U.B., Blackstone, E.H., Lauer, M.S.: Random survival forests. Ann. Appl. Stat. 2(3), 841–860 (2008). https://doi.org/10.1214/08-aoas169
9. Kocak, B., Ates, E., Durmaz, E.S., Ulusan, M.B., Kilickesmez, O.: Influence of segmentation margin on machine learning-based high-dimensional quantitative CT texture analysis: a reproducibility study on renal clear cell carcinomas. Eur. Radiol. 29(9), 4765–4775 (2019)
10. Leseur, J., et al.: Pre-and per-treatment 18F-FDG PET/CT parameters to predict recurrence and survival in cervical cancer. Radiother. Oncol. 120(3), 512–518 (2016)
11. Lundberg, S.M., Lee, S.I.: A unified approach to interpreting model predictions. In: Guyon, I., et al. (eds.) Advances in Neural Information Processing Systems 30, pp. 4765–4774. Curran Associates, Inc. (2017). http://papers.nips.cc/paper/7062-a-unified-approach-to-interpreting-model-predictions.pdf
12. Meinshausen, N., Bühlmann, P.: Stability selection. J. Roy. Stat. Soc. Ser. B (Stat. Methodol.) 72(4), 417–473 (2010)
13. Oreiller, V., et al.: Head and neck tumor segmentation in PET/CT: the HECKTOR challenge. Med. Image Anal. 77, 102336 (2021)
14. Paidpally, V., Chirindel, A., Lam, S., Agrawal, N., Quon, H., Subramaniam, R.M.: FDG-PET/CT imaging biomarkers in head and neck squamous cell carcinoma. Imaging Med. 4(6), 633 (2012)

15. Pedregosa, F., et al.: scikit-learn: machine learning in Python. J. Mach. Learn. Res. **12**, 2825–2830 (2011)
16. Pölsterl, S.: scikit-survival: a library for time-to-event analysis built on top of scikit-learn. J. Mach. Learn. Res. **21**(212), 1–6 (2020). http://jmlr.org/papers/v21/20-729.html
17. Pulte, D., Brenner, H.: Changes in survival in head and neck cancers in the late 20th and early 21st century: a period analysis. Oncologist **15**(9), 994 (2010)
18. Sabatini, M.E., Chiocca, S.: Human papillomavirus as a driver of head and neck cancers. Br. J. Cancer **122**(3), 306–314 (2020)
19. Vallieres, M., et al.: Radiomics strategies for risk assessment of tumour failure in head-and-neck cancer. Sci. Rep. **7**(1), 1–14 (2017)
20. Van Griethuysen, J.J., et al.: Computational radiomics system to decode the radiographic phenotype. Can. Res. **77**(21), e104–e107 (2017)
21. Xie, H., Zhang, X., Ma, S., Liu, Y., Wang, X.: Preoperative differentiation of uterine sarcoma from leiomyoma: comparison of three models based on different segmentation volumes using radiomics. Mol. Imag. Biol. **21**(6), 1157–1164 (2019)

An Ensemble Approach for Patient Prognosis of Head and Neck Tumor Using Multimodal Data

Numan Saeed[✉], Roba Al Majzoub, Ikboljon Sobirov, and Mohammad Yaqub

Mohamed bin Zayed University of Artificial Intelligence, Abu Dhabi, UAE
{numan.saeed,roba.majzoub,ikboljon.sobirov,mohammad.yaqub}@mbzuai.ac.ae
https://mbzuai.ac.ae/biomedia

Abstract. Accurate prognosis of a tumor can help doctors provide a proper course of treatment and, therefore, save the lives of many. Traditional machine learning algorithms have been eminently useful in crafting prognostic models in the last few decades. Recently, deep learning algorithms have shown significant improvement when developing diagnosis and prognosis solutions to different healthcare problems. However, most of these solutions rely solely on either imaging or clinical data. Utilizing patient tabular data such as demographics and patient medical history alongside imaging data in a multimodal approach to solve a prognosis task has started to gain more interest recently and has the potential to create more accurate solutions. The main issue when using clinical and imaging data to train a deep learning model is to decide on how to combine the information from these sources. We propose a multimodal network that ensembles deep multi-task logistic regression (MTLR), Cox proportional hazard (CoxPH) and CNN models to predict prognostic outcomes for patients with head and neck tumors using patients' clinical and imaging (CT and PET) data. Features from CT and PET scans are fused and then combined with patients' electronic health records for the prediction. The proposed model is trained and tested on 224 and 101 patient records respectively. Experimental results show that our proposed ensemble solution achieves a C-index of 0.72 on The HECKTOR test set that saved us the first place in prognosis task of the HECKTOR challenge. The full implementation based on PyTorch is available on https://github.com/numanai/BioMedIA-Hecktor2021.

Keywords: Cancer prognosis · Head and neck tumor · CT scans · PET scans · Multimodal data · Mutli-task logistic regression · Cox proportional hazard · Deep learning · Convolutional neural network

1 Introduction

Each year, 1.3 million people are diagnosed with head and neck (H&N) cancer worldwide on average [16]. However, the mortality rate can be lowered to 70%

Team name: MBZUAI-BioMedIA

© Springer Nature Switzerland AG 2022
V. Andrearczyk et al. (Eds.): HECKTOR 2021, LNCS 13209, pp. 278–286, 2022.
https://doi.org/10.1007/978-3-030-98253-9_26

with early detection of H&N tumor [16]. Therefore, diagnosis and prognosis are the two primary practices involved in most medical treatment pipelines, especially for cancer-related diseases. After determining the presence of cancer, a doctor tries to prescribe the best course of treatment yet with limited information, it is very challenging. An early survival prediction can help doctors pinpoint a specific and suitable treatment course. Different biomarkers from radiomics field can be used to predict and prognose medical cases in a non-invasive fashion [7]. It is used in oncology to help with cancer prognosis, allowing patients to plan their lives and actions in their upcoming days. In addition, it enables doctors to better plan for the time and mode of action followed for treatment [11]. This is necessary to make more accurate predictions, which, in turn, is likely to lead to better management by the doctors.

Many other research fields also strive to assist medical doctors, at least to a point of alleviating their work process. One of the most common statistical frameworks used for the prediction of the survival function for a particular unit is the Cox proportional hazard model (CoxPH), proposed by Cox [5] in 1972. It focuses on developing a hazard function, i.e., an age-specific failure rate. Nevertheless, CoxPH comes with specific issues, such as the fact that the proportion of hazards for any two patients is constant or that the time for the function is unspecified. Yu et al. [17] proposed an alternative to CoxPH - multi-task logistic regression (MTLR). MTLR can be understood as a sequence of logistic regression models created at various timelines to evaluate the probability of the event happening. Fotso [6] improved the MTLR model by integrating neural networks to achieve nonlinearity, yielding higher results.

Deep learning (DL) has gained a considerable amount of attention in classification, detection, and segmentation tasks of the medical research field. Furthermore, their use in far more complicated tasks such as prognosis and treatment made DL even more popular, as it can handle data in large amounts and from different modalities, both tabular and visual.

Many studies have been conducted to perform prognosis of cancer using D.L. Sun et al. [14] propose a deep learning approach for the segmentation of brain tumor and prognosis of survival using multimodal MRI images. 4524 radiomic features are extracted from the segmentation outcome, and further feature extraction is performed with a decision tree and cross-validation. For survival prediction, they use a random forest model. In a similar task done by Shboul et al. [13], a framework for glioblastoma and abnormal tissue segmentation and survival prediction is suggested. The segmentation results, along with other medical data, are combined to predict the survival rate. Tseng et al. [15] develop a multiclass deep learning model to analyze the historical data of oral cancer cases. They achieve superior results compared to traditional logistic regression.

A few studies have been conducted for the prognosis of H&N cancer [9]. The prognosis studied in [12] shows that they achieve the area under the curve (AUC) of 0.69 for their best-performing dataset for H&N tumor, while for the rest of the datasets, they achieve AUC between 0.61 and 0.68. Furthermore, Kazmierski et al. [9] use electronic health record (EHR) data and pre-treatment radiological images to develop a model for survival prediction in H&N cancer. Out of the many trials they experimented with, a non-linear, multitask approach that uses the EHR data and tumor volume produced the highest result for prognosis.

Such results to this task are unlikely to motivate clinicians to use machine learning models in clinical practice; therefore, more accurate prognosis is critical to help solve this problem. In this paper, we are proposing a multimodal machine learning algorithm that, without prior information on the exact location of the tumor, utilizes both tabular and imaging data for the prognosis of Progression Free Survival (PFS) for patients who have H&N oropharyngeal cancer. This work is carried out to address the prognosis task of the MICCAI 2021 Head and Neck Tumor segmentation and outcome prediction challenge (HECKTOR) [1,2].

2   Materials and Methods

2.1   Dataset

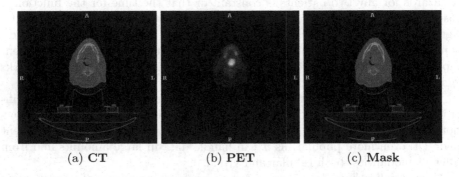

(a) **CT**   (b) **PET**   (c) **Mask**

Fig. 1. Examples from the training set of HECKTOR 2021 [1]. (a) shows the CT slice, (b) shows the PET slice, and (c) shows the mask (in red) superimposed on the CT image. (Color figure online)

The HECKTOR committee [1,2] provided CT and PET scans, manual segmentation masks and electronic health record (EHR) dataset. The ground truth segmentation masks for H&N oropharyngeal cancer was manually delineated by oncologists as shown in Fig. 1. EHR contains data about a patient's age, gender, weight, tumor stage, tobacco and alcohol consumption, chemotherapy experience, presence of human papillomavirus (HPV) and other data. A clinically relevant endpoint was provided in the training set to predict PFS for each patient. The data is multi-centric, collected from four centers in Canada, one center in Switzerland and another one in France. The total number of patients involved in the study was 325, out of which 224 were training cases and 101 test cases. However, some data points are missing in the EHR, e.g., tobacco

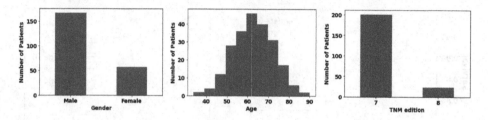

Fig. 2. EHR data visualization. The graph on the left shows the male-to-female ratio. The chart in the middle depicts the distribution of age. The chart on the right shows the TNM edition for all patients.

and alcohol data were only provided by one of the centers mentioned above in the training set. The dataset has 75% of the patients censored, who might have stopped following up or changed the clinic.

Visualization of the EHR data was performed to observe the distribution of patients in terms of gender and age, as shown in Fig. 2. Most of the patients are males, and the age ranges between 35 to 90 years with the peak at around 60 years. The TNM (T: tumor, N: nodes, and M: metastases) edition 7 is used for most patients to describe the volume and spread of cancer in a patient's body, while the rest of the patients' cases were represented using TNM edition 8.

2.2 Data Analysis and Image Preprocessing

We initially analyzed the EHR data from the training dataset using the CoxPH model by splitting them into training, validation and testing sets to experiment on different hyperparameters and configurations of the solution. Then, we tried to observe the effects of different covariates on the survival rate using the trained CoxPH model. In Fig. 3 (a), the gender covariate is varied by assigning males and females the values 0 and 1 respectively. The results show that the survival rate of males is less than that of females. Similarly, to observe the effect of metastasis (M), M1, M2, and Mx are assigned the values 0, 1, and 2 respectively. The analysis shows that the patients with cancer spread to other parts of their bodies (M1) have lower survival rates, which is in line with the medical science. Some data points for tobacco and alcohol were not available, so we tried to impute them through assigning values of 1 for consumers, −1 for non-consumers and 0 for patients with missing tobacco and alcohol consumption data. Then, we tried another approach where we dropped the use of incomplete data features and trained the model on other available data features. The obtained results of the cross validation revealed that dropping incomplete data points leads to better prognosis results than imputing them.

As for the image dataset, PET and CT scans were preprocessed using the bounding box information available in the provided csv file to obtain $144 \times 144 \times 144$ cropped images. To prepare the image data for our model input, we normalized the two images to the same scale and a fused image was created

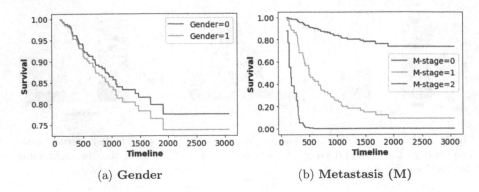

(a) Gender   (b) Metastasis (M)

Fig. 3. Partial effects on outcome when covariate(s) is varied. (a) shows the survival rate for males and females. The baseline we compared to and males' survival prediction curves are superimposed in the figure. (b) depicts the metastasis effects on the survival of patients.

by averaging the two scans for each patient. To further reduce the volume, the fused output image is cropped again based on a specific distance away from the center of the $144 \times 144 \times 144$ cube as shown in Fig. 4. Two possible crop resolutions were considered: $50 \times 50 \times 50$ and $80 \times 80 \times 50$; however, the latter was adopted since it resulted in a better outcome.

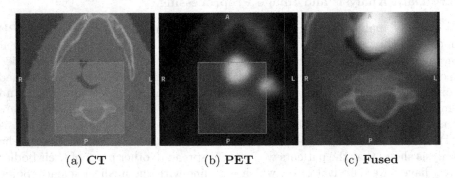

(a) **CT**   (b) **PET**   (c) **Fused**

Fig. 4. An example of combining CT and PET scans. (a) depicts the $(144 \times 144 \times 144)$ CT scan with a rectangle to show the region to be cropped. (b) shows the $(144 \times 144 \times 144)$ PET scan and the cropping region. (c) shows the $(80 \times 80 \times 50)$ fused image in the cropped form.

2.3 Network Architecture

The baseline we proposed has an MTLR model as its cornerstone and was later expanded and developed by integrating and optimizing features obtained from different inputs. First, from the EHR data provided in the training dataset, a prognostic model was developed. Then, we optimized the vanilla MTLR by experimenting with different hyperparameters such as learning rates, the depth and width of the feedforward layers, and the constant C in l_2 regularization term of the loss function of MTLR from [8] depicted in the following equation:

$$\min_{\Theta} \frac{C}{2} \sum_{j=1}^{m} \|\boldsymbol{\theta}_j\|^2 - \sum_{i=1}^{n} \left[\sum_{j=1}^{m} y_j(s_i)(\boldsymbol{\theta}_j \cdot \boldsymbol{x}_i + b_j) - \log \sum_{k=0}^{m} \exp f_\Theta(\boldsymbol{x}_i, k) \right]$$

The smoothness of the predicted survival curves depends on the change between consecutive timepoints and is controlled by C.

Next, we investigated the effect of multimodality on the performance of the model by integrating the available image data. Features were extracted from the fused crops through the use of a 3D convolutional neural network (CNN) adopted from [10] named Deep-CR. Unlike [10], we optimized the CNN architecture using OPTUNA framework [3] to obtain the best hyperparameters, including the kernel sizes and the number of layers. These features were concatenated with our tabular data and fed into two fully connected layers. Finally, the risk was calculated using MTLR, and the output was averaged with CoxPH model risk output.

For our best-performing model, inspired by the Deep-CR approach, we developed a network as shown in Fig. 5 to predict the individual risk scores. An optimized Deep-CR network with two blocks was developed. Each block consists of two 3D convolutional, ReLU activation and batch normalization layers. The 3D CNN blocks are followed by 3D MaxPooling layers. The kernel sizes of the 3D CNN layers in each block are 3 and 5. The number of output channels of the 3D CNN layers are 32, 64, 128 and 256 respectively as shown in the Fig. 5. The number of neurons in the two feed forward layers are 256 each. The batch size, learning rate, and dropout were experimentally set to 16, 0.016, and 0.2 respectively for the training. The model was trained for 100 epochs using Adam optimizer on a single GPU NVIDIA RTX A6000 (48 GB). We refer to this network as Deep Fusion, and we implement two variants of it; (V1) with three 3D CNN paths that take three types of image inputs (CT, PET, and fused images) and (V2), which includes one 3D CNN path with a single image input (the fused data) as shown in Fig. 5. Each CNN outputs a feature vector of length 256. In Deep Fusion V1, the 3 feature vectors are concatenated. The feature vector was then combined with EHR data to train a fully connected layer followed by two MTLR layers to estimate the risk. Finally, the risk predictions from the MTLR model and the CoxPH results are averaged to calculate the final risk scores.

3   Results

We trained the models using all the training data before applying it on the HECKTOR test set. The concordance index (C-index), one of the metrics used to measure the performance of a prognosis model [4], was used to report the results on the HECKTOR test dataset on each of the different models previously mentioned as shown in Table 1. The baseline model which only uses MTLR to estimate the risk has 0.66 C-index. Slight improvement on C-index was achieved

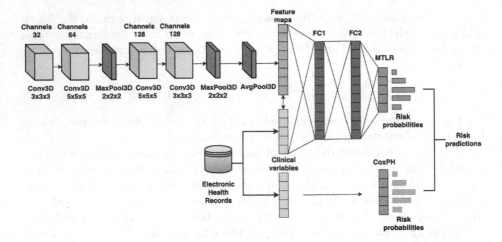

Fig. 5. Overall Architecture of Deep Fusion V2. Features are extracted from fused CT and PET scans using the CNN network and concatenated with the EHR features. Then, the output is passed to the FC layers before MTLR. Lastly, risk scores from MTLR and CoxPH models are averaged to get the final risk predictions.

when combining image features and EHR in the MTLR + Deep-CR model (C-index of 0.67). The results obtained using Deep Fusion (V1) has also achieved a C-index score of 0.67. However, the best estimation of risk was obtained using Deep Fusion (V2) with C-index of 0.72.

Table 1. C-index scores obtained on the HECKTOR 2021 testing dataset.

Models	C-index
MTLR (Baseline)	0.66
MTLR + Deep-CR [10]	0.67
MTLR + CoxPH + Deep Fusion (V1)	0.67
MTLR + CoxPH + Deep Fusion (V2)	**0.72**

4   Discussion and Conclusion

The results of the Deep Fusion (V1) suffered a low score of 0.67 C-index compared to the (V2). This augmentation-like approach of feeding CT, PET and Fused version into individual CNN architectures, combining the outputs and forwarding them into MTLR, then finally concatenating them with CoxPH results was hypothesized to yield better results. However, C-index of 0.67 is achieved compared to 0.72 in (V2). Multiple possibilities could have contributed to this discrepancy. First, the training of 3 different CNNs was not optimized to generate a well representative feature vector. This may have led to misleading feature vectors that make it hard to train a discriminative MTLR model. Second, the final aggregation of the output was merely concatenating the three feature vectors. Introducing more sophisticated aggregation of these feature vectors such as attention mechanism may improve the representation power in the latent space.

In this paper, we introduced a machine learning model which uses EHR of patients along with a fused CT and PET images to perform prognosis for H&N tumor patients. The model has achieved 0.72 C-index using an optimized network. This was part of the HECKTOR 2021 challenge which has only allowed a small number of submissions during the challenge period. This has limited this work on exploring other possible experiments and optimizations to the proposed models. Nevertheless, investigating different CNN architectures as well as ways to combine EHR and imaging data has the potential to improve the model.

References

1. Orciller, V., et al.: Head and neck tumor segmentation in PET/CT: the HECKTOR challenge. Med. Image Anal. **77**, 102336 (2022)
2. Andrearczyk, V., et al.: Overview of the HECKTOR challenge at MICCAI 2021: automatic head and neck tumor segmentation and outcome prediction in PET/CT images. In: Andrearczyk, V., Oreiller, V., Hatt, M., Depeursinge, A. (eds.) HECKTOR 2021. LNCS, vol. 13209, pp. 1–37. Springer, Cham (2022)
3. Akiba, T., Sano, S., Yanase, T., Ohta, T., Koyama, M.: Optuna: a next-generation hyperparameter optimization framework. CoRR abs/1907.10902 (2019). http://arxiv.org/abs/1907.10902
4. Allende, A.S.: Concordance index as an evaluation metric (October 2019). https://medium.com/analytics-vidhya/concordance-index-72298c11eac7
5. Cox, D.R.: Regression models and life-tables. J. Roy. Stat. Soc. Ser. B (Methodol.) **34**(2), 187–202 (1972). https://doi.org/10.1111/j.2517-6161.1972.tb00899.x
6. Fotso, S.: Deep neural networks for survival analysis based on a multi-task framework (2018)
7. Gillies, R.J., Kinahan, P.E., Hricak, H.: Radiomics: images are more than pictures, they are data. Radiology **278**(2), 563–577 (2016). https://doi.org/10.1148/radiol.2015151169
8. Jin, P.: Using survival prediction techniques to learn consumer-specific reservation price distributions (2015)
9. Kazmierski, M., et al.: A machine learning challenge for prognostic modelling in head and neck cancer using multi-modal data (2021)
10. Kim, S., Kazmierski, M., Haibe-Kains, B.: Deep-CR MTLR: a multi-modal approach for cancer survival prediction with competing risks (2021)
11. Mackillop, W.J.: The Importance of Prognosis in Cancer Medicine. American Cancer Society (2006). https://doi.org/10.1002/0471463736.tnmp01.pub2
12. Parmar, C., Grossmann, P., Rietveld, D., Rietbergen, M.M., Lambin, P., Aerts, H.J.W.L.: Radiomic machine-learning classifiers for prognostic biomarkers of head and neck cancer. Front. Oncol. **5**, 272 (2015). https://doi.org/10.3389/fonc.2015.00272
13. Shboul, Z.A., Alam, M., Vidyaratne, L., Pei, L., Elbakary, M.I., Iftekharuddin, K.M.: Feature-guided deep radiomics for glioblastoma patient survival prediction. Front. Neurosci. **13**, 966 (2019). https://doi.org/10.3389/fnins.2019.00966

14. Sun, L., Zhang, S., Chen, H., Luo, L.: Brain tumor segmentation and survival prediction using multimodal MRI scans with deep learning. Front. Neurosci. **13**, 810 (2019). https://doi.org/10.3389/fnins.2019.00810

15. Tseng, W.T., Chiang, W.F., Liu, S.Y., Roan, J., Lin, C.N.: The application of data mining techniques to oral cancer prognosis. J. Med. Syst. **39**(5), 1–7 (2015). https://doi.org/10.1007/s10916-015-0241-3

16. Wang, X., Li, B.: Deep learning in head and neck tumor multiomics diagnosis and analysis: review of the literature. Front. Genet. **12**, 42 (2021). https://doi.org/10. 3389/fgene.2021.624820. https://www.frontiersin.org/article/10.3389/fgene.2021. 624820

17. Yu, C.N., Greiner, R., Lin, H.C., Baracos, V.: Learning patient-specific cancer survival distributions as a sequence of dependent regressors. In: Shawe-Taylor, J., Zemel, R., Bartlett, P., Pereira, F., Weinberger, K.Q. (eds.) Advances in Neural Information Processing Systems. vol. 24. Curran Associates, Inc. (2011). https:// proceedings.neurips.cc/paper/2011/file/1019c8091693ef5c5f55970346633f92- Paper.pdf

Progression Free Survival Prediction for Head and Neck Cancer Using Deep Learning Based on Clinical and PET/CT Imaging Data

Mohamed A. Naser$^{(\boxtimes)}$ ⓘ, Kareem A. Wahid ⓘ, Abdallah S. R. Mohamed ⓘ,
Moamen Abobakr Abdelaal ⓘ, Renjie He ⓘ, Cem Dede ⓘ, Lisanne V. van Dijk ⓘ,
and Clifton D. Fuller ⓘ

Department of Radiation Oncology, The University of Texas MD Anderson Cancer, Houston,
TX 77030, USA
manaser@mdanderson.org

Abstract. Determining progression-free survival (PFS) for head and neck squamous cell carcinoma (HNSCC) patients is a challenging but pertinent task that could help stratify patients for improved overall outcomes. PET/CT images provide a rich source of anatomical and metabolic data for potential clinical biomarkers that would inform treatment decisions and could help improve PFS. In this study, we participate in the 2021 HECKTOR Challenge to predict PFS in a large dataset of HNSCC PET/CT images using deep learning approaches. We develop a series of deep learning models based on the DenseNet architecture using a negative log-likelihood loss function that utilizes PET/CT images and clinical data as separate input channels to predict PFS in days. Internal model validation based on 10-fold cross-validation using the training data (N = 224) yielded C-index values up to 0.622 (without) and 0.842 (with) censoring status considered in C-index computation, respectively. We then implemented model ensembling approaches based on the training data cross-validation folds to predict the PFS of the test set patients (N = 101). External validation on the test set for the best ensembling method yielded a C-index value of 0.694, placing 2nd in the competition. Our results are a promising example of how deep learning approaches can effectively utilize imaging and clinical data for medical outcome prediction in HNSCC, but further work in optimizing these processes is needed.

Keywords: PET · CT · Head and neck cancer · Oropharyngeal cancer · Deep learning · Progression-free survival · Outcome prediction model

1 Introduction

Head and neck squamous cell carcinomas (HNSCC) are among the most prevalent cancers in the world. Approximately 890,000 new cases of HNSCC are diagnosed a year, and rates are projected to continue to increase [1]. While prognostic outcomes for HNSCC, particularly for oropharyngeal cancer (OPC), have improved over recent years, patients still have a significant probability of disease recurrence or death [2]. Determination of

V. Andrearczyk et al. (Eds.): HECKTOR 2021, LNCS 13209, pp. 287–299, 2022.
https://doi.org/10.1007/978-3-030-98253-9_27

progression-free survival (PFS) in HNSCC is a highly challenging task since the ultimate healthcare outcomes of patients are driven by a complex interaction between a large number of variables, including clinical demographics, treatment approaches, and underlying disease physiology. While risk prediction models based on clinical demographics for HNSCC have been developed in the past [3], these methods may lack prediction potential due to their use of a small number of simple variables or inherently linear nature.

PET/CT imaging provides an avenue to combine anatomical with functional imaging to gauge the phenotypic properties of tumors. Distinct morphologic and metabolic information derived from PET/CT has been linked to the underlying pathology of HNSCC tumors and is invaluable in diagnosis, staging, and therapeutic assessment [4]. Therefore, PET/CT imaging data has been as an attractive biomarker in developing clinical outcome prediction models for HNSCC. Radiomic analysis of PET/CT images in HNSCC has been heavily investigated in the past, with models based on statistical, geometric, and texture information contained within regions of interest showing promise in predicting PFS [5, 6]. However, radiomic methods rely on hand-crafted features and pre-defined regions of interest, which can introduce bias into analysis pipelines [7]. Therefore, deep learning approaches, which do not require a-priori definition of regions of interest and where features are learned during the model training process, have been touted as attractive alternatives in the medical imaging domain [7–10]. However, while deep learning methods for clinical prediction are promising, they are often bottlenecked by small homogenous training and evaluation datasets that limit the generalizability of models [11]. Therefore, developing and validating deep learning models for HNSCC outcome prediction on large multi-institutional datasets is of paramount importance.

The 2021 HECKTOR Challenge provides an opportunity to utilize large heterogenous training and testing PET/CT datasets with matched clinical data to further unlock the power of clinical prediction models for HNSCC. This manuscript describes the development and evaluation of a deep learning prediction model based on the DenseNet architecture that can implement PET/CT images and clinical data to predict PFS for HNSCC patients.

2   Methods

We developed and trained a deep learning model Sect. 2.3 through a cross-validation procedure Sect. 2.4 for predicting the progression-free survival (PFS) endpoint of OPC patients (based on censoring status and time-to-event between PET/CT scan and event) using co-registered ^{18}F-FDG PET and CT imaging data Sect. 2.1 and associated clinical data Sect. 2.2. The performance of the trained model for predicting PFS was validated using an internal cross-validation approach and applied to a previously unseen testing set Sect. 2.5.

2.1   Imaging Data

The data set used in this manuscript consists of co-registered ^{18}F-FDG PET and CT scans for 325 OPC patients (224 patients used for training and 101 patients used for testing).

All training and testing data were provided in Neuroimaging Informatics Technology Initiative (NIfTI) format and were released through AIcrowd [12] for the HECKTOR Challenge at MICCAI 2021 [13–15].

All images (i.e., PET and CT) were cropped to fixed bounding box volumes, provided with the imaging data by the HECKTOR Challenge organizers [12], of size 144 × 144 × 144 mm^3 in the x, y and z dimensions and then resampled to a fixed image resolution of 1 mm in the x, y, and z dimensions using spline interpolation of order 3. The cropping and resampling codes used were based on the code provided by the HECKTOR Challenge organizers (https://github.com/voreille/hecktor). The CT intensities were truncated to the range of [–200, 200] Hounsfield Units (HU) to increase soft tissue contrast and then were normalized to a [–1, 1] scale. The intensities of PET images were normalized with z-score normalization. These normalization schemes are customary in previous deep learning studies [16]. We used the Medical Open Network for AI (MONAI) [17] software transformation packages to rescale and normalize the intensities of the PET/CT images. Image processing steps used in this manuscript are displayed in Fig. 1A and 1B.

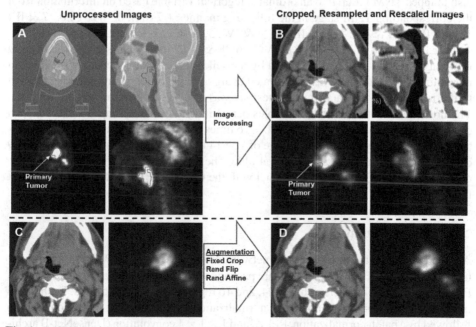

Fig. 1. An illustration of the workflow used for image processing. (A) Overlays of the provided ground truth tumor segmentation mask (used to highlight tumor location in the image but not used in any analysis done in the paper) and the original CT (top) and PET (bottom) images. (B) Illustration of data augmentation that included cropping, random flips, and affine transformations used during the model training.

2.2 Clinical Data

Clinical data for patients were provided in.csv files, which included PFS outcome information and clinical variables. As per the HECKTOR website, "progression is defined based on RECIST criteria: either a size increase of known lesions (change of T and or N), or appearance of new lesions (change of N and/or M). Disease-specific death is also considered a progression event for patients previously considered stable." We used all the clinical variables provided by the HECKTOR Challenge with complete fields (no NaN values present) for all patients. These variables included Center ID, Gender, Age, TNM edition, chemotherapy status, TNM group, T-stage, N-stage, and M-stage (9 variables in total). We elected not to use variables with incomplete fields (NaN values present), i.e., performance status, smoking status, etc., to avoid issues in the model building process. We mapped the individual T-, N-, and M- staging information to ordinal categorical variables in the following manner: T-stage: ('Tx': 0, 'T1': 1, 'T2': 2, 'T3': 3, 'T4': 4, 'T4a': 5, 'T4b': 6) – 7 classes; N-stage ('N0': 0, 'N1': 1, 'N2': 'N2a': 3, 'N2b': 4, 'N2c': 5, 'N3': 6) – 7 classes; M-stage ('Mx': 0, 'M0': 0, 'M1': 1) – 2 classes. We also mapped TNM group into an ordinal categorical variable based on information from the corresponding TNM stage in the following manner: ('7 & I': 0, '8 & I': 0, '7 & II': 1, '8 & II': 1, '7 & III': 2, '8 & III': 2, '7 & IV': 3, '8 & IV': 3, '7 & IVA': 4, '8 & IVA': 4, '7 & IVB': 5, '8 & IVB': 5, '7 & IVC': 6, '8 & IVC': 6) – 7 classes. We then used a min-max rescaling approach provided by the scikit-learn Python package [18] for data normalization. The scaler was instantiated using only the training data and then applied to the test set, i.e., no data from the test set was allowed to leak into the training set. The normalized clinical data (9 variables per patient) was then reshaped such that each clinical variable is repeated $144 \times 144 \times 16$ times in the x, y, and z dimensions, respectively and therefore a volumetric image of size $144 \times 144 \times 144$ was formed from the 9 concatenated images of each clinical variable. The volumetric image generated from the clinical data was used as a third channel with the CT and PET images to the DenseNet model Sect. 2.3.

2.3 Model Architecture

A deep learning convolutional neural network model based on the DenseNet121 architecture included in the MONAI software package was used for the analysis. As shown in Fig. 2, the network consists of 6, 12, 24, and 16 repetitions of dense blocks. Each dense block contained a pre-activation batch normalization, ReLU, and $3 \times 3 \times 3$ convolution followed by a batch normalization, ReLU, and $1 \times 1 \times 1$ convolution (DenseNet-B architecture). The network implementation in PyTorch used was provided by MONAI [17]. The model has 3 input channels each of size $144 \times 144 \times 144$ and 20 output channels representing different time intervals of the predicted survival probabilities Sect. 2.4.

2.4 Model Implementation

We used a 10-fold cross-validation approach where the 224 patients from the training data were divided into 10 non-overlapping sets. Each set (22 patients) was used for model validation while the remaining 202 patients in the remaining sets were used for training,

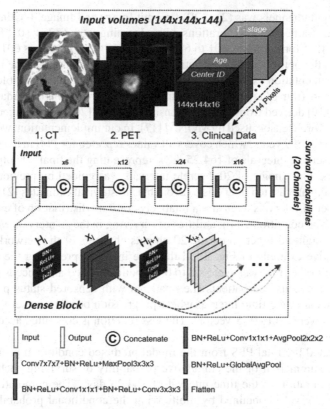

Fig. 2. Schematic of the Densenet121 architecture used for the prediction model and the 3 input volumetric images representing CT, PET and clinical data. Clinical data is reformatted into a volumetric input for the model. The number of repeated dense blocks of (6, 12, 24, and 16) are given above each group of blocks.

i.e., each set was used once for testing and 9 times for training. The processed PET and CT images Sect. 2.1 were cropped to a fixed size of (144, 144, 144) per image per patient. The clinical data was provided as a third channel to the model. We implemented additional data augmentation to the CT and PET channels, including random horizonal flips of 50% and random affine transformations with an axial rotation range of 12° and a scale range of 10%. The image processing and data augmentation Sect. 2.1, and network architecture Sect. 2.3 were used from the software packages provided by the MONAI framework [17]; code for these packages can be found at "https://github.com/Project-MONAI/".

We implemented two main approaches for model building which utilizes imaging data only (Image), i.e., 2 channel input – CT/PET, or using imaging plus clinical data (Image + Clinical), i.e., 3 channel input – CT/PET + clinical data. We used a batch size of 2 patients' images and clinical data and, therefore, the shape of the input tensor provided to the network Sect. 2.3 for the three-channel inputs was (2, 3, 144, 144, 144) and for the two-channel inputs was (2, 2, 144, 144, 144). The shape of the output tensor for the 20 output channels was (2, 1, 20) for both Image and Image + Clinical models. The model was trained for 800 iterations with a learning rate of (2×10^{-4} for iterations 0 to 300, 1×10^{-4} for iterations 301 to 600, and 5×10^{-5} for iterations 601 to 800). We used Adam as the optimizer and a negative log-likelihood function as the loss function [19]. An implementation example of the log-likelihood function compatible with Keras was provided in (https://github.com/chestrad/survival_prediction/blob/master/loss.py) by Kim et al. [19] derived from work by Gensheimer et al. [20]. We modified the code to match with PyTorch tensors used by MONAI [17]. In our implementation, we divided the total time interval of 9 years, which covers all values reported for PFS in the training data set, into 20 discreate intervals of 164.25 days, representing the final 20 output channels of the network. Our choice of 20 discrete intervals was guided by previous heuristics suggesting using at least ten intervals to avoid bias in survival estimates [21] and a large enough number of intervals to ensure a relatively constant distribution of events in each interval [20]. The conditional probabilities of surviving in these intervals were obtained by applying a sigmoid function on the 20 outputs channels of the network. As shown in the illustrative example of Fig. 3, all the time intervals preceding the events were set to 1, and all other intervals were set to 0 for both censored and uncensored patients (S vector). For uncensored patients (i.e., patients with censored status p = 1, where progression occurs), the time interval where progression occurs was set to 1 while all other intervals were set to 0 ($\bar{S}$ vector). The loss function is computed according to the equation shown in Fig. 3.

We estimated the final PFS from the model predicted conditional probabilities by obtaining the summation of the cumulative probability of surviving each time interval times the duration of the time interval of 164.25 days. The cumulative probability of a time interval is obtained by multiplying the conditional probabilities of surviving all preceding time intervals times the time interval duration (i.e., $\text{PFS}^{pred} = \sum_{k=1}^{20} \left(\prod_{i=1}^{k} S_i^{pred} \right) *$ (time interval duration $= 164.25$)).

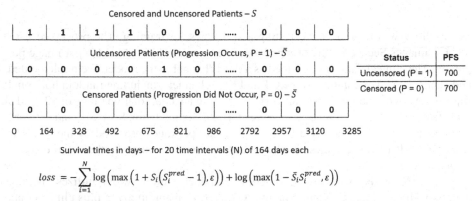

$$loss = -\sum_{i=1}^{N} \log\left(\max\left(1 + S_i\left(S_i^{pred} - 1\right), \varepsilon\right)\right) + \log\left(\max\left(1 - \bar{S}_i S_i^{pred}, \varepsilon\right)\right)$$

Fig. 3. A graphical illustration of the negative log likelihood loss function used by the DenseNet model.

2.5 Model Validation

For each validation fold (i.e., 22 patients), we trained the DenseNet121 architecture Sect. 2.3 on the remaining 202 patients. Therefore, we obtained 10 different trained models from 10-fold cross-validation. We evaluated the performance of each separate model on the corresponding validation set using concordance index (C-index) function of the lifelines package (https://github.com/CamDavidsonPilon/lifelines) which accounts for both censored and uncensored data through the indication of the time events occurrence of progression (i.e., progression occurs – True or False). Broadly, the C-index represents the ability of an algorithm to correctly provide a ranking of survival times based on individual risk scores [22]. We estimated the mean C-index by averaging all the C-index values obtained from each fold. It is also possible to measure the C-index by ignoring the events observed. Therefore, as an alternative metric, we also measured this modified C-index in reporting results.

For the test set (101 patients), we implemented two different model ensembling approaches to estimate the PFS. In the first approach, we estimated the PFS for each patient by each model and then obtained the mean value of the 10 predicted PFS values (AVERAGE approach). In the second approach, we first estimated the mean conditional probability survival vector by getting the mean value for each time interval. Then we computed the cumulative survival probability for each interval to estimate the consensus PFS values from the 10 models (CONSENSUS approach). These ensembling approaches were chosen due to their simplicity and ease of implementation.

3 Results

The performance (predicted PFS predictions vs. ground truth) of each set of the 10-fold cross-validation procedure for the Image model and Image + Clinical model are shown in Fig. 4 and Fig. 5, respectively. For both models, most individual fold predictions were not significantly different from the ground truth PFS ($p > 0.05$), except sets 6 and 8 for the Image model and sets 2 and 6 for the Image + Clinical model. The cumulative mean performance measured over all folds for both models are shown in Table 1. Both models offered similar performance, regardless of the C-index calculation method.

When evaluated through external validation (test set), our model predictions from each cross-validation fold were ensembled using two methods (AVERAGE and CONSENSUS), leading to a total of four tested models (Image AVERAGE, Image CONSENSUS, Image + Clincial AVERAGE, Image + Clinical CONSENSUS). The results on the test set for these separate models are shown in Table 2. The highest performing model overall was the Image + Clinical CONSENSUS model with a C-index of 0.694, which achieved a 2nd place ranking in the competition.

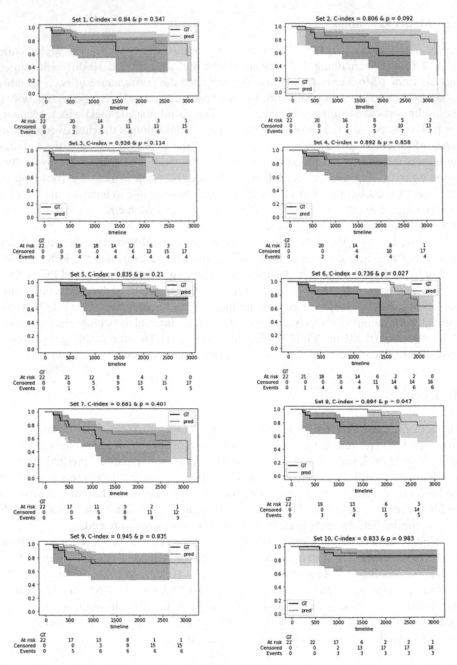

Fig. 4. Kaplan Meier plots showing survival probabilities as a function of time in days for the ground truth (GT) PFS and the predicted PFS by the model using only imaging data (i.e., CT and PET) for each validation set of 22 patients. The C-index and the p-value of the logrank test for the GT and predicted PFS are shown above each subplot.

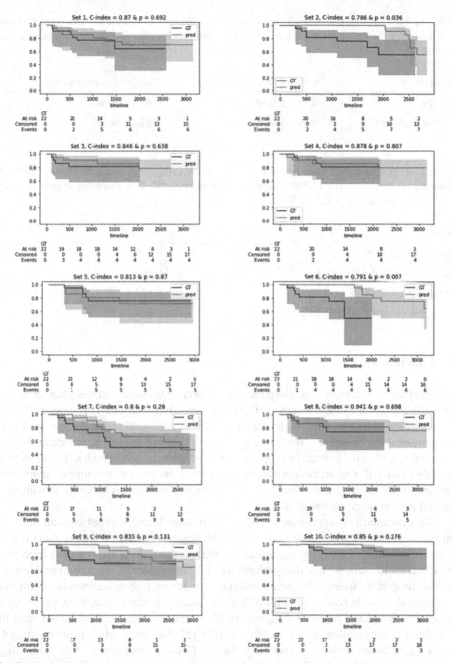

Fig. 5. Kaplan Meier plots showing survival probabilities as a function of time in days for the ground truth (GT) PFS and the predicted PFS by the model using both imaging and clinical data for each validation set of 22 patients. The C-index and the p-value of the logrank test for the GT and predicted PFS are shown above each subplot.

Table 1. Mean C-index across all cross-validation folds for evaluated models (Image, Image + Clinical). C-index can be calculated with and without observed events; both calculations are reported.

Model	C-index (events used)	C-index (no events used)
Image	0.842 ± 0.080	0.620 ± 0.082
Image + clinical	0.841 ± 0.045	0.622 ± 0.067

Table 2. Progression-free survival prediction measured using the C-index for ensemble models submitted to the 2021 HECKTOR Challenge evaluation portal. Two ensemble approaches were tested (AVERAGE, CONSENSUS).

Model	C-index
Image AVERAGE	0.645
Image + clinical AVERAGE	0.689
Image CONSENSUS	0.651
Image + clinical CONSENSUS	0.694

4 Discussion

In this study, we have utilized deep learning approaches based on DenseNet applied to PET/CT images and clinical data of HNSCC patients to predict PFS. The determination of prognostic outcomes is an unmet need for HNSCC patients that could improve clinical decision-making processes. While the performance of our models is not ideal, they are reasonable within the context of prognostic prediction, which is known to be notoriously complex task [6]. Our main innovation stems from the use of ensembling approaches applied to predictions from internal cross-validation, which seems to reasonably improve overall performance on unseen data.

We evaluated two approaches (PET/CT inputs with and without clinical data) for internal validation through a 10-fold cross-validation approach in the training set to gauge performance before applying the models to the test set. When investigating individual cross-validation sets, we find that performance for both models is often comparable to the ground truth PFS. Specifically, when utilizing C-index with observed events information considered, models could reach fairly high prediction performance, with values up to 0.842 for the Image model. Compared to PFS prediction models evaluated with the C-index in other studies, this is notably higher [5]. However, when implementing C-index calculations without considering observed events information, this value drops precipitously for both models. Specifically, a maximum value of 0.622 is achieved, which is more consistent with previously reported values for similar tasks. Interestingly, the addition of clinical data generally did not make any noticeable improvements in model performance, regardless of the C-index metric calculation method, possibly indicating the majority of informative data was contained within the PET/CT images. Moreover, it may be possible that secondary to the cross-validation evaluation scheme, the limited number of patients in the evaluation sets did not have significant signal in their clinical variables to improve performance. This observation runs counter to results seen in other clinical prediction models, where the addition of clinical data to imaging information often improves performance [6].

For the external validation (testing set) we further investigate the effect of specific model ensembling techniques through the AVERAGE and CONSENSUS methods. Generally, the CONSENSUS method (consensus from cumulative survival probability derived from mean conditional probability survival vectors) seems to offer performance gains over the AVERAGE method. Importantly, model performance on the test set is improved compared to the training set (assuming C-index calculation did not incorporate the observed events). Specifically, we achieve C-index gains of approximately 0.07 for test set performance. This may be due to the ensemble approach adding generalizability acquired from the combined inference capabilities shared across multiple cross-validation sets. Interestingly, while adding clinical data did not make noticeable differences in the training set, this was not the case in the test set, which offered substantial C-index gains of approximately 0.04. The discrepancy may be explained by clinical data being more relevant when used in conjunction with model ensembling. Model ensembling is known to be a powerful technique in machine learning generally. This has been reiterated in other tasks for the HECKTOR Challenge, where ensembling provided impressive performance gains for tumor segmentation [23]. Moreover, our ensembling approaches demonstrated encouraging results in the 2021 HECKTOR Challenge leaderboard, placing 2nd overall and achieving a C-index only 0.03 lower than the winning entry (BioMedIA). Therefore, we emphasize that model ensembling techniques may also be relevant for deep learning prediction models in HNSCC.

While we have taken steps to ensure a robust analysis, our study contains several limitations. Firstly, while we have included clinical data in our model-building process, we have not included all the data initially provided by the HECKTOR Challenge. Specifically, we have not included clinical data with incomplete fields (NaNs) for any patients, such as tobacco status, alcohol status, and human papillomavirus status. Importantly, it is precisely these variables that are often the most highly correlated to prognosis in HNSCC [2, 24]. Therefore, data imputation techniques or related methods should be implemented in future studies to fully realize available clinical variables' discriminative capabilities. A second limitation of our approach is we have used an out-of-the-box DenseNet architecture that has not been specifically optimized for imaging data combined with tabular clinical data. Moreover, to ensure that we could utilize our DenseNet architecture with an additional channel input for clinical data, we have concatenated clinical data into a 3D volume for model input. While using an open-source commonly available approach improves reproducibility, further studies should determine how architectural modifications can be made to optimize performance on this specific task. Additionally, in our model design process, we have selected discrete intervals based on observations in the training data. However, this assumes the training data fully encapsulates the PFS landscape, which may not necessarily be the case. While previous studies have demonstrated these models are relatively robust to the choice of specific discrete intervals [20], further work should be performed to determine if there are more optimal intervals that can be implemented for this type of data. Finally, since deep learning ensembling applied to PFS prediction is relatively understudied, we have approached model ensembling of individual cross-validation folds through relatively simple methods that involve linear combinations of individual model predictions. More complex methods for deep learning ensembling [25] may provide additional predictive power for PFS tasks.

5 Conclusion

Using PET/CT and clinical data inputs, we developed, trained, and validated a series of deep learning models that could predict PFS in HNSCC patients in large-scale datasets provided by the 2021 HECKTOR Challenge. Cross-validation performance on the training set achieved mean C-index values of up to 0.622 (0.842 with alternative C-index calculations). Simple model ensembling approaches improved this performance further with reported C-index values up to 0.694 on the testing set. While our models offer modest performance on test data, these methods can be enhanced through additional clinical inputs, improved architectural modifications, and alternative ensembling approaches.

Acknowledgements. M.A.N. is supported by a National Institutes of Health (NIH) Grant (R01 DE028290-01). K.A.W. is supported by a training fellowship from The University of Texas Health Science Center at Houston Center for Clinical and Translational Sciences TL1 Program (TL1TR003169), the American Legion Auxiliary Fellowship in Cancer Research, and a NIDCR F31 fellowship (1 F31 DE031502-01). C.D.F. received funding from the National Institute for Dental and Craniofacial Research Award (1R01DE025248-01/R56DE025248) and Academic-Industrial Partnership Award (R01 DE028290), the National Science Foundation (NSF), Division of Mathematical Sciences, Joint NIH/NSF Initiative on Quantitative Approaches to Biomedical Big Data (QuBBD) Grant (NSF 1557679), the NIH Big Data to Knowledge (BD2K) Program of the National Cancer Institute (NCI) Early Stage Development of Technologies in Biomedical Computing, Informatics, and Big Data Science Award (1R01CA214825), the NCI Early Phase Clinical Trials in Imaging and Image-Guided Interventions Program (1R01CA218148), the NIH/NCI Cancer Center Support Grant (CCSG) Pilot Research Program Award from the UT MD Anderson CCSG Radiation Oncology and Cancer Imaging Program (P30CA016672), the NIH/NCI Head and Neck Specialized Programs of Research Excellence (SPORE) Developmental Research Program Award (P50 CA097007) and the National Institute of Biomedical Imaging and Bioengineering (NIBIB) Research Education Program (R25EB025787). He has received direct industry grant support, speaking honoraria and travel funding from Elekta AB.

References

1. Johnson, D.E., Burtness, B., Leemans, C.R., Lui, V.W.Y., Bauman, J.E., Grandis, J.R.: Head and neck squamous cell carcinoma. Nat. Rev. Dis. Prim. **6**, 1–22 (2020)
2. Chow, L.Q.M.: Head and neck cancer. N. Engl. J. Med. **382**, 60–72 (2020)
3. Budach, V., Tinhofer, I.: Novel prognostic clinical factors and biomarkers for outcome prediction in head and neck cancer: a systematic review. Lancet Oncol. **20**, e313–e326 (2019)
4. Goel, R., Moore, W., Sumer, B., Khan, S., Sher, D., Subramaniam, R.M.: Clinical practice in PET/CT for the management of head and neck squamous cell cancer. Am. J. Roentgenol. **209**, 289–303 (2017)
5. Haider, S.P., Burtness, B., Yarbrough, W.G., Payabvash, S.: Applications of radiomics in precision diagnosis, prognostication and treatment planning of head and neck squamous cell carcinomas. Cancers Head Neck **5**, 1–19 (2020)

6. Chinnery, T., et al.: Utilizing artificial intelligence for head and neck cancer outcomes prediction from imaging. Can. Assoc. Radiol. J. **72**, 73–85 (2021)
7. Hosny, A., Aerts, H.J., Mak, R.H.: Handcrafted versus deep learning radiomics for prediction of cancer therapy response. Lancet Digit. Health **1**, e106–e107 (2019)
8. Sun, Q., et al.: Deep learning vs. radiomics for predicting axillary lymph node metastasis of breast cancer using ultrasound images: don't forget the peritumoral region. Front. Oncol. **10**, 53 (2020)
9. Avanzo, M., et al.: Machine and deep learning methods for radiomics. Med. Phys. **47**, e185–e202 (2020)
10. Hosny, A., et al.: Deep learning for lung cancer prognostication: a retrospective multi-cohort radiomics study. PLoS Med. **15**, e1002711 (2018)
11. Willemink, M.J., et al.: Preparing medical imaging data for machine learning. Radiol. **295**, 4–15 (2020)
12. AIcrowd MICCAI 2020: HECKTOR Challenges
13. Andrearczyk, V.: Overview of the HECKTOR challenge at MICCAI 2020: automatic head and neck tumor segmentation in PET/CT. In: Andrearczyk, V., Oreiller, V., Depeursinge, A. (eds.) HECKTOR 2020. LNCS, vol. 12603, pp. 1–21. Springer, Cham (2021). https://doi.org/10.1007/978-3-030-67194-5_1
14. Andrearczyk, V., et al.: Overview of the HECKTOR challenge at MICCAI 2021: automatic head and neck tumor segmentation and outcome prediction in PET/CT images. In: Andrearczyk, V., Oreiller, V., Hatt, M., Depeursinge, A. (eds.) HECKTOR 2021. LNCS, vol. 13209, pp. 1–37. Springer, Cham (2022)
15. Valentin, O., et al.: Head and neck tumor segmentation in PET/CT: the HECKTOR challenge. Med. Image Anal. **25**(77), 102336 (2021)
16. Naser, M.A., Dijk, L.V., He, R., Wahid, K.A., Fuller, C.D.: Tumor segmentation in patients with head and neck cancers using deep learning based-on multi-modality PET/CT images. In: Andrearczyk, V., Oreiller, V., Depeursinge, A. (eds.) HECKTOR 2020. LNCS, vol. 12603, pp. 85–98. Springer, Cham (2021). https://doi.org/10.1007/978-3-030-67194-5_10
17. The MONAI Consortium: Project MONAI (2020). https://doi.org/10.5281/zenodo.4323059
18. Pedregosa, F., et al.: Scikit-learn: machine learning in Python. J. Mach. Learn. Res. **12**, 2825–2830 (2011)
19. Kim, H., Goo, J.M., Lee, K.H., Kim, Y.T., Park, C.M.: Preoperative CT-based deep learning model for predicting disease-free survival in patients with lung adenocarcinomas. Radiol. **296**, 216–224 (2020). https://doi.org/10.1148/radiol.2020192764
20. Gensheimer, M.F., Narasimhan, B.: A scalable discrete-time survival model for neural networks. PeerJ **7**, e6257 (2019)
21. Breslow, N., Crowley, J.: A large sample study of the life table and product limit estimates under random censorship. Ann. Stat. **2**, 437–453 (1974)
22. Uno, H., Cai, T., Pencina, M.J., D'Agostino, R.B., Wei, L.-J.: On the C-statistics for evaluating overall adequacy of risk prediction procedures with censored survival data. Stat. Med. **30**, 1105–1117 (2011)
23. Iantsen, A., Visvikis, D., Hatt, M.: Squeeze-and-excitation normalization for automated delineation of head and neck primary tumors in combined PET and CT images. In: Andrearczyk, V., Oreiller, V., Depeursinge, A. (eds.) HECKTOR 2020. LNCS, vol. 12603, pp. 37–43. Springer, Cham (2021). https://doi.org/10.1007/978-3-030-67194-5_4
24. Leemans, C.R., Snijders, P.J.F., Brakenhoff, R.H.: The molecular landscape of head and neck cancer. Nat. Rev. Cancer. **18**, 269–282 (2018)
25. Ganaie, M.A., Hu, M.: Ensemble deep learning: a review. arXiv Prepr. arXiv:2104.02395 (2021)

Combining Tumor Segmentation Masks with PET/CT Images and Clinical Data in a Deep Learning Framework for Improved Prognostic Prediction in Head and Neck Squamous Cell Carcinoma

Kareem A. Wahid(✉) 🔟, Renjie He 🔟, Cem Dede 🔟, Abdallah S. R. Mohamed 🔟,
Moamen Abobakr Abdelaal 🔟, Lisanne V. van Dijk 🔟, Clifton D. Fuller 🔟,
and Mohamed A. Naser 🔟

Department of Radiation Oncology, The University of Texas MD Anderson Cancer, Houston,
TX 77030, USA
{kawahid,manaser}@mdanderson.org

Abstract. PET/CT images provide a rich data source for clinical prediction models in head and neck squamous cell carcinoma (HNSCC). Deep learning models often use images in an end-to-end fashion with clinical data or no additional input for predictions. However, in the context of HNSCC, the tumor region of interest may be an informative prior in the generation of improved prediction performance. In this study, we utilize a deep learning framework based on a DenseNet architecture to combine PET images, CT images, primary tumor segmentation masks, and clinical data as separate channels to predict progression-free survival (PFS) in days for HNSCC patients. Through internal validation (10-fold cross-validation) based on a large set of training data provided by the 2021 HECKTOR Challenge, we achieve a mean C-index of 0.855 ± 0.060 and 0.650 ± 0.074 when observed events are and are not included in the C-index calculation, respectively. Ensemble approaches applied to cross-validation folds yield C-index values up to 0.698 in the independent test set (external validation), leading to a 1st place ranking on the competition leaderboard. Importantly, the value of the added segmentation mask is underscored in both internal and external validation by an improvement of the C-index when compared to models that do not utilize the segmentation mask. These promising results highlight the utility of including segmentation masks as additional input channels in deep learning pipelines for clinical outcome prediction in HNSCC.

Keywords: PET · CT · Head and neck cancer · Oropharyngeal cancer · Deep learning · Progression-free survival · Outcome prediction model · Segmentation mask

1 Introduction

Deep learning has been studied extensively for medical image segmentation and clinical outcome prediction in head and neck squamous cell carcinoma (HNSCC) [1]. While

© Springer Nature Switzerland AG 2022
V. Andrearczyk et al. (Eds.): HECKTOR 2021, LNCS 13209, pp. 300–307, 2022.
https://doi.org/10.1007/978-3-030-98253-9_28

utilizing medical images in an end-to-end framework with no additional data streams has become commonplace, previous studies have noted the importance of adding regions of interest in a deep learning workflow to improve predictive performance [2, 3]. These regions of interest may help deep learning models localize to areas that harbor more relevant information for the downstream prediction task of interest. This may be particularly salient for HNSCC patients, where the prognostic status is often informed by the re-appearance of tumor volumes in or near the original region of interest (recurrence/treatment failure) [4]. Moreover, PET/CT images provide a rich source of information for the tumor region of interest in HNSCC [5] that may be combined with deep learning models to improve prognostic prediction, such as for progression-free survival (PFS). Therefore, the development of PET/CT-based deep learning approaches that can effectively combine previously segmented tumor regions of interest with existing architectures is an important component in exploring novel and effective HNSCC outcome prediction models.

In this study, we develop and evaluate a deep learning model based on the DenseNet architecture that combines PET/CT images, primary tumor segmentation masks, and clinical data to predict PFS in HNSCC patients provided by the 2021 HECKTOR Challenge. By combining these various information streams, we demonstrate reasonable performance on internal and external validation sets, achieving a top ranking in the competition.

2   Methods

We developed a deep learning model for PFS prediction of HNSCC patients using co-registered ^{18}F-FDG PET and CT imaging data, ground truth primary tumor segmentation masks, and associated clinical data. The censoring status and time-to-event between PET/CT scan and event, imaging data, segmentation masks, and clinical data were used to train the model. The performance of the trained model for predicting PFS was validated using a 10-fold cross-validation approach. An ensemble model based on the predictions from the 10-fold cross-validation models was packaged into a Docker container to evaluate models on unseen testing data. Abbreviated methodology salient to the use of segmentation masks in our models is described below. For further details, we refer the reader to our parallel study on Task 2 of the HECKTOR Challenge, i.e., PFS prediction without using segmentation mask information [6].

Data from 224 HNSCC patients from multiple institutions was provided in the 2021 HECKTOR Challenge [7–10] training set. Data for these patients included co-registered ^{18}F-FDG PET and CT scans, clinical data (Table 1), and ground truth manual segmentations of primary tumors derived from clinical experts. Images and masks were cropped to a bounding box of $144 \times 144 \times 144$ mm^3 and resampled to a resolution of 1 mm. The CT intensities were truncated to the range of [−200, 200] and then normalized to a [−1, 1] scale, while PET intensities were normalized with a z-score. Clinical data were mapped to ordinal categorical variables if applicable (Table 1) and then min-max rescaled. The clinical data was reshaped into a $144 \times 144 \times 144$ volume by concatenating clinical variables repeatedly ($144 \times 144 \times 16$ matrix of repeating elements for each variable) to act as a volumetric input to the deep learning model.

Table 1. Clinical variables used in this study and corresponding categorical mappings (if applicable) used in model building. Note: No variables with empty values (NaNs) in any entries were used in this study. * TNM group ordinal categorical variable was also based on information from the corresponding TNM stage.

Clinical variable	Categorical mapping
Center ID	None – used as is, 0,1,2,3,4,5
Age	None – used as is, continuous
Gender	None – used as is, Male:0, Female:1
TNM edition	None – used as is, 7:0, 8:1
Chemotherapy status	None – used as is, No:0, Yes:1
TNM group*	'7 & I': 0, '8 & I': 0, '7 & II': 1, '8 & II': 1, '7 & III': 2, '8 & III': 2, '7 & IV': 3, '8 & IV': 3, '7 & IVA': 4, '8 & IVA': 4, '7 & IVB': 5, '8 & IVB': 5, '7 & IVC': 6, '8 & IVC': 6
T-stage	'Tx': 0, 'T1': 1, 'T2': 2, 'T3': 3, 'T4': 4, 'T4a': 5, 'T4b': 6
N-stage	'N0': 0, 'N1': 1, 'N2': 2, 'N2a': 3, 'N2b': 4, 'N2c': 5, 'N3': 6
M-stage	'Mx': 0, 'M0': 0, 'M1': 1

A deep learning convolutional neural network model based on the DenseNet121 [11] architecture included in the MONAI Python package [12] was used for the analysis (Fig. 1); further details on the architecture can be found in the MONAI documentation. Each dense block was comprised of a pre-activation batch normalization, ReLU, and $3 \times 3 \times 3$ convolution followed by a batch normalization, ReLU, and $1 \times 1 \times 1$ convolution, where dense blocks were repeated in a (6, 12, 24, 16) pattern. Processed PET/CT, tumor mask, and clinical data volumes were used as separate input channels to the model. We used data augmentation by MONAI [12] during training which was applied to the PET/CT and tumor mask images. The augmentation included random horizontal flips of 50% and random affine transformations with an axial rotation range of 12° and a scale range of 10%. We used a batch size of 2 patients' images, masks, and clinical data. The model was trained for 800 iterations with a learning rate of 2×10^{-4} for iterations 0 to 300, 1×10^{-4} for iterations 301 to 600, and 5×10^{-5} for iterations 601 to 800. We used an Adam optimizer and a negative log-likelihood loss function. The negative log-likelihood loss function was adapted from previous studies [13, 14] which can account for both PFS and censoring information; a detailed description of our implementation is given in [6].

As described on the HECKTOR Challenge website, progression was defined based on RECIST criteria: either a size increase of known lesions (change of T and or N), or appearance of new lesions (change of N and/or M), where disease-specific death was also considered a progression event. To model PFS we divided the total time interval of 9 years, which covers all values reported for PFS in the training data set, into 20 discrete intervals of 164.25 days, representing the final 20 output channels of the network. The discretization scheme was adapted from previous studies [13, 14]. The conditional probabilities of surviving in the discretized intervals were obtained by applying a sigmoid

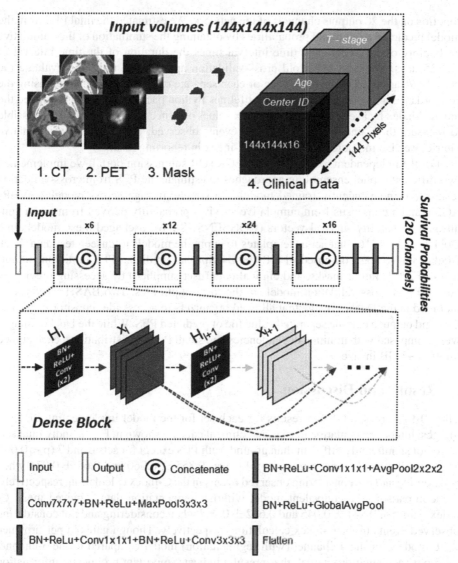

Fig. 1. Schematic of the Densenet121 architecture used for the prediction model and the four channel input volumetric images representing CT, PET, tumor mask, and clinical data. Ground truth tumor segmentation masks for model training were provided by the 2021 HECKTOR Challenge and accessed via a Docker framework for test set evaluation. The number of repeated dense blocks of (6, 12, 24, and 16) are given above each group of blocks.

function on the 20 outputs channels of the network. We estimated the final PFS from the model predicted conditional probabilities by obtaining the summation of the cumulative probability of surviving each time interval times the duration of the time interval of 164.25 days. We used a 10-fold cross-validation approach to train and evaluate the model. We assessed the performance of each separate cross-validation model using the concordance index (C-index) from the lifelines Python package [15]. We estimated the mean C-index by averaging all the C-index values obtained from each fold. It is possible to measure the C-index by ignoring the events observed; therefore, as an alternative metric, we also measured this modified C-index in reporting results.

For the independent test set on the HECKTOR submission portal, we implemented two different model ensembling approaches to estimate the PFS from cross-validation results (10 independently trained models): a simple average across models (AVERAGE), and a consensus from cumulative survival probability derived from mean conditional probability survival vectors (CONSENSUS). We packaged these models into Docker images [16] that act as templates to apply the models to unseen test data. Each Docker image was composed of a Python script that predicted PFS (anti-concordant) from PET/CT, tumor masks, and clinical data (user supplied) by accessing previously built 10-fold cross validation models in the form of.pth files. Two BASH scripts were included in the Docker image that allowed organizers to access file repositories for test data and output a comma separated value file of predicted PFS. While the Docker images were composed with minimal dependencies and small Python distributions, each zipped file was ~3 GB in size.

3   Results and Discussion

The 10-fold cross-validation results for each set for the model implementing PET/CT images, tumor segmentation masks, and clinical data are shown in Fig. 2. Individual sets were not significantly different than ground truth PFS except for sets 6 and 7 ($p < 0.05$). The overall mean C-index across all folds was 0.855 ± 0.060 and 0.65 ± 0.074 when considering and not considering observed events in the C-index calculation, respectively. For comparison, the equivalent model without segmentation data achieved mean C-index values of 0.841 ± 0.045 and 0.622 ± 0.067 when considering and not considering observed events in the C-index calculation, respectively. Though gains in performance were modest for the 4-channel (with segmentation) model compared to the 3-channel (without segmentation) model, these results indicate important prognostic information can further be teased out of a region of interest to improve the performance of a deep learning clinic prediction model. Upon submitting our ensemble models on external test set validation, the AVERAGE method demonstrates a C-index of 0.696 while the CONSENSUS method demonstrates a C-index of 0.698. Both these methods improve upon the analogous submissions in Task 2 of the HECKTOR Challenge, i.e. 0.689 and 0.694 for the AVERAGE and CONSENSUS models without segmentation masks, respectively [6]. Compared to the internal validation results, the increase in external validation performance for these models is likely secondary to improved generalization introduced by the ensembling approaches. Importantly, our best model was ranked 1st in the HECKTOR Challenge Task 3 leaderboard, beating the 2nd place model's C-index by 0.0272 (civodlu). Previous studies have shown that combining deep learning

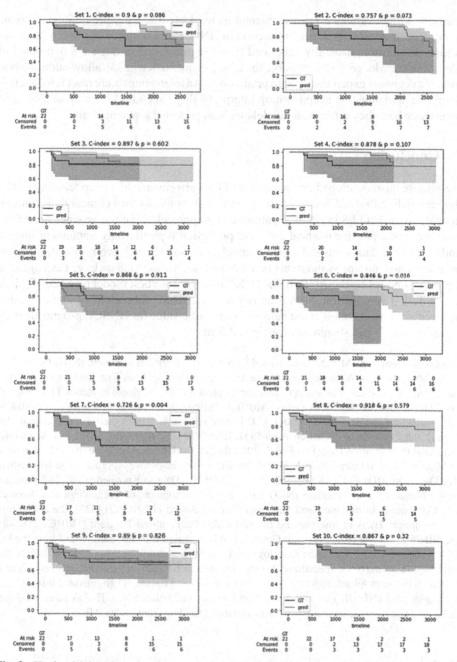

Fig. 2. Kaplan Meier plots showing survival probabilities as a function of time in days for the ground truth (GT) PFS and the predicted PFS by the four channel (PET, CT, tumor segmentation mask, clinical) model. The C-index and the p-value of the logrank test for the GT and predicted PFS are shown above each subplot. The tables below each plot correspond to the number of patients experiencing GT PFS at each time-step.

methods with information from segmentation masks (particularly pre-defined radiomic features) can improve outcome prediction in HNSCC [2], which is consistent with our results. In addition, manually generated tumor segmentation masks are a byproduct of physician knowledge, so it is possible the added segmentation masks allow our models to impart a degree of expert insight into predictions. An interesting future research direction would be implementing deep learning interpretability methods [17] on these models to investigate how they differ from models without provided segmentations.

4   Conclusion

Herein, we have developed and investigated the performance of a deep learning model that can utilize PET/CT images, tumor segmentation masks, and clinical data simultaneously to predict PFS in HNSCC patients. This approach is innovative since most deep learning techniques for medical outcome prediction typically only implement images and/or clinical data as model input channels. Our models achieve reasonable performance through internal-validation and external test set validation on large heterogenous datasets provided by the 2021 HECKTOR Challenge; our best model received 1st place in the Task 3 competition. Moreover, our results indicate subregions of interest acting as separate input channels could help add prognostic value for HNSCC prognostic deep learning models and should be investigated further.

Acknowledgements. M.A.N. is supported by a National Institutes of Health (NIH) Grant (R01 DE028290-01). K.A.W. is supported by a training fellowship from The University of Texas Health Science Center at Houston Center for Clinical and Translational Sciences TL1 Program (TL1TR003169), the American Legion Auxiliary Fellowship in Cancer Research, and a NIDCR F31 fellowship (1 F31 DE031502-01). C.D.F. received funding from the National Institute for Dental and Craniofacial Research Award (1R01DE025248-01/R56DE025248) and Academic-Industrial Partnership Award (R01 DE028290), the National Science Foundation (NSF), Division of Mathematical Sciences, Joint NIH/NSF Initiative on Quantitative Approaches to Biomedical Big Data (QuBBD) Grant (NSF 1557679), the NIH Big Data to Knowledge (BD2K) Program of the National Cancer Institute (NCI) Early Stage Development of Technologies in Biomedical Computing, Informatics, and Big Data Science Award (1R01CA214825), the NCI Early Phase Clinical Trials in Imaging and Image-Guided Interventions Program (1R01CA218148), the NIH/NCI Cancer Center Support Grant (CCSG) Pilot Research Program Award from the UT MD Anderson CCSG Radiation Oncology and Cancer Imaging Program (P30CA016672), the NIH/NCI Head and Neck Specialized Programs of Research Excellence (SPORE) Developmental Research Program Award (P50 CA097007) and the National Institute of Biomedical Imaging and Bioengineering (NIBIB) Research Education Program (R25EB025787). He has received direct industry grant support, speaking honoraria and travel funding from Elekta AB.

References

1. Wang, X., Li, B.: Deep learning in head and neck tumor multiomics diagnosis and analysis: review of the literature. Front. Genet. **12**, 42 (2021)
2. Diamant, A., Chatterjee, A., Vallières, M., Shenouda, G., Seuntjens, J.: Deep learning in head & neck cancer outcome prediction. Sci. Rep. **9**, 1–10 (2019)

3. Afshar, P., Mohammadi, A., Plataniotis, K.N., Oikonomou, A., Benali, H.: From handcrafted to deep-learning-based cancer radiomics: challenges and opportunities. IEEE Signal Process. Mag. **36**, 132–160 (2019)
4. Mohamed, A.S.R., et al.: Methodology for analysis and reporting patterns of failure in the Era of IMRT: head and neck cancer applications. Radiat. Oncol. **11**, 1–10 (2016)
5. Paidpally, V., Chirindel, A., Lam, S., Agrawal, N., Quon, H., Subramaniam, R.M.: FDG-PET/CT imaging biomarkers in head and neck squamous cell carcinoma. Imaging Med. **4**, 633 (2012)
6. Naser, M.A., et al.: Progression free survival prediction for head and neck cancer using deep learning based on clinical and PET-CT imaging data. medRxiv **6**, 1 (2021). https://doi.org/10.1101/2021.10.14.21264955
7. Andrearczyk, V., et al.: Overview of the HECKTOR challenge at MICCAI 2020: automatic head and neck tumor segmentation in PET/CT. In: Andrearczyk, V., Oreiller, V., Depeursinge, A. (eds.) HECKTOR 2020. LNCS, vol. 12603, pp. 1–21. Springer, Cham (2021). https://doi.org/10.1007/978-3-030-67194-5_1
8. Andrearczyk, V., et al.: Automatic segmentation of head and neck tumors and nodal metastases in PET-CT scans. In: Medical Imaging with Deep Learning. pp. 33–43. PMLR (2020)
9. Andrearczyk, V., et al.: Overview of the HECKTOR challenge at MICCAI 2021: automatic head and neck tumor segmentation and outcome prediction in PET/CT images. In: Andrearczyk, V., Oreiller, V., Hatt, M., Depeursinge, A. (eds.) HECKTOR 2021. LNCS, vol. 13209, pp. 1–37. Springer, Cham (2022)
10. Valentin, O., et al.: Head and neck tumor segmentation in PET/CT: the HECKTOR challenge. Med. Image Anal. **44**, 177–195 (2021)
11. Iandola, F., Moskewicz, M., Karayev, S., Girshick, R., Darrell, T., Keutzer, K.: Densenet: implementing efficient convnet descriptor pyramids. arXiv Prepr. arXiv:1404.1869 (2014)
12. The MONAI Consortium: Project MONAI, (2020). https://doi.org/10.5281/zenodo.4323059
13. Kim, H., Goo, J.M., Lee, K.H., Kim, Y.T., Park, C.M.: Preoperative CT-based deep learning model for predicting disease-free survival in patients with lung adenocarcinomas. Radiol. **296**, 216–224 (2020). https://doi.org/10.1148/radiol.2020192764
14. Gensheimer, M.F., Narasimhan, B.: A scalable discrete-time survival model for neural networks. PeerJ **7**, e6257 (2019)
15. Davidson-Pilon, C.: lifelines: survival analysis in Python. J. Open Source Softw. **4**, 1317 (2019)
16. Merkel, D.: Docker: lightweight linux containers for consistent development and deployment. Linux J. **2014**, 2 (2014)
17. Singh, A., Sengupta, S., Lakshminarayanan, V.: Explainable deep learning models in medical image analysis. J. Imaging **6**, 52 (2020)

Self-supervised Multi-modality Image Feature Extraction for the Progression Free Survival Prediction in Head and Neck Cancer

Baoqiang Ma$^{(\boxtimes)}$ ⓘ, Jiapan Guo$^{(\boxtimes)}$ ⓘ, Alessia De Biase$^{(\boxtimes)}$ ⓘ,
Nikos Sourlos$^{(\boxtimes)}$ ⓘ, Wei Tang ⓘ, Peter van Ooijen ⓘ, Stefan Both,
and Nanna Maria Sijtsema ⓘ

University Medical Center Groningen (UMCG),
Groningen 9700, RB, Netherlands
{b.ma,j.guo,a.de.biase,n.sourlos,w.tang,p.m.a.van.ooijen,
s.both,n.m.sijtsema}@umcg.nl

Abstract. Long-term survival of oropharyngeal squamous cell carcinoma patients (OPSCC) is quite poor. Accurate prediction of Progression Free Survival (PFS) before treatment could make identification of high-risk patients before treatment feasible which makes it possible to intensify or de-intensify treatments for high- or low-risk patients. In this work, we proposed a deep learning based pipeline for PFS prediction. The proposed pipeline consists of three parts. Firstly, we utilize the pyramid autoencoder for image feature extraction from both CT and PET scans. Secondly, the feed forward feature selection method is used to remove the redundant features from the extracted image features as well as clinical features. Finally, we feed all selected features to a DeepSurv model for survival analysis that outputs the risk score on PFS on individual patients. The whole pipeline was trained on 224 OPSCC patients. We have achieved a average C-index of 0.7806 and 0.7967 on the independent validation set for task 2 and task 3. The C-indices achieved on the test set are 0.6445 and 0.6373, respectively. It is demonstrated that our proposed approach has the potential for PFS prediction and possibly for other survival endpoints.

Keywords: Progression free survival prediction · OPSCC ·
DeepSurv · Autoencoder · PET-scans and CT-scans

1 Introduction

Almost 60,000 US patients are diagnosed with head and neck (H&N) cancer every year, causing 13,000 deaths annually [1]. The treatment strategies of H&N cancer

Aicrowd Group Name: "umcg"

V. Andrearczyk et al. (Eds.): HECKTOR 2021, LNCS 13209, pp. 308–317, 2022.
https://doi.org/10.1007/978-3-030-98253-9_29

such as Oropharyngeal squamous cell carcinoma (OPSCC) are usually nonsurgical such as chemotherapy, radiotherapy, and combinations of these. Although loco-regional control of most OPSCC is good, five-year OS for OPSCC have ranged from 46% to 85% including all stages, and 40–85% in advanced stage cohorts [2]. It would be beneficial to be able to identify patients with an expected worse treatment response before start of treatment. When prediction models for tumor related endpoints and complications would be available it would become possible to select the most optimal treatment method (with the optimal balance between predicted tumor control and complications) for individual patients. E.g. a more intensive treatment regimen could be considered for patients with a predicted high-risk for tumor recurrence, whereas a de-intensified treatment regimen could be an option for patients with a low risk for tumor recurrence, to limit the risk of complications like swallowing problems and xerostomia [3]. Therefore, we have developed a PFS prediction mode using clincial data and image data.

Radiomics [4] - quantitative imaging features from high throughput extraction - has been successfully applied to outcome prediction of H&N cancers [5–8]. However, its clinical application is restricted due to its dependence on manual segmentation and handcrafted features [9]. Deep learning-based methods includes algorithms and techniques that identify more complex patterns than radiomics in large image data sets without handcrafted feature extraction, and they have been employed in various medical image fields [10–12] as well as H&N cancer outcome prediction [13–16]. In our method, we select Autoencoders as the basic architecture for image feature extraction.

Features significantly relating to PFS prediction can be obtained through features selection process. The obtained features can be used to create a survival analysis model, such as Cox proportional hazard model (CPHM) [17], random survival forests (RFS) [18] and DeepSurv [19] (a Cox proportional hazards deep neural network). We chose DeepSurv as our PFS prediction model because it can successfully model increasingly complex relationships between a patient's covariates and their risk of failure.

Our aim is building a DeepSurv model with the capability of predicting PFS prior to treatment using available clinical data and image features of CT and PET extracted by the Autoencoder. The work described in this paper was used to participate in the task 2 and task 3 of HECKOR 2021 challenge [20, 21] .

2   Materials

2.1   Dateset Description

The training set includes 224 head and neck cancer patients from 5 hospitals (CHGJ, CHUS, CHMR, CHUM and CHUP). There are co-registered 3D CT and FDG-PET images and GTVt images (primary Gross Tumor Volume label) for each patient. The voxels sizes are nearly 1.0 mm in the x and y directions and vary between 1.5 to 3.0 mm along the z direction. A bounding box is provided for each patient around the oropharyngeal primary tumors. Clinical data of all patients can be found in a csv file. The testing set consists of 101 patients treated in 2 hospitals (CHUP and CHUV). Co-registered CT- and PET-scans, bounding box and clinical data are available for patients in the test set, but not the GTV contour.

2.2 Data Preparation

We selected Gender, Age, T-stage, N-stage, TNM group, HPV status and Chemotherapy as the potential predictive data. Age is normalized by dividing by 100, and other clinical data are used as categorical variables. Through KM-survival analysis, some categories values with similar survival curves were combined. A detailed description of the definition of the categorical values is summarized in Table 1.

Table 1. The summary of classification of values in each category variables of clinical data.

Category variable	Value classification
Gender	(Male = 0), (Female = 1)
T-stage	(T1 T2 T3 = 0), (T4 T4a T4b = 1)
N-stage	(N0 N1 = 0), (N2 N2a N2b N2c = 1), (N3 = 2)
TNM group	(I = 0), (II III IV = 1), (IVA IVB IVC = 2)
HPV status	(negative = 0), (positive = 1), (unknown: 2)
Chemotherapy	(not = 0), (yes = 1)

The bounding box region image of CT, PET and GTVt (the mask image of gross tumor volume of the primary tumor) of each patient in training set and testing set are first cropped and extracted. Then, these CT and PET 3D images were resampled to $1 \times 1 \times 1 \, mm^3$ pixel spacing with trilinear interpolation. The GTVt masks were resampled to the same resolution with $1 \times 1 \times 1 \, mm^3$ CT but using nearest interpolation. CT region image pixel values are truncated to $[-200, 200]$ and then normalized to $[0,1]$ by the max-min value method. The pixel values of the PET region image smaller than 0 are set to 0. PET region images are normalized by first z-score and then the max-min value method. Finally, the normalized CT and PET region images of each patient are summed up to form a new combined image named CT/PET image. The pre-processed CT, PET and CT/PET with the size of $1 \times 144 \times 144 \times 144$ are the input of the autoencoder in task 2 (not using GTVt). To get the input of autoencoder in task 3 (using GTVt), we first dilated GTVt with size of 5 voxels, and use the dilated GTVt to multiply with CT and PET, then extracting the GTVt-region CT, PET and CT/PET images by two methods. The first method is GTVt center cropping (to a size of $64 \times 64 \times 64$). The second method is first cropping a sub-cube of GTVt according to the border positions in three directions of the tumor, then resampling the sub-cube to a size of $64 \times 64 \times 64$ voxels. Method 1 gives GTVt images with the same pixel spacing across cases whereas method 2 results in GTVt images with varying pixel spacing across cases depending on the tumor size. By combining both GTVt images information about the tumor size as well as tumor images with an optimal resolution, tumor information are effectively used in the training process. Images from the two methods are concatenated together to a size of $2 \times 64 \times 64 \times 64$, and they are the input of the task 3 autoencoder.

We adopt two different strategies to divide the provided dataset (224 patients) into a training and validation set. The first one is to use leave-one-center-out, in which 4 centers are used as the training set while one as the validation set. The second one is that we randomly selected 179 patients as the training while 45 as the validation set. Thus, we could perform 6-fold cross validation and ensemble results of different models on the final test set (101 patients).

3   Methods

The success of deep learning methods in computer vision tasks has brought its wide applications to the medical image analysis. They, however, require a large amount of labeled samples. The model performances are also biased by the manually provided labels. In this work, we proposed a deep learning based pipeline that adopts unsupervised learning approach for image feature extraction in the prediction of progression free survival (PFS) for head and neck cancer. We utilized a self-supervised deep learning approach for the extraction of tumor characteristics from both CT and PET scans. The extracted image features and clinical parameters were then used to train a DeepSurv model for time-to-event prediction on the PFS. Figure 1 illustrates the proposed pipeline.

For each image set of CT, PET, CT/PET, CT-GTVt, PET-GTVt and CT/PET-GTVt, we trained 6 models using 6-fold-cross-validation. The train/validation set ratio of each fold is different, because we performed leave one center out cross validation. For each fold, feature selection is performed in all image features that were identified by the autoencoders. The selected image features (around 2–6 features) and selected clinical data (Age, T-stage and HPV status) are used to train a DeepSurv model. We trained 30 DeepSurv models using the training set in each fold, and finally selected 3 models with the highest validation set C-index. In total 18 DeepSurv models (each fold has 3 DeepSurv models) are obtained, and their predicted risk scores on the test set are averaged to obtain the final result in the test set.

3.1   Autoencoder

Autoencoders are used to extract high-level features through reconstructing the input. In our method, we used CT, PET and CT/PET images to train three autoencoders, separately. An autoencoder consists of an encoder and a decoder. The encoder compresses the input image to high-level features, then the decoder reconstructs input image from these features. Those high-level features from the last layer of the encoder are chosen for the feature selection process.

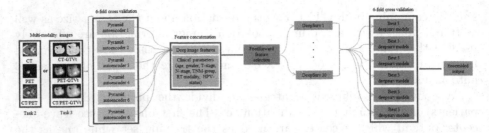

Fig. 1. The whole pipeline of the proposed method. 6-fold cross validation is applied in training autoencoder and DeepSurv models. Each modality image (CT, PET and CT/PET of task 2 or CT-GTVt, PET-GTVt and CT/PET-GTVt of task 3) is used to train one autoencoder to extract image features. All extracted image features and clinical data are selected using feedforward selection. The selected features are applied to train 30 DeepSurv models for each fold. Finally, 18 models (3 models for each fold) with highest validation C-index are used for testing, and their output on the test set are ensembled.

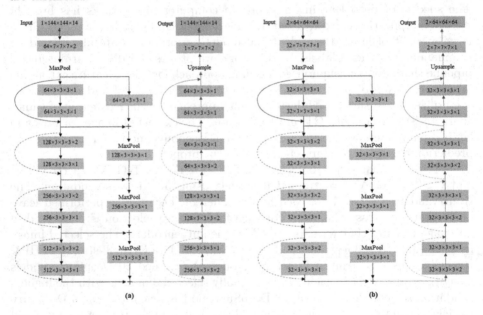

Fig. 2. The 3D ResNet-like architecture of autoencoders in task 2 (a) and task 3 (b). In task 2, the input is CT, PET or CT/PET patches in boxing regions with size $144 \times 144 \times 144$. In task 3, the input is CT, PET or CT/PET patches in primary tumor region only with size $2 \times 64 \times 64 \times 64$. Identity (solid arrows) and projection (dashed arrows) shortcuts are shown in residual blocks. Yellow and green rectangle stand for convolution and deconvolution, seperately (Color figure online)

The architecture of autoencoders in task 2 and task 3 are displayed in Fig. 2. Our autoencoder is built upon on 3D ResNet [22] with the use of a pyramid architecture between convolution blocks. We used NxHxWxDxS to descript the convolution or transpose convolution kernel number (N), size in three directions (H, W, D) and the stride (S) in Fig. 2. The encoder consists of one convolution layer, a maxpooling layer and 4 convolution residual blocks and a pyramid architecture to combine different level images features. The stride of all maxpooling layers is 2. At the end of the encoder, 512 high-level feature maps with size of $5 \times 5 \times 5$ (task 2) or 32 high-level feature maps with size of $4 \times 4 \times 4$ (task 3) are obtained. Four continuous transposed convolution residual blocks, an upsampling layer with sampling factor 2 and a final transposed convolution layer form the decoder. A Relu function and Batch Normalization layer follow all convolution and transposed convolution layers of the autoencoder.

A combined loss function including L1 loss, mean square error (MSE) and Structual Similarity (SSIM) is employed in the autoencoder training process. The L1 loss for one training example can be written as:

$$L_{L1} = ||A_{(x)} - x||_1 \qquad (1)$$

the MSE is defined as:

$$L_{MSE} = ||A_{(x)} - x||_2 \qquad (2)$$

For SSIM the loss function L_{SSIM} as described in [23] was used: SSIM is designed by modeling any image distortion as a combination of three factors that are loss of correlation, luminance distortion and contrast distortion. The combined loss function is:

$$L_{combined} = L_{L1} + L_{MSE} + 0.5 * L_{SSIM} \qquad (3)$$

3.2 Feature Selection

We first selected clinical data which are known to be related to PFS prediction. Gender, Age, T-stage, N-stage, TNM group, HPV status and Chemotherapy are kept for selection. We used the SequentialFeatureSelector of scikit-learn as feature selector (set direction as forward). The estimator of the selector is set as CPHM model. We ran the feature selector 1000 times using a random subset of the training set every time. Finally, the most frequently selected 3 features (Age, T-stage, HPV status) are reserved to perform PFS prediction.

We use the same feature selection method for image features selection. First each 3D feature map extracted from autoencoders is changed to single value feature by maxpooling. Then these 512 CT, 512 PET, 512 CT/PET, 32 CT-gtv, 32 PET-gtv and 32 CT/PET-gtv features are input to feature selector. All those image features are ranked according to their selected frequency. The 2–6 features with highest rankings are chosen.

3.3 DeepSurv

DeepSurv [19] is a deep learning based survival analysis model. We do not elaborate on it here and refer the interested readers to [19]. We set the DeepSurv architecture as two fully connected layers with 50 nodes, Relu and Batch Normalization. The DeepSurv outputs the risk score of PFS. The loss function is the average negative log partial likelihood.

4   Experiment

4.1   Training Details

The Autoencoders were trained using the Adam optimizer with the initial learning rate 0.001 in Tesla V100 GPU. The total number of training epochs is set to 80. The learning rate will decrease by multiplying by 0.1 if the training loss doesn't reduce in 10 consecutive epochs. Flipping and random rotation are used for data augmentation.

The official DeepSurv (https://github.com/jaredleekatzman/DeepSurv) code setting is applied to train our DeepSurv models. The total training steps are 5000, the validation set is used to select the best C-index model.

4.2   Results

This section shows the reconstructed images of the autoencoder and the C-index on the training set, validation set and test set.

The input images and the reconstructed images by autoencoders from one patient in the test set are displayed in Fig. 3. The reconstructed PET image is very similar to the input one. And we can recognize the highlighted tumor region in the reconstructed CT/PET image. Autoencoders successfully restored the shape of the tumor from the high-level features when we compare the input and output of CT-GTVt, PET-GTVt and CT/PET-GTVt. These results show that the high-level features extracted from autoencoders are representative and relevant.

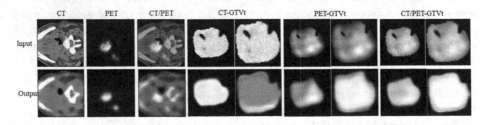

Fig. 3. The input and output of autoencoders. Tumor region images are successfully reconstructed in the output PET and CT/PET images. Tumor shape information are recovered on the output of CT-GTVt, PET-GTVt and CT/PET-GTVt images.

We summarized the training set and validation set C-index values of the DeepSurv model with highest validation C-index of each fold in Table 2. In task 2, we used selected clinical features and images features of CT, PET and CT/PET to train the DeepSurv models. In task 3, in addition to the features used in task 2, we added image features from CT-GTVt, PET-GTVt and CT/PET-GTVt. However, compared with C-index values on the validation set in fold 1, 4 and 6 of task2, we did not obtain a higher C-index value in task 3 when adding image features from CT-GTVt, PET-GTVt and CT/PET-GTVt. Therefore, we used the image features from task 2 also in task 3 in these three folds in where task 2 and task3 had the same C-indexes results. In fold 2, 3 and 5, task 3 obtained higher C-index values on both training and validation sets than task 2 because of adding features from CT-GTVt, PET-GTVt and CT/PET-GTVt.

From the Table 3, we can see that our method achieved good C-index values in the independent test set (0.6445 in task2 and 0.6373 in task 3). Although the C-index values of the validation set in task 3 were little higher than that in task 2, the C-index values of the test set were a little lower in task 3. This is possible due to the noise when training. The experimental results showed that our method does not need GTVt to locate the tumor. Our autoencoder can automatically extract image features in the tumor region, which is demonstrated by the reconstructed PET and CT/PET images in Fig. 3, in which the highlighted tumor image was constructed successfully from high-level features extracted by the encoder.

Table 2. The C-index values of training set and validation set of task 2 and task 3, using the best DeepSurv model with highest validation C-index.

Task name	Set name	fold1	fold2	fold3	fold4	fold5	fold6
Task2	Training	0.5939	0.6418	0.4389	0.7233	0.5723	0.7009
Task2	Validation	0.8073	0.7052	0.7644	0.8360	0.7383	0.8324
Task3	Training	0.5939	0.8025	0.4852	0.7233	0.5723	0.7009
Task3	Validation	0.8073	0.7063	0.8506	0.8360	0.7477	0.8324

Table 3. The C-index values on test set of task 2 and task 3.

Task name	C-index
Task2	0.6445
Task3	0.6373

5   Discussion and Conclusion

We have shown that our method was able to predict PFS with relative high average C-indexes of 0.7806 and 0.7967 for task 2 and 3 respectively in the validation set of all folds. However, the C-index of the test set is much lower than that of the validation sets (>0.7), which shows that our DeepSurv models overfit on the validation set. For example in fold 3, the C-index on the training set were very low (0.4389 in task2 and 0.4852 in task 3) but much higher (0.7644 in task2 and 0.8506 in task 3) on the validation set. The reason may be that the validation set (only 18 patients) has a very different feature distribution from the training set (206 patients). In our experiment, we selected the DeepSurv models with highest C-index values in the validation set for testing purpose, but they might perform worse in both the training and testing set. Thus, these models performing worse in the training set will decrease the final test set C-index value.

In order to improve the result on the test set in the future, we plan to change the methods in three aspects. Firstly, we will only select a part of DeepSurv models have good C-index in both the training and validation set for using on the test set, such as only using models in fold 2,4 and 6. Secondly, splitting the training set and validation set in a another way to make them have similar feature distribution. Finally when we retrain DeepSurv models of each fold in the future, we should save the model with high validation C-index on the condition of a high training C-index value instead of selecting models with only highest validation C-index.

We proposed a method that used a 3D pyramid autoencoder to extract high-level image features for PFS prediction. Obtained images features and clinical data are selected to acquire PFS-prediction related features. These selected features are applied to train a DeepSurv model for PFS prediction. Experimental results demonstrated that whether using GTVt or not, we could obtain a good C-index value on the test set. The proposed method has the potential for PFS prediction and possibly for other survival endpoints.

References

1. Siegel, R.L., Miller, K.D., Jemal, A.: Cancer statistics, 2016. CA Cancer J. Clin. **66**, 7–30 (2016)
2. Clark, J.M., et al.: Long-term survival and swallowing outcomes in advanced stage oropharyngeal squamous cell carcinomas. Papillomavirus Res. **7**, 1–10 (2019)
3. Tolentino, E.S., et al.: Oral adverse effects of head and neck radiotherapy: literature review and suggestion of a clinical oral care guideline for irradiated patients. J. Appl. Oral Sci. Revista FOB **19**, 448–54 (2011)
4. Kumar, V., et al.: Radiomics: the process and the challenges. Magn. Reson. Imaging **30**(9), 1234–1248 (2012)
5. Cheng, N.M., Fang, Y.D., Tsan, D.L., Lee, L.Y., Chang, J.T., Wang, H.M., et al.: Heterogeneity and irregularity of pretreatment (18)F-fluorodeoxyglucose positron emission tomography improved prognostic stratification of p16-negative high-risk squamous cell carcinoma of the oropharynx. Oral Oncol. **78**, 156–62 (2018)
6. Haider SP., Zeevi T., Baumeister P.: Potential added value of PET/CT radiomics for survival prognostication beyond AJCC 8th edition staging in oropharyngeal squamous cell carcinoma. Cancers (Basel) **12**(7) (2020). https://doi.org/10.3390/cancers12071778

7. Leijenaar, R.T., Carvalho, S., Hoebers, F.J., Aerts, H.J., van Elmpt, W.J., Huang, S.H., et al.: External validation of a prognostic CT-based radiomic signature in oropharyngeal squamous cell carcinoma. Acta Oncol. **54**(9), 1423–9 (2015). https://doi.org/10.3109/0284186x.2015.1061214

8. Wu, J., et al.: Tumor subregion evolution based imaging features to assess early response and predict prognosis in oropharyngeal cancer. J. Nucl. Med. **61**(3), 327–36 (2020). https://doi.org/10.2967/jnumed.119.230037

9. Bi, W.L., et al.: Artificial intelligence in cancer imaging: clinical challenges and applications. CA Cancer J. Clin. **69**(2), 127–57 (2019). https://doi.org/10.3322/caac.21552

10. Ma, B., Zhao, Y., Yang, Y., et al.: MRI image synthesis with dual discriminator adversarial learning and difficulty-aware attention mechanism for hippocampal subfields segmentation. Comput. Med. Imaging Graph. **86**, 101800 (2020)

11. Zhao, Y., Ma, B., Jiang, P., Zeng, D., Wang, X., Li, S.: Prediction of Alzheimer's disease progression with multi-information generative adversarial network. IEEE J. Biomed. Health Inform. **25**(3), 711–719 (2020)

12. Zeng, D., Li, Q., Ma, B., Li, S.: Hippocampus segmentation for preterm and aging brains using 3D densely connected fully convolutional networks. IEEE Access **8**, 97032–97044 (2020)

13. Diamant, A., Chatterjee, A., Vallières, M., Shenouda, G., Seuntjens, J.: Deep learning in head & neck cancer outcome prediction. Sci. Rep. **9**(1), 1–10 (2019)

14. Kann, B.H., et al.: Pretreatment identification of head and neck cancer nodal metastasis and extranodal extension using deep learning neural networks. Sci. Rep. **8**(1), 1–11 (2018)

15. Fujima, N., et al.: Prediction of the local treatment outcome in patients with oropharyngeal squamous cell carcinoma using deep learning analysis of pretreatment FDG-PET images. BMC Cancer **21**(1), 1–13 (2021)

16. Cheng, N.M., et al.: Deep learning for fully automated prediction of overall survival in patients with oropharyngeal cancer using FDG-PET imaging. Clin. Cancer Res. **27**, 3948–3959 (2021)

17. Cox, D.R.: Regression models and life-tables. In: Kotz, S., Johnson, N.L. (eds.) Breakthroughs in Statistics. SSS, pp. 527–541. Springer, New York (1992). https://doi.org/10.1007/978-1-4612-4380-9_37

18. Ishwaran, H., Kogalur, U.B., Blackstone, E.H., Lauer, M.S.: Random survival forests. Ann. Appl. Stat. **2**(3), 841–860 (2008)

19. Katzman, J.L., Shaham, U., Cloninger, A., Bates, J., Jiang, T., Kluger, Y.: DeepSurv: personalized treatment recommender system using a Cox proportional hazards deep neural network. BMC Med. Res. Methodol. **18**(1), 1–12 (2018)

20. Andrearczyk, V., et al.: Overview of the HECKTOR challenge at MICCAI 2021: automatic head and neck tumor segmentation and outcome prediction in PET/CT images. In: Andrearczyk, V., Oreiller, V., Hatt, M., Depeursinge, A. (eds.) HECKTOR 2021. LNCS, vol. 13209, pp. 1–37. Springer, Cham (2022)

21. Oreiller, V., et al.: Head and neck tumor segmentation in PET/CT: the HECKTOR challenge. Med. Image Anal. **77**, 102336 (2022)

22. Hara, K., Kataoka, H., Satoh, Y.: Learning spatio-temporal features with 3D residual networks for action recognition. In: Proceedings of the IEEE International Conference on Computer Vision Workshops, pp. 3154–3160 (2017)

23. Wang, Z., Bovik, A.C., Sheikh, H.R., Simoncelli, E.P.: Image quality assessment: from error visibility to structural similarity. IEEE Trans. Image Process. **13**(4), 600–612 (2004)

Comparing Deep Learning and Conventional Machine Learning for Outcome Prediction of Head and Neck Cancer in PET/CT

Bao-Ngoc Huynh[1], Jintao Ren[2], Aurora Rosvoll Groendahl[1],
Oliver Tomic[1], Stine Sofia Korreman[2]($\boxtimes$),
and Cecilia Marie Futsaether[1]($\boxtimes$)

[1] Faculty of Science and Technology, Norwegian University of Life Sciences,
Universitetstunet 3, 1433 Ås, Norway
cecilia.futsaether@nmbu.no
[2] Department of Clinical Medicine, Aarhus University,
Nordre Ringgade 1, 8000 Aarhus, Denmark
stine.korreman@oncology.au.dk

Abstract. Prediction of cancer treatment outcomes based on baseline patient characteristics is a challenging but necessary step towards more personalized treatments with the aim of increased survival and quality of life. The HEad and neCK TumOR Segmentation Challenge (HECK-TOR) 2021 comprises two major tasks: auto-segmentation of GTVt in FDG-PET/CT images and outcome prediction for oropharyngeal head and neck cancer patients. The present study compared a deep learning regressor utilizing PET/CT images to conventional machine learning methods using clinical factors and radiomics features for the patient outcome prediction task. With a concordance index of 0.64, the conventional machine learning approach trained on clinical factors had the best test performance. Team: Aarhus_Oslo

Keywords: Head and neck cancer · Machine learning · Deep learning · Gross tumor volume · Radiomics · Outcome prediction

1 Introduction

Head and neck cancer (HNC) is one of the most common cancers worldwide [1]. The long-term survival of HNC patients is generally poor, with a 50% rate of 5-year survival [2], usually due to the development of distant metastasis or second primary cancer [3,4]. Traditionally, head and neck cancer has been diagnosed in older patients with heavy smoking and alcohol consumption. However, the incidence of human papillomavirus (HPV)-related oropharyngeal cancer is increasing, predominantly among younger persons in Western countries

---

B.-N. Huynh and J. Ren—Authors contributed equally.

ⓒ Springer Nature Switzerland AG 2022
V. Andrearczyk et al. (Eds.): HECKTOR 2021, LNCS 13209, pp. 318–326, 2022.
https://doi.org/10.1007/978-3-030-98253-9_30

[5]. HPV-related oropharyngeal cancers have better prognosis than other head and neck cancers. However, up to 25% of HPV-related oropharyngeal cancer patients will experience a relapse within 5-years of treatment [6]. Developing methods to separate between low- and high-risk patients before treatment is, therefore, essential to provide more personalized treatment for increased survival and quality of life.

The MICCAI 2021 - HECKTOR2021 (HEad and neCK TumOR segmentation and outcome prediction in PET/CT images 2021) challenge consisted of three tasks: 1. Primary tumor (GTVt) automatic segmentation from Positron Emission Tomography/Computed Tomography (PET/CT) images. 2. Progression-Free Survival (PFS) outcome prediction from PET/CT images and available clinical data. 3. PFS outcome prediction, similar to task 2, from PET/CT images and clinical data, with the addition of GTVt delineation contours [7,8]. Thanks to this challenge, participants from different backgrounds worldwide could test and compare their approaches with the same data and evaluation criteria. The present study focuses on task 2 - PFS outcome prediction from PET/CT images and clinical information.

Radiotherapy, chemotherapy, and surgery are the most common treatment options for HNC. Depending on the characteristics of the patient's disease, such as the cancer site and stage, different cancer treatments might be the standard of care. However, different treatment outcomes still occur even for patients having similar disease characteristics [9]. Therefore, more information than just traditional clinical risk factors are likely needed for more accurate prediction of treatment outcome prognosis.

As an addition to the clinical information, radiomics has attracted a great deal of attention. Radiomics features are information extracted from medical images, including first-order statistical features and textural features [10]. Since the introduction of the term "radiomics" in 2012, many studies have addressed the potential of including radiomics features in cancer treatment prognosis [11–13].

Deep learning is yet another approach for extracting information from medical images. Deep learning has a reputation for outperforming most other methods when tackling the same problem, making it an excellent choice for outcome prediction. Furthermore, unlike radiomics, which uses hand-crafted features, deep learning, particularly convolutional neural networks (CNNs), can extract information or features that are linked to clinical outcomes [13]. In addition, this approach can potentially learn other patient characteristics such as T and N categories from medical images.

Therefore, in Task 2 of HECKTOR2021, we used two approaches: 1. Conventional machine learning models trained on clinical factors or both clinical factors (CFs) and radiomics features (RFs) extracted from PET/CT images and 2. Deep learning models trained on PET/CT images. As a result, we could evaluate the effect of three different input modalities on PFS outcome prediction.

2   Materials and Methods

2.1   Data

The input data were fluorodeoxyglucose (FDG)-PET/CT images and patient clinical information of 224 oropharyngeal cancer patients from five clinical centers provided by HECKTOR2021. The PET image intensities had been converted to Standardized Uptake Values (SUV). Available patient clinical information consisted of age, gender, tobacco and alcohol consumption, performance status, HPV status, treatment type, tumor (T), node (N) and metastasis (M) category, TNM stage and TNM edition. In addition, HECKTOR2021 provided a $144 \times 144 \times 144$ mm^3 bounding box to crop the relevant head and neck region. No delineation of gross tumour volume was included in the input data.

The prediction target PFS was time-to-event in days. The censoring data was based on the RECIST criteria [14], where either a size increase of known lesions or the appearance of new lesions was counted as a progression event.

HECKTOR2021 provided an external test set of 101 patients from two centers, of which one was an external center not present in the training data.

2.2   Data Preprocessing

Clinical Factors. While HECKTOR2021 provided over ten clinical factors, not all were used when training due to missing data or low variance. The chosen columns were gender, age, T category, N category, TNM stage and TNM edition, all of which were transformed into numeric data when necessary.

Radiomics Features. Each FDG-PET/CT image was resampled to a 1mm^3 isotropic grid using spline interpolation, then cropped into a $144 \times 144 \times 144$ mm^3 volume using the provided bounding box. Subsequently, these cropped images were discretized using a fixed bin width of 25 before extracting over 100 radiomics features (RFs) following the Image Biomarker Standardization Initiative (IBSI) standard [15]. The implementation of the radiomics feature extraction pipeline is available in GitHub[1], which utilizes the PyRadiomics open-source python package.

As the extracted features can be highly correlated, feature selection was necessary to reduce the input data dimension. However, due to limited time, we manually selected five radiomics features based on similar studies [16]. Selected features were (i) the non-uniformity of the gray level size zone matrix (glszm) extracted from the CT image, (ii) skewness of the PET image, (iii) cluster shade of the gray level co-occurrence matrix, (iv) the small area emphasis of the glszm, and (v) the long run high gray level emphasis of the run length matrix; features (iii)-(v) were extracted from the PET images.

---

[1] https://github.com/NMBU-Data-Science/imskaper.

PET/CT Images for Deep Learning. Similar to the process before extracting radiomics features, the PET/CT images were resampled and cropped into $144 \times 144 \times 144 \, \text{mm}^3$ volumes.

The primary tumor (GTVt) and involved lymph nodes (GTVn) often have comparably high SUV, which may affect the ability of the deep learning model to extract different characteristics from the GTVt and GTVn. Therefore, we applied a normalization scheme, namely PET-sin, to the cropped PET images. The PET-sin images were the results of mapping each PET voxel intensity x_n (in SUV) to $y_n = \sin(x_n)$, where y_n is the new voxel intensity of the transformed PET-sin images. The resulting PET-sin images contained voxels with intensities in the range $[-1, 1]$, where the GTVn can be distinguished from the GTVt by having a multi-layer structure resembling the ellipsoidal onion in the axial view (Fig. 1).

Fig. 1. An example image slice from patient CHUM054. From left to right: CT with blended PET image, original PET, and PET-sin. GTVn on the PET-sin image (right) shows an onion-like structure.

In addition, for the CNN trained on PET/CT images, the prediction target - the PFS in days - was first clipped into $[0, 3000]$ and then normalized into the range in $[0, 1]$ to reduce the high range of the original data.

2.3   Experiments

Baseline Model and Evaluation Metrics. Since the primary goal of HECKTOR2021 - Task 2 was survival outcome prediction, a simple Cox Proportional Hazard model (CoxPH) [17] with clinical factors as inputs was used as the baseline model. The concordance index (CI) was the evaluation metrics used to compare different model performances.

Conventional Machine Learning Approach. While there are various machine learning models for regression problems, ensemble models are more likely to outperform the other methods. Therefore, we used grid-search to tune the best hyper-parameters for two ensemble models: a random forest regressor and a gradient boosting regressor. Tuned hyper-parameters were the number of trees, maximum depth for each tree, and learning rate. In the case of gradient boosting, loss function tuning was also included. Each combination of hyper-parameters was trained and evaluated using five-fold cross-validation. The best combination of model and tuned hyper-parameters was trained on all training data to predict the survival outcome of the external test set.

Deep Learning Approach. In the deep learning approach, we did not include the patient information but only image data, namely CT and PET-sin. For CNNs, we used the encoder part of the SE Norm UNet model introduced by Iantsen et al. [18] with three fully connected layers (4096, 512, and 1 units) added to the top. Five-fold cross-validation was also applied in this approach. Each model was trained for 150 epochs using the Adam optimizer on an NVIDIA RTX Quadro 8000 (48 GB) GPU with a batch size of 4. The initial learning rate was set to 3e−6.

Choosing a suitable loss function for the CNN model was challenging. As the PFS prediction task is a regression problem, the mean squared errors, mean absolute errors and Huber loss are good candidates. Since the Huber loss is less sensitive to outliers [19], which are expected for the PFS of censored data, Huber loss is the best choice among the suggested loss functions. On the other hand, because the CI focuses on the order of elements, the Canberra distance loss, which minimizes the differences between two ranked lists [20], is also a good option for the loss function. Therefore, we chose to use the fusion of the Canberra distance loss and Huber loss ($\delta = 1$) as the loss function to train the CNN model. In this fused loss function, the weight of the Huber loss was twice the Canberra distance.

3   Results

3.1   Cross-validation Results

When comparing cross-validation results, there were no substantial differences in terms of CI between the CoxPH model and the Random Forest Regressor (Table 1). Adding radiomics features did not contribute to the overall performance of these two models. On the other hand, gradient boosting obtained the best cross-validation results, with radiomics features playing an important role in improving the model performance.

In addition, unlike other models which minimized the error rate (for instance, mean absolute error), the best model from gradient boosting used the quantile loss function [21], which predicted the lower boundary (5–10 percentile) of the target instead.

With only the PET/CT images as input data, the deep learning approach obtained a moderate cross-validation CI (Table 1). However, the obtained CI was the lowest among all experiments, including the baseline model.

Table 1. The CI for each fold of the best-tuned models trained on clinical factors (CFs) or both CFs and radiomics features (RFs) and the deep learning model.

Model	Inputs	Fold 0	Fold 1	Fold 2	Fold 3	Fold 4	Average
CoxPH	CFs	0.56	0.64	0.67	0.57	0.60	0.61
CoxPH	CFs + RFs	0.60	0.61	0.67	0.46	0.60	0.59
Random Forest	CFs	0.61	0.69	0.63	0.51	0.66	0.62
Random Forest	CFs + RFs	0.61	0.63	0.65	0.56	0.59	0.61
Gradient Boosting	CFs	0.61	0.68	0.76	0.70	0.62	0.67
Gradient Boosting	CFs + RFs	0.66	0.76	0.62	0.78	0.68	0.70
Deep learning	PET/CT	0.48	0.58	0.59	0.53	0.67	0.57

3.2 Test Set Submission

Since the CI depends on censoring data, it is interesting to see the external test set performance of a model trained only on observed data. In addition, the best models using different inputs should be evaluated on the test set. Therefore, the following models were selected for test set submissions:

1. Gradient boosting model with best hyper-parameters combination trained on clinical factors of observed (uncensored) data (GB-CFs).
2. Ensemble based on mean predicted values of five-fold deep learning models trained on FDG-PET/CT images (Deep learning).
3. Gradient boosting model with best hyper-parameters combination trained on clinical factors of all training data (GB-CFs-All).
4. Gradient boosting model with best hyper-parameters combination trained on both clinical factors and radiomics features of all training data (GB-CFs-RFs-All).

Table 2. Test set submission results.

Model	GB-CFs	Deep learning	GB-CFs-All	GB-CFs-RFs-All
Concordance index	0.57	0.57	0.64	0.49

Results of the test set submissions are shown in Table 2. The performance of the deep learning approach remained unchanged compared to the cross-validation results, with a CI similar to the model trained on only observed data (GB-CFs). However, the CI of the gradient boosting regressors trained on CFs and both CFs and RFs dropped drastically. In particular, the CI of the model trained on CFs and RFs was approximately equal to random predictions. Although there was a drop in CI on the test set, the gradient boosting regressor trained on clinical factors of all training data (GB-CFs-All) achieved the best performance of the submitted models.

4   Discussion and Conclusion

From cross-validation results, models whose goal was to minimize the differences between the predicted results and the actual survival outcomes were prone to overfitting. This led to the low cross-validation results of the random forest regressor and the CNN. In addition, censoring data were not considered while training these models despite the importance of censoring data for the CI. In addition, only 25% of training data have observed events, which means these machine learning models encountered considerable noise before converging.

In the case of gradient boosting with the quantile loss function, predicting the lower bound of the target seems to benefit from the proportion of observed data. In a dummy simulation of survival data, where only 25% of the data were observed, predicting the lower or upper bound of the target can achieve higher CI than the best-fit prediction.

While the CNNs can potentially extract some clinical factors directly from PET/CT images such as TNM staging, other clinical factors and patient characteristics may be missing, for example, patients' gender and age. This missing information can explain why the CNN model could not achieve higher CI than the other models. Therefore, combining clinical factors and medical image data using CNNs may improve the performance.

While radiomics features were shown to significantly impact outcome prediction models in previous studies [12,16], including these features decreased the cross-validation performance of the Cox model and the random forest model. Also, the best-tuned gradient boosting regressor trained on both clinical factors and radiomics features (GB-CFs-RFs-All) failed to predict the external test set. This poor performance may be because of the manual selection of radiomics features instead of a feature selection model, where the most relevant features can be used for training. In addition, the radiomics features were extracted from a large bounding box, which may entail a risk of feature aggregation problems [22]. The improvement of CI when including radiomics features of the GB-CFs-RFs-All model during cross-validation can result from overfitting, which can also explain the CI drop in the test set submission.

Although the gradient boosting regressor trained on only clinical factors of all training data (GB-CFs-All) achieved the highest CI among the submitted models, there was still a CI drop in the test set. This decrease of CI may be due to the different proportions of observed data between the training and test sets. Also, overfitting could be one of the reasons for the lower test set performance.

Despite having the lowest cross-validation CI, the deep learning model performed stably, as it obtained similar CI on both cross-validation and test set results. This consistent behavior may indicate that the deep learning approach can generalize to data from different centers. However, no conclusions can be drawn at the moment as the deep learning model is currently underfitting.

Surprisingly, the gradient boosting regressor (GB-CFs) trained only on observed data (only 25% of training data) achieved similar performance to the CNN. Therefore, it is essential to integrate censoring information when training survival models.

In this study, different PFS prediction models were built using conventional machine and deep learning approaches with different clinical factors, radiomics features and PET/CT images as input data. A conventional machine learning model trained on clinical factors only achieved moderate performance while adding radiomics features or using medical image features did not improve test set performances.

Acknowledgments. This work is supported by Aarhus University Research Foundation (No. AUFF-F-2016-FLS-8-4); Danish Cancer Society (No. R231-A13856, No. R191-A11526) and DCCC Radiotherapy - The Danish National Research Center for Radiotherapy.

References

1. Economopoulou, P., Psyrri, A.: Head and neck cancers: Essentials for clinicians, chap. 1. ESMO Educational Publications Working Group (2017)
2. Yeh, S.A.: Radiotherapy for head and neck cancer. In: Seminars in Plastic Surgery, vol. 24, pp. 127–136. Thieme Medical Publishers (2010)
3. Ferlito, A., Shaha, A.R., Silver, C.E., Rinaldo, A., Mondin, V.: Incidence and sites of distant metastases from head and neck cancer. ORL **63**(4), 202–207 (2001)
4. Baxi, S.S., Pinheiro, L.C., Patil, S.M., Pfister, D.G., Oeffinger, K.C., Elkin, E.B.: Causes of death in long-term survivors of head and neck cancer. Cancer **120**(10), 1507–1513 (2014)
5. Chow, L.Q.: Head and neck cancer **382**(1), 60–72 (2020). https://doi.org/10.1056/nejmra1715715
6. Culié, D., et al.: Oropharyngeal cancer: first relapse description and prognostic factor of salvage treatment according to p16 status, a GETTEC multicentric study **143**, 168–177 (2021). https://doi.org/10.1016/j.ejca.2020.10.034
7. Oreiller, V., et al.: Head and neck tumor segmentation in PET/CT: the HECKTOR challenge. Med. Image Anal. **77**, 102336 (2021)
8. Andrearczyk, V., et al.: Overview of the HECKTOR challenge at MICCAI 2021: automatic head and neck tumor segmentation and outcome prediction in PET/CT images. In: Andrearczyk, V., Oreiller, V., Hatt, M., Depeursinge, A. (eds.) HECKTOR 2021. LNCS, vol. 13209, pp. 1–37. Springer, Cham (2022)
9. Caudell, J.J., et al.: The future of personalised radiotherapy for head and neck cancer. Lancet Oncol. **18**(5), e266–e273 (2017)
10. Lambin, P., et al.: Radiomics: extracting more information from medical images using advanced feature analysis. Eur. J. Cancer **48**(4), 441–446 (2012)
11. Aerts, H.J., et al.: Decoding tumour phenotype by noninvasive imaging using a quantitative radiomics approach. Nat. Commun. **5**(1), 4006 (2014)
12. Vallieres, M., et al.: Radiomics strategies for risk assessment of tumour failure in head-and-neck cancer. Sci. Rep. **7**(1), 10117 (2017)
13. Liu, Z., et al.: The applications of radiomics in precision diagnosis and treatment of oncology: opportunities and challenges. Theranostics **9**(5), 1303–1322 (2019)
14. Therasse, P., et al.: New guidelines to evaluate the response to treatment in solid tumors. J. Natl. Cancer Inst. **92**(3), 205–216 (2000)
15. Zwanenburg, A., et al.: The image biomarker standardization initiative: standardized quantitative radiomics for high-throughput image-based phenotyping. Radiology **295**(2), 328–338 (2020)

16. Groendahl, A.R., et al.: PO-0967 prediction of treatment outcome for head and neck cancers using radiomics of PET/CT images. Radiother. Oncol. **133**, S526 (2019). https://www.sciencedirect.com/science/article/pii/S0167814019313878. ESTRO 38, 26–30 April 2019, Milan, Italy

17. David, C.R., et al.: Regression models and life tables (with discussion). J. Roy. Stat. Soc. **34**(2), 187–220 (1972)

18. Iantsen, A., Visvikis, D., Hatt, M.: Squeeze-and-excitation normalization for automated delineation of head and neck primary tumors in combined PET and CT images. In: Andrearczyk, V., Oreiller, V., Depeursinge, A. (eds.) HECKTOR 2020. LNCS, vol. 12603, pp. 37–43. Springer, Cham (2021). https://doi.org/10.1007/978-3-030-67194-5_4

19. Huber, P.J.: Robust Statistics, vol. 523. Wiley, Hoboken (2004)

20. Jurman, G., Riccadonna, S., Visintainer, R., Furlanello, C.: Canberra distance on ranked lists. In: Proceedings of Advances in Ranking NIPS 2009 Workshop, pp. 22–27. Citeseer (2009)

21. Koenker, R., Hallock, K.F.: Quantile regression. J. Econ. Perspect. **15**(4), 143–156 (2001)

22. Fontaine, P., Acosta, O., Castelli, J., De Crevoisier, R., Müller, H., Depeursinge, A.: The importance of feature aggregation in radiomics: a head and neck cancer study. Sci. Rep. **10**(1), 1–11 (2020)

Author Index

Printed in the United States
by Baker & Taylor Publisher Services